Numerical M[illegible]thod[illegible]eers

A PROGRAM[illegible]

Frontispiece (opposite title page). Train crash at Montparnasse Station in 1895. Numerical modelling of *short duration* events such as this impact problem could be tackled using an explicit *one-step* method to solve the differential equations of motion (see Chapter 7). Full scale physical modelling on the other hand, can be rather time-consuming and expensive. (Photograph courtesy of Éditions Gendre, Paris.)

DE L'OUEST
OUEST
721

Numerical Methods for Engineers

A PROGRAMMING APPROACH

D.V. Griffiths and I.M. Smith

Both of the University of Manchester

OXFORD

BLACKWELL SCIENTIFIC PUBLICATIONS

LONDON EDINBURGH BOSTON

MELBOURNE PARIS BERLIN VIENNA

To Valerie, Janet and Catherine

Editorial offices:
Osney Mead, Oxford OX2 0EL
25 John Street, London WC1N 2BL
23 Ainslie Place, Edinburgh EH3 6AJ
3 Cambridge Center, Cambridge
Massachusetts 02142, USA
54 University Street, Carlton
Victoria 3053, Australia

First published 1991

Set by Interprint Ltd, Malta
Printed and bound in Great Britain
by Hartnolls Ltd, Bodmin

DISTRIBUTORS

Marston Book Services Ltd
PO Box 87
Oxford OX2 0DT
(*Orders*: Tel: 0865 791155
Fax: 0865 791927
Telex: 837515)

USA
Blackwell Scientific Publications, Inc.
3 Cambridge Center
Cambridge, MA 02142
(*Orders*: Tel: 800 759-6102)

Canada
Oxford University Press
70 Wynford Drive
Don Mills
Ontario M3C 1J9
(*Orders*: Tel: (416) 441-2941)

Australia
Blackwell Scientific Publications
(Australia) Pty Ltd
54 University Street
Carlton, Victoria 3053
(*Orders*: Tel: (03) 347-0300)

British Library
Cataloguing in Publication Data

Griffiths, D.V. (Denwood Vaughan), *1953–*
Numerical methods for engineers: a programming approach.
1. Engineering. Analysis. Numerical methods
I. Title II. Smith, I.M. (Ian Moffat), *1940–*
620

ISBN 0-632-02751-7
ISBN 0-632-02753-3 pbk

Library of Congress
Cataloging in Publication Data

Griffiths, D.V.
Numerical methods for engineers: a programming approach/D.V. Griffiths, I.M. Smith.
p. cm.
Includes bibliographical references.
ISBN 0-632-02751-7. — ISBN 0-632-02753-3 (pbk.)
1. Engineering mathematics—Data processing.
2. Numerical analysis—Data processing. I. Smith, I.M. II. Title.
TA335.G75 1991
620′.001′51—dc20

Contents

List of Programs

Chapter 6

Chapter 7

Preface

This book is based on a lecture course in numerical methods in engineering given by the authors to second year undergraduates (sophomores/juniors) at the University of Manchester. The course has evolved over a number of years into its present blend of formal lectures and practical 'hands-on' experience in problem solving. For the majority of students, it will constitute the only formal training they will receive in the numerical techniques which underpin so much of engineering analysis today. As such it forms a bridge between the classical mathematics and heavily computer code-based techniques such as the finite element and boundary element methods. For a smaller number of students, it will constitute an introduction to the deeper study of these techniques and methods.

There are two main reasons for compiling what may seem to be just yet another introductory book on numerical methods. Firstly, the authors found that existing texts did not cover the range of material they needed. For example, few books incorporate methods of weighted residuals in their treatment of the solution of ordinary differential equations, although these methods are central to the understanding of finite element techniques. In order to achieve breadth of coverage, the authors have inevitably been forced to be selective. This book, therefore, describes only the aspects of numerical analysis likely to be of most use to engineers, and which can be embodied in a single year's course. No attempt is made to catalogue every known method for solving simultaneous equations or ordinary differential equations.

Secondly, although many books now incorporate computer code, this is not done in a systematic or didactic way. In contrast, the authors propose a philosophy of program-writing based on the use of subprograms. The computer language employed is FORTRAN 77, and so the subprograms are FORTRAN FUNCTIONs and SUBROUTINEs. In this way, the book serves as an introduction to the use of powerful mathematical subroutine libraries. Every aspect of numerical analysis which is described in the book is illustrated by a compact program which makes use of a simple library. All the FORTRAN programs and subroutines described in this book can be obtained from the publishers in the form of a diskette.

The authors are actively involved in numerical modelling, usually using finite elements, both at research level and in engineering consultancy. Although based in England, both authors have considerable experience of engineering education in North America, being engineering graduates of UC Berkeley, and frequent visitors to academic symposia. The first author recently spent fifteen months at Princeton University teaching and conducting research in the area of finite element modelling of the response of dams to earthquake loading.

D.V. Griffiths/I.M. Smith

Acknowledgements

The authors are indebted to the following reviewers for their helpful and constructive comments regarding the contents of this book:

Professor W.F. Ames	Georgia Tech
Professor Z.P. Bazant	Northwestern University
Dr D.N. Fenner	King's College, London
Professor G.M.L. Gladwell	University of Waterloo
Professor J.G. Hartley	Georgia Tech
Professor W. Miller Jr.	University of Minnesota
Professor N. Whitaker	University of Massachusetts
Professor R. White	North Carolina State University

Thanks also go to Navin Sullivan of Blackwell Scientific Publications for his encouragement in this project, and particularly for keeping things ticking over while the two authors were living on opposite sides of the Atlantic.

Finally, the authors would like to thank the secretarial staff of Manchester University Engineering Department who all had a part to play in the preparation of the manuscript. Particular thanks go to Barbara Michniowski and Doreen Dye who did the majority of the typing, and to Sally Henshall for the figures.

Numerical Methods for Engineers
A PROGRAMMING APPROACH

1

Introduction and Programming Preliminaries

1.1 Introduction

There are many existing texts aimed at introducing engineers to the use of numerical methods. Increasingly, these contain computer code which enables a particular problem to be solved. In the present book, this idea is taken to a logical conclusion, namely that the natural way of learning how numerical methods work is to code them for a digital computer, and then to do numerical experiments. In order to achieve this aim efficiently, it is argued that it pays to adopt a style of programming which facilitates transfer of expertise from one numerical analysis task to another.

1.2 Software and hardware

This book contains forty-two computer programs and thirty-three 'library' sub-programs, which are written in FORTRAN 77. While this language is not ideal, as will be shown, it is by far the most common in use in engineering practice today, a situation which shows no sign of changing. A new version of FORTRAN, entirely compatible with FORTRAN 77, will remedy many of its deficiencies and will soon become widely available.

As to hardware, the only prerequisite is a machine capable of compiling and running FORTRAN 77 programs. This embraces the vast majority of computers now available commercially, from the 'PC' range upwards through medium-sized computers to 'supercomputers'.

1.3 Mathematical subroutine libraries

Nearly all computer manufacturers supply some kind of 'library' of mathematical software with the machine. Thus, while it would be perfectly feasible for each user to write programs to work out the sine of an angle, for example by summing a series, this is a task which occurs so frequently that it makes sense to have the code permanently stored in every machine of that make. In FORTRAN machines, the language provides for about forty 'intrinsic' or 'library' functions: ABS, ACOS, AIMAG, AINT, ANINT, ASIN, ATAN, etc., which enable users to carry out simple but commonly occurring mathematical tasks.

To supplement these, most manufacturers will also supply a 'mathematical subroutine library' which enables rather more complex operations to be undertaken – solving linear simultaneous equations is a common example – without the user having to write the coding. While such libraries can be helpful, particularly when they make best use of some special hardware capabilities of a particular machine, they tend to be limited in extent and often are not 'portable', i.e. transferrable from one machine to another.

Table 1.1. Contents of NAG mathematical subroutine library

	'Chapter'	Subject area	Nature
	A02	Complex arithmetic	Utility routine
(3)	* C02	Zeros of polynomials	
	C05	Roots of one or more transcendental equations	
	C06	Summation of series	
(6)	* D01	Quadrature	
(7)	* D02	Ordinary differential equations	
(8)	* D03	Partial differential equations	
(5)	* D04	Numerical differentiation	Determination
	D05	Integral equations	Numerical
(5)	* E01	Interpolation	analysis
(5)	* E02	Curve and software fitting	
	E04	Maximising or minimising a function	
(2)	* F01	Matrix operations including inversion	
(4)	* F02	Eigenvalues and eigenvectors	
(2)	* F03	Determinants	
(2)	* F04	Simultaneous linear equations	
	F05	Orthogonalisation	
	G01	Simple calculations on statistical data	
	G02	Correlation and regression analysis	
	G04	Analyses of variance	
	G05	Random number generators	Statistical
	G07	Univariate estimation	analysis
	G08	Nonparametric statistics	
	G11	Contingency table analysis	
	G13	Time series analysis	
	H	Operations research	
	M01	Sorting	
	P01	Error Trapping	
	S	Approximations of Special Functions	Utility
	*X*01	Mathematical Constants	routines
	*X*02	Machine Constants	
	*X*03	Innerproducts	
	*X*04	Input/Output Utilities	

* Area covered in the present text

( ) Chapter in this book dealing with this topic

For this reason, organisations have evolved which support mathematical subroutine libraries available for use on many different computers. A good example is the NAG mathematical subroutine library, containing over 600 subroutines. These attempt to cover a wide area of numerical applications, and are organised into 'chapters' which are listed in Table 1.1.

These chapters can be very broadly classified into 'deterministic' numerical analyses, statistical analyses and utility routines. The present book deals only with the first of these classes and even then with a subset, asterisked in the table. Preceding each asterisk, in parentheses, is the chapter in the present book which forms an introduction to the same topic.

Other libraries with which students may be familiar are IMSL, MATLAB and LINPACK and EISPACK, the last two being sub-libraries dealing specifically with linear algebra and eigenvalue analyses respectively.

It can be seen that the majority of deterministic analyses methods will be dealt with in the following chapters. The selection is governed by limitations of space and of teaching time in typical courses. The book is also directed towards coverage of probably the most important area of numerical analysis concerning engineers, namely the solution of differential equations.

In the chapters that follow, we shall therefore illustrate how simple mathematical subroutines are constructed and how these are then assembled to form small computer programs to address various numerical analysis tasks. This will serve as an introduction for students and engineers to the use of much more comprehensive software such as the NAG mathematical subroutine library.

Before writing any programs, we must first illustrate how subroutines can best be constructed to be 'portable' between one user program and another.

1.4 Functions and subroutines

In the previous section, we remarked that the FORTRAN language contains a list of 'intrinsic' or 'library functions' for carrying out simple tasks like taking sines and cosines of angles. A FUNCTION in FORTRAN is a special subprogram having a specific structure, and any number of these can be appended by programmers to their 'main' programs.

Suppose, for example, we wished to refer many times in a 'main' program to the function $f(x)=3x^2+15x-23$. To save writing all of the algebra out each time, we could create a FORTRAN FUNCTION, as follows:

```
      FUNCTION F(X)
C
C      THIS IS AN EXAMPLE OF A FUNCTION CALLED F
C      THE INPUT IS X
C
      F=3.0*X*X+15.0*X-23.0
      RETURN
      END
```

Then in the main program, as long as the 'argument' of F has a value, we can refer to F as many times as we like, for example in statements like A = F(X1)*F(X2)/F(X3). The FUNCTION subprogram is particularly simple since it 'returns' a single quantity (the value of the function $f(x)$) to the main program each time it is 'called' from that main program.

If more than one quantity is to be returned to the main program from a subprogram call (even if the argument, although a single name, is an array) then a more complicated type of subprogram called a SUBROUTINE must be used. This also has a name, but that name cannot be used in expressions. Instead, several arguments or 'parameters' are used to pass information into and out of the subprogram.

As an example, our FUNCTION subprogram could be written as a SUBROUTINE in the following way:

```
      SUBROUTINE FUN(F,X)
C
C      THIS IS AN EXAMPLE OF A SUBROUTINE CALLED FUN
C      THE INPUT IS X AND THE OUTPUT IS F
C
      F=3.0*X*X+15.0*X-23.0
      RETURN
      END
```

In the main program, we would then have to 'call' the SUBROUTINE with its different arguments in order to work out the previously determined A, for example:

```
CALL FUN(F1,X1)
CALL FUN(F2,X2)
CALL FUN(F3,X3)
A=F1*F2/F3
```

In Chapters 2 to 7 which follow, the use of a SUBROUTINE rather than a FUNCTION will be the norm, since our parameters will often be arrays. The use of arrays in this way causes some additional complications in FORTRAN 77 which are now described.

1.4.1 Arrays as parameters in subroutines

In FORTRAN 77 main programs, arrays must have sizes which are known before the program is compiled. Thus a typical main program might have a structure

```
      PROGRAM DECLAR
C
C      DECLARATION OF ARRAY SIZES
C
      REAL A(100),B(1000,50),C(50,100)
      ..........
      ..........
      ..........
      STOP
      END
```

In order to avoid the need to be forever altering lots of parameters of arrays for problems of different sizes, the language allows a PARAMETER statement which alleviates this difficulty somewhat. Using this statement, the above program can be written:

```
      PROGRAM DECLAR
C
C      DECLARATION OF ARRAY SIZES USING PARAMETER STATEMENT
C
      PARAMETER(IA=100,IB=1000,IC=50)
C
      REAL A(IA),B(IB,IC),C(IC,IA)
      ...........
      ...........
      ...........
      STOP
      END
```

In SUBROUTINEs which use arrays as parameters, it is necessary for these arrays to be compatibly 'mapped' onto the main program fixed storage area. And yet the purpose of a SUBROUTINE is that it should be 'portable', that is it should be able to be transferred freely between user programs without any restrictions related to these programs' main storage strategy.

Consider first a SUBROUTINE involving only one-dimensional arrays. We shall see that a common task is to find the scalar or 'dot' product DOTPR of two vectors, say V1 and V2. Within the SUBROUTINE, V1 and V2 have to be 'declared', but since all the subprogram has to do to carry out the mapping is to locate the first element in store of each of the vectors, i.e. V1(1) and V2(1), FORTRAN 77 allows us to declare them as V1(*) and V2(*) respectively. Thus the SUBROUTINE can be written:

```
      SUBROUTINE VDOTV(V1,V2,DOTPR,N)
C
C              DOT PRODUCT V1*V2
C
      REAL V1(*),V2(*)
      DOTPR = 0.0
      DO 1 I = 1,N
    1 DOTPR = DOTPR + V1(I)*V2(I)
      RETURN
      END
```

The input parameters are the names of the vectors to be multiplied, V1 and V2, together with the length of the vectors, N. The output consists of the dot product, named as the single parameter DOTPR. Note that the actual size of V1 and V2 in the main program is irrelevant as long as it is greater than, or equal to N.

The situation becomes rather more complicated when arrays processed by subprograms have two or more dimensions, as shown in Fig. 1.1. In FORTRAN 77, arrays are stored by columns. That is, for a two-dimensional array A, the elements are stored in the order A(1, 1), A(2, 1), ... A(M, 1), ... A(IA, 1), A(1, 2), A(2, 2) ... A(M, 2) ... A(IA, 2), etc. Thus, for appropriate mapping between main program and subprogram, the SUBROUTINE has to pick up not just the first element of array A, namely A(1, 1), but also

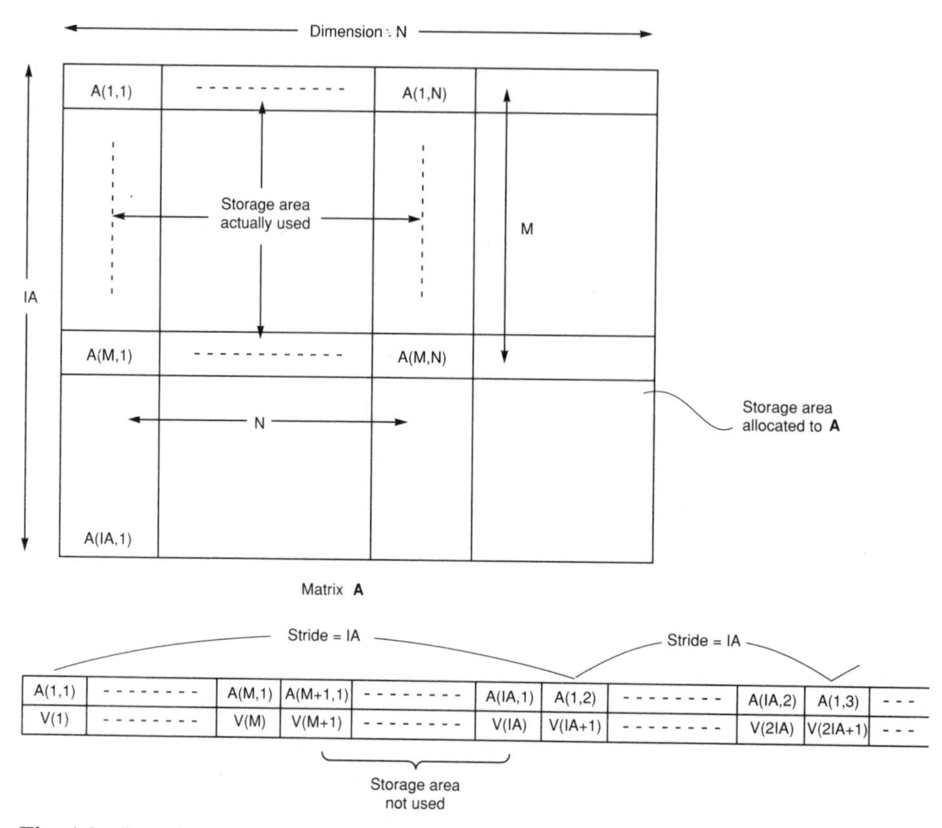

Fig. 1.1 Two-dimensional array storage in FORTRAN

has to find the correct locations in store of A(1, 2), A(1, 3), etc., the first elements in the second, third, etc. columns of A. To do this, the subroutine needs to know the number of elements in each column of A, called the 'stride', and denoted by IA in the figure. Thus, even though the actual column size of an array A being processed by the subroutine might be M, as long as IA is greater than, or equal to, M, we can locate A(1, 2) as being the number in store immediately following A(IA, 1).

Since the stride is enough information for all elements in A to be located, FORTRAN 77 allows us to declare A within a SUBROUTINE as A(IA,*).

As an illustration, consider the computation of the vector or cross product of V1 and V2, of lengths M and N respectively, which will now be an array A(M, N). The appropriate SUBROUTINE could be written:

```
      SUBROUTINE VVMULT(V1,V2,A,IA,M,N)
C
C      FORMS A VECTOR PRODUCT
C
      REAL V1(*),V2(*),A(IA,*)
      DO 1 I = 1,M
          DO 1 J = 1,N
    1 A(I,J) = V1(I)*V2(J)
      RETURN
      END
```

Thus, IA must be greater than or equal to M, and the second dimension of A, as declared in the main program, must be greater than or equal to N for this subprogram to work.

Some libraries, such as the NAG mathematical subroutine library, would add a further subprogram parameter, say JA, which would be the second dimension of A as declared in the main program. Although this has the benefit that checks can be made to make sure that $N \leq JA$, the increase in number of parameters does not justify its inclusion in the simple programs described in this book.

1.5 Errors

In the chapters that follow, it is assumed that calculations in numerical methods will be made using a digital computer. It will be very rare for computations to be made in exact arithmetic and, in general, 'real' numbers represented in 'floating point' form will be used. This means that the number 3 will tend to be represented by 2.99999 ... 9 or 3.0000 ... 1. Therefore, all calculations will, in a strict sense, be erroneous and what will concern us is that these errors are within the range that can be tolerated for engineering purposes. The most significant measure of error is not the 'absolute' error, but the 'relative' error. For example, if x_0 is exact and x an approximation to it, then what is significant is not $x-x_0$ (absolute) but $(x-x_0)/x_0$ (relative). In engineering calculations it must always be remembered that excessive computational accuracy may be unjustified if the data (for example some physical parameters) are not known to within a relative error far larger than that achievable in the calculations.

Modern computers (even 'hand' calculators) can represent numbers to a high precision – perhaps to eight or more places of decimals. Nevertheless, in extreme cases, errors can arise in calculations due to three main sources, called 'roundoff', 'truncation' and 'cancellation' respectively.

1.5.1 Roundoff

Consider the following examples of decimal manipulations to two decimal places:

$0.56 \times 0.65 = 0.36$ and so $(0.56 \times 0.65) \times 0.54 = 0.19$

whereas

$0.65 \times 0.54 = 0.35$ and so $(0.56 \times (0.65 \times 0.54)) = 0.20$

Alternatively, consider the sum:

$51.4 \times 23.25 - 50.25 \times 22.75 + 1.25 \times 10.75 - 1.2 \times 9.8 = 53.54$ (exactly)

If this calculation is worked to two decimal places, four significant figures, zero decimal places and three significant figures respectively, adding forwards (F) or backwards (B), adding and rounding (R) or rounding each time (C), the following answers are found:

	FR	FC	BR	BC
2 decimal places	53.54	53.54	53.54	53.54
4 significant figures	53.68	53.67	54	54
0 decimal places	53	54	53	54
3 significant figures	61.6	51.7	60	60.

These represent extremes by modern standards, since calculations tend to be done on computers, but illustrate the potential pitfalls in numerical work.

In doing floating point arithmetic, computers hold a fixed number of digits in the 'mantissa', that is the digits following the decimal point and preceding the 'exponent' $\times 10^6$ in the number 0.243875×10^6. The effect of roundoff in floating point calculations can be seen in the alternative calculations to six digits of accuracy:

$$(0.243875 \times 10^6 + 0.41648 \times 10^1) - 0.243826 \times 10^6 = 0.530000 \times 10^2$$

and

$$(0.243875 \times 10^6 - 0.243826 \times 10^6) + 0.412648 \times 10^1 = 0.531265 \times 10^2$$

It should however be emphasised that many modern computers represent floating point numbers by 64 binary digits, which retain about 15 significant figures, and so roundoff errors are less of a problem than they used to be. For engineering purposes, sensitive calculations on machines with only 32-bit numbers should always be done in 'double precision'.

1.5.2 Truncation

Errors due to this source occur when an infinite process is replaced by a finite one. For example, it was mentioned earlier in the context of intrinsic functions in FORTRAN that trigonometric functions are computed by summing series.

Thus $S = \sum_{i=0}^{\infty} a_i x^i$ is replaced by the finite sum $\sum_{i=0}^{N} a_i x^i$

For example

$$\exp(x) \simeq 1 + \frac{x}{1!} + \frac{x^2}{2!} + \frac{x^3}{3!} + \frac{x^4}{4!} + \cdots$$

is summed to N terms. Suppose we wish to calculate $\exp(\frac{1}{3})$. We might begin by creating an error by specifying exp(0.3333) so that

$$\begin{aligned}\epsilon_1 = \text{propagated error} &= e^{0.3333} - e^{1/3} \\ &= -0.0000465196.\end{aligned}$$

Then we might truncate the series after five terms, leading to

$$\epsilon_2 = \text{truncation error} = -\left(\frac{0.3333^5}{5!} + \frac{0.3333^6}{6!} + \cdots\right)$$

$$= -0.0000362750.$$

Finally we might sum with rounded values:

$1 + 0.3333 + 0.0555 + 0.0062 + 0.0005 = 1.3955$

where the propagated error from roundings is -0.0000296304, leading to a final total error of -0.0001124250.

1.5.3 Cancellation

The quadratic equation $x^2 - 2ax + \epsilon = 0$ has the roots $x_1 = a + \sqrt{a^2 - \epsilon}$ and $x_2 = a - \sqrt{a^2 - \epsilon}$. If a is 100 and ϵ is 1, $x_2 = 100 - \sqrt{10000 - 1}$. In extreme cases, this could lead to dangerous cancellation and it would be safer to write $x_2 = \epsilon/(a + \sqrt{a^2 - \epsilon})$.

Mathematical subroutine libraries often contain useful routines which help users to appreciate the possibilities of numerical errors in digital computations. For example, the NAG library (see Table 1.1) contains utility routines enabling the following 'machine constants' to be inspected and used:

1 Smallest positive ϵ such that $1.0 + \epsilon > 1.0$
2 Smallest representable positive real number
3 Largest representable positive real number
4 Largest negative permissible argument for EXP
5 Largest positive permissible argument for EXP
6 Smallest representable positive real number whose reciprocal is also representable
7 Largest permissible argument for SIN and COS
8 Base of floating point arithmetic
9 Largest representable integer
10 Largest positive integer power to which 2.0 can be raised without overflow
11 Largest negative integer power to which 2.0 can be raised without underflow
12 Maximum number of decimal digits that can be represented

In the above, 'overflow' (a number is too big to be represented) usually causes calculations to be terminated while 'underflow' (a number is too small to be represented) often causes that number to be replaced by 'zero'. Wide variations in these machine 'constants' will be found amongst various different machines.

1.6 Conclusions

A style of programming using 'portable' SUBROUTINEs and FUNCTIONs in FORTRAN 77 has been outlined. Using this strategy, we shall see in subsequent chapters that the same subprograms find use in many different contexts of numerical analysis.

Attention has to be devoted to the possibility and limitation of calculation errors, which will vary from machine to machine.

Chapters 2 to 7 go on to describe small programs built up using subprograms wherever possible. These chapters cover a subset of, and form an introduction to, a more comprehensive subroutine library such as the NAG mathematical subroutine library.

Chapter 2 deals with the numerical solution of sets of linear simultaneous equations, while Chapter 3 considers roots of nonlinear equations. In Chapter 4, eigenvalue problems are considered, while Chapter 5 deals with interpolation and curve fitting. Chapter 6 is devoted to 'quadrature', that is to numerical evaluation of integrals, while Chapter 7 introduces the solution of ordinary differential equations by numerical means. Chapter 8 is an introduction to the solution of partial differential equations.

In all chapters, mathematical ideas and definitions are introduced as they occur, and most numerical aspects discussed are illustrated by a small computer program. The 'library' routines are described and listed in Appendices 1 and 2 respectively.

1.7 Further reading

Dijkstra, E.W. (1976). *A Discipline of Programming*, Prentice-Hall, Eaglewood Cliffs, New Jersey.

Dongarra, J., Bunch, J., Moler, G. and Stewart, G. (1979). *LINPACK User's Guide*, SIAM Pub., Philadelphia.

Ford, B. and Sayers, D.K. (1976). Developing a Single Numerical Algorithms Library for Different Machine Ranges, *ACM Trans. Math. Software*, **2**, p. 115.

Garbow, B., Boyle, J., Dongarra, J. and Moler, C. (1977). Matrix Eigensystem Routines – EISPACK Guide Extension, *Lecture Notes on Computer Science*, **51**, Springer-Verlag, New York.

Moré, J., Garbow, B. and Hillstrom, K. (1980). *User Guide for MINPACK-1*. Argonne Nat. Lab. Report ANL-80-74, Chicago, Ill.

Phillips, J. (1987). *The NAG Library – A Beginner's Guide*, Clarendon Press, Oxford.

Rice, J. (1983). *Numerical Methods, Software and Analysis*, McGraw-Hill, New York.

Smith, B., Boyle, J., Garow, B. Ikebe, Y., Klema, V. and Moler, C. (1976) Matrix Eigensystem Routines – EISPACK Guide, 2nd edn, *Lecture Notes in Computer Science* **6**, Springer-Verlag, New York.

Smith, I.M. and Griffiths, D.V. (1988). *Programming the Finite Element Method*, 2nd edn, Wiley, Chichester.

Wilkinson, J.H. (1963). *Rounding Errors in Algebraic Processes*, HMSO, London.

2

Linear Algebraic Equations

2.1 Introduction

One of the commonest numerical tasks facing engineers is the solution of sets of linear algebraic equations of the form

$$
\begin{aligned}
a_{11}x_1 + a_{12}x_2 + a_{13}x_3 &= b_1 \\
a_{21}x_1 + a_{22}x_2 + a_{23}x_3 &= b_2 \\
a_{31}x_1 + a_{32}x_2 + a_{33}x_3 &= b_3
\end{aligned}
\tag{2.1}
$$

commonly written

$$\mathbf{Ax} = \mathbf{b} \tag{2.2}$$

where $\mathbf{A}$ is a 'matrix' and $\mathbf{x}$ and $\mathbf{b}$ are 'vectors'.

In these equations the a_{ij} are constant known quantities, as are the b_i. The problem is to determine the unknown x_i. In this chapter we shall consider two different solution techniques, usually termed 'direct' and 'iterative' methods. The direct methods are considered first and are based on row by row 'elimination' of terms, a process usually called 'Gaussian elimination'.

2.2 Gaussian elimination

We begin with a specific set of equations:

$$
\begin{aligned}
10x_1 + x_2 - 5x_3 &= 1 \quad \text{(a)} \\
-20x_1 + 3x_2 + 20x_3 &= 2 \quad \text{(b)} \\
5x_1 + 3x_2 + 5x_3 &= 6 \quad \text{(c)}
\end{aligned}
\tag{2.3}
$$

To 'eliminate' terms, we could, for example, multiply equation (a) by two and add to equation (b). This would produce an equation from which the term in x_1 had been eliminated. Similarly, we could multiply equation (a) by 0.5 and subtract from equation (c). This would also eliminate the term in x_1 leaving an equation in (at most) x_2 and x_3.

We could formally write this process

$$\begin{aligned}
&\text{(b)} - \left(\frac{-20}{10}\right) \times \text{(a)} \rightarrow 5x_2 + 10x_3 = 4 \quad \text{(d)} \\
&\text{(c)} - \left(\frac{5}{10}\right) \times \text{(a)} \rightarrow 2.5x_2 + 7.5x_3 = 5.5 \quad \text{(e)}
\end{aligned} \tag{2.4}$$

One more step of the same procedure would be

$$\text{(e)} - \left(\frac{2.5}{5}\right) \times \text{(d)} \rightarrow 2.5x_3 = 3.5 \tag{2.5}$$

Thus, for sets of N simultaneous equations, however big N might be, after N steps of this process a single equation involving only the unknown x_N would remain. Working backwards from eq. 2.5 – a procedure usually called 'back-substitution' – x_3 can first be found as 3.5/2.5 or 1.4. Knowing x_3, substitution in eq. 2.4(d) gives x_2 as -2.0 and finally substitution in eq. 2.3(a) gives x_1 as 1.0. Writing the back-substitution process in terms of matrices and vectors, we have

$$\begin{bmatrix} 10 & 1 & -5 \\ 0 & 5 & 10 \\ 0 & 0 & 2.5 \end{bmatrix} \begin{Bmatrix} x_1 \\ x_2 \\ x_3 \end{Bmatrix} = \begin{Bmatrix} 1 \\ 4 \\ 3.5 \end{Bmatrix} \tag{2.6}$$

or,

$$\mathbf{U}\mathbf{x} = \mathbf{y} \tag{2.7}$$

The matrix **U** is called an 'upper triangular matrix' and it is clear that such matrices will be very convenient in linear equation work.

In a similar way, if we had the system of equations

$$\begin{bmatrix} l_{11} & 0 & 0 \\ l_{21} & l_{22} & 0 \\ l_{31} & l_{32} & l_{33} \end{bmatrix} \begin{Bmatrix} x_1 \\ x_2 \\ x_3 \end{Bmatrix} = \begin{Bmatrix} y_1 \\ y_2 \\ y_3 \end{Bmatrix} \tag{2.8}$$

or,

$$\mathbf{L}\mathbf{x} = \mathbf{y} \tag{2.9}$$

It would be relatively easy to calculate **x** given **L** and **y**. The matrix **L** is called a 'lower triangular matrix', and the process of finding **x** in eqs 2.9 is called 'forward-substitution'. The direct methods we shall discuss all involve, in some way or another, matrices like **L** and **U**.

Example 2.1. Hand solution of three equations

Use Gaussian elimination to solve the following set of equations:

$$\begin{aligned} 2x_1-3x_2+x_3&=7\\ x_1-x_2-2x_3&=-2\\ 3x_1+x_2-x_3&=0 \end{aligned}$$

Solution 2.1

Eliminate the first column

$$\begin{aligned} 2x_1-3x_2+x_3&=7\\ 0.5x_2-2.5x_3&=-5.5\\ 5.5x_2-2.5x_3&=-10.5 \end{aligned}$$

Eliminate the second column

$$\begin{aligned} 2x_1-3x_2+x_3&=7\\ 0.5x_2-2.5x_3&=-5.5\\ 25x_3&=50 \end{aligned}$$

Back-substitute

$$\begin{aligned} x_3&=2\\ x_2&=(-5.5+2.5(2))/0.5=-1\\ x_1&=(7-2+3(-1))/2=1 \end{aligned}$$

Program 2.1. Gaussian elimination
For our first computer program in this chapter, let us construct a program for Gaussian elimination following the steps in eqs 2.4 and 2.5. The code is listed as Program 2.1 and, apart from simple counters, involves the following nomenclature:

Simple variables
N Number of equations to be solved

Variable length arrays
A Coefficient matrix of the equations
B right-hand side vector

PARAMETER *restriction*
IN ≥ N

```
      PROGRAM P21
C
C      PROGRAM 2.1 GAUSSIAN ELIMINATION
C
C      ALTER NEXT LINE TO CHANGE PROBLEM SIZE
C
      PARAMETER (IN=20)
C
      REAL A(IN,IN),B(IN)
C
      READ (5,*) N
```

```
      READ (5,*) ((A(I,J),J=1,N),I=1,N)
      READ (5,*) (B(I),I=1,N)
      WRITE(6,*) ('************** GAUSSIAN ELIMINATION ************')
      WRITE(6,*)
      WRITE(6,*) ('COEFFICIENT MATRIX')
      CALL PRINTA(A,IN,N,N,6)
      WRITE(6,*)
      WRITE(6,*) ('RIGHT HAND SIDE VECTOR')
      CALL PRINTV(B,N,6)
      WRITE(6,*)
C      CONVERT TO UPPER TRIANGULAR FORM
      DO 1 K = 1,N - 1
          IF (ABS(A(K,K)).GT.1.E-6) THEN
              DO 2 I = K + 1,N
                  X = A(I,K)/A(K,K)
                  DO 3 J = K + 1,N
                      A(I,J) = A(I,J) - A(K,J)*X
    3             CONTINUE
                  B(I) = B(I) - B(K)*X
    2         CONTINUE
          ELSE
              WRITE (6,*) ('ZERO PIVOT FOUND IN LINE:')
              WRITE (6,*) K
              STOP
          END IF
    1 CONTINUE
      WRITE(6,*) ('MODIFIED MATRIX')
      CALL PRINTA(A,IN,N,N,6)
      WRITE(6,*)
      WRITE(6,*) ('MODIFIED RIGHT HAND SIDE VECTOR')
      CALL PRINTV(B,N,6)
      WRITE(6,*)
C         BACK SUBSTITUTION
      DO 5 I = N,1,-1
          SUM = B(I)
          IF (I.LT.N) THEN
              DO 6 J = I + 1,N
                  SUM = SUM - A(I,J)*B(J)
    6         CONTINUE
          END IF
          B(I) = SUM/A(I,I)
    5 CONTINUE
C         PRINT THE RESULTS
      WRITE(6,*) ('SOLUTION VECTOR')
      CALL PRINTV(B,N,6)
      STOP
      END
```

In eqs 2.4 and 2.5 it can be seen that the elements of the original arrays **A** and **b** are progressively altered during the calculation. In Program 2.1, once the terms in **x** have been calculated, they are stored in **b** since the original **b** has been lost anyway.

In passing it should be noted that eqs 2.4 involve division by the coefficient a_{11} (equal to 10 in this case) while eqs 2.5 involve division by the *modified* coefficient a_{22} (equal to 5 in this case).

These coefficients a_{kk} where $1 \leq k < n$ are called the 'pivots' and it will be clear that they might be zero, either at the beginning of the elimination process, or during it. We shall return to this problem later, but for the moment shall merely check whether a_{kk} is or has become zero and stop the calculation if this is so. Input details are shown in Fig. 2.1(a) with output in Fig. 2.1(b).

The program begins by reading in N, **A** and **b**. In the section commented 'convert to upper triangular form' the check is first made to see if a_{kk} is greater than 'zero' (a small number in this case). Rows 2 to N are then processed according to eqs 2.4 and 2.5 and the modified **A** and **b** printed out for comparison with eqs 2.6. Note that only the numbers in the upper triangular

Number of equations	N		
	3		
Coefficient matrix	(A(I, J), J = 1, N), I = 1, N		
	10.0	1.0	−5.0
	−20.0	3.0	20.0
	5.0	3.0	5.0
Right-hand side	B(I), I = 1, N		
	1.0	2.0	6.0

Fig. 2.1(a) Input data for Program 2.1

```
************** GAUSSIAN ELIMINATION ************

COEFFICIENT MATRIX
    .1000E+02   .1000E+01  -.5000E+01
   -.2000E+02   .3000E+01   .2000E+02
    .5000E+01   .3000E+01   .5000E+01

RIGHT HAND SIDE VECTOR
    .1000E+01   .2000E+01   .6000E+01

MODIFIED MATRIX
    .1000E+02   .1000E+01  -.5000E+01
   -.2000E+02   .5000E+01   .1000E+02
    .5000E+01   .2500E+01   .2500E+01

MODIFIED RIGHT HAND SIDE VECTOR
    .1000E+01   .4000E+01   .3500E+01

SOLUTION VECTOR
    .1000E+01  -.2000E+01   .1400E+01
```

Fig. 2.1(b) Results from Program 2.1

part of **A** are of any significance. The back-substitution calculation is then performed, leaving the original unknowns **x** stored in **b** which is printed out.

2.2.1 Observations on the elimination process

In Program 2.1, some terms below the diagonal in matrix **U** (eq. 2.7) have been computed even although they are known in advance to be 'zero'. This is obviously work which need not be done. Further, during the conversion of **A** to upper triangular form, it was necessary to operate also on **b**. Therefore if equations with the same coefficients **A** have to be solved for different **b**, which are not known in advance as is often the case, the conversion of **A** to triangular form would be necessary for every **b** using this method.

We therefore seek a way of implementing Gaussian elimination so that multiple right-hand side **b** vectors can be processed after only a single 'decomposition' of **A** to triangular form. Such methods involve 'factorisation' of **A** into triangular matrix components. For example, it can be shown that matrix **A** can always be written as the product

$$\mathbf{A} = \mathbf{L}\mathbf{U} \tag{2.10}$$

where **L** is a lower triangular matrix and **U** an upper triangular matrix, in the forms

$$\mathbf{L}=\begin{bmatrix} l_{11} & 0 & 0 \\ l_{21} & l_{22} & 0 \\ l_{31} & l_{32} & l_{33} \end{bmatrix} \tag{2.11}$$

and

$$\mathbf{U}=\begin{bmatrix} u_{11} & u_{12} & u_{13} \\ 0 & u_{22} & u_{23} \\ 0 & 0 & u_{33} \end{bmatrix} \tag{2.12}$$

The numbers l_{kk} and u_{kk} are arbitrary except that their product is known. For example

$$l_{11}u_{11}=a_{11} \tag{2.13}$$

It is conventional to assume that either l_{kk} *or* u_{kk} are unity, hence, typically

$$\begin{bmatrix} a_{11} & a_{12} & a_{13} \\ a_{21} & a_{22} & a_{23} \\ a_{31} & a_{32} & a_{33} \end{bmatrix}=\begin{bmatrix} 1 & 0 & 0 \\ l_{21} & 1 & 0 \\ l_{31} & l_{32} & 1 \end{bmatrix}\begin{bmatrix} u_{11} & u_{12} & u_{13} \\ 0 & u_{22} & u_{23} \\ 0 & 0 & u_{33} \end{bmatrix} \tag{2.14}$$

is a usual statement of **LU** factorisation which is elaborated in the next section.

2.3 Equation solution using LU factorisation

When the triangular factors **L** and **U** in eqs 2.10 and 2.14 have been computed, equation solution proceeds as follows:

$$\mathbf{Ax}=\mathbf{b} \tag{2.15}$$

or $\quad \mathbf{LUx}=\mathbf{b}$

We now let

$$\mathbf{Ux}=\mathbf{y} \tag{2.16}$$

Hence

$$\mathbf{Ly}=\mathbf{b} \tag{2.17}$$

Since **L** and **b** are known, and **L** does not depend on **b**, this process is simply the 'forward-substitution' we saw in eq. 2.9. Once eq. 2.17 has been solved for **y**, eq. 2.16 is then the 'back-substitution' described by eq. 2.7. A solution algorithm will therefore consist of three phases, namely a factorisation (eq. 2.15) followed by a forward-substitution (eq. 2.17) and a back-substitution (eq. 2.16). The procedures of factorisation and forward- and back-substitution will be used in other contexts in this book and elsewhere, so that it makes sense to code them as library subroutines. They are called LUFAC, SUBFOR and SUBBAC respectively and their actions and parameters are described in Appendix 1 with full listings in Appendix 2.

Equations 2.14 are evaluated as follows:

Row 1 $u_{11}=a_{11}$, $u_{12}=a_{12}$, $u_{13}=a_{13}$.

This shows that with unity on the diagonal of **L**, the first row of **U** is simply a copy of the first row of **A**. Subroutine LUFAC therefore begins by nulling **L** (called LOWTRI) and **U** (called UPTRI) and by copying the first row of **A** into UPTRI.

Row 2 $l_{21}u_{11}=a_{21}$ $\therefore l_{21}=\dfrac{a_{21}}{u_{11}}$

Having found l_{21}, u_{22} and u_{23} can be computed since

$l_{21}u_{12}+u_{22}=a_{22}$ $\therefore u_{22}=a_{22}-l_{21}u_{12}$

and

$l_{21}u_{13}+u_{23}=a_{23}$ $\therefore u_{23}=a_{23}-l_{21}u_{13}$

Row 3 $l_{31}u_{11}=a_{31}$ $\therefore l_{31}=\dfrac{a_{31}}{u_{11}}$

$l_{31}u_{12}+l_{32}u_{22}=a_{32}$ $\therefore l_{32}=\dfrac{a_{32}-l_{31}u_{12}}{u_{22}}$

Having found l_{31} and l_{32}, u_{33} can be computed from

$l_{31}u_{13}+l_{32}u_{23}+u_{33}=a_{33}$ $\therefore u_{33}=a_{33}-l_{31}u_{13}-l_{32}u_{23}$

Subroutine LUFAC carries out these operations in two parts, commented 'lower triangular components' and 'upper triangular components' respectively. A 'zero' pivot is tested for in the same way as was done in Program 2.1.

Example 2.2

Use **LU** factorisation to solve the following set of equations:

$$
\begin{aligned}
2x_1 - 3x_2 + x_3 &= 7\\
x_1 - x_2 - 2x_3 &= -2\\
3x_1 + x_2 - x_3 &= 0
\end{aligned}
$$

Solution 2.2

Factorise the coefficient matrix into upper and lower triangular matrices, hence

$\mathbf{A}=\mathbf{LU}$

$$
\begin{bmatrix} 2 & -3 & 1\\ 1 & -1 & -2\\ 3 & 1 & -1 \end{bmatrix} = \begin{bmatrix} 1 & 0 & 0\\ l_{21} & 1 & 0\\ l_{31} & l_{32} & 1 \end{bmatrix} \begin{bmatrix} u_{11} & u_{12} & u_{13}\\ 0 & u_{22} & u_{23}\\ 0 & 0 & u_{33} \end{bmatrix}
$$

solving for l_{ij} and u_{ij} gives

$$\mathbf{L}=\begin{bmatrix} 1 & 0 & 0 \\ 0.5 & 1 & 0 \\ 1.5 & 11 & 1 \end{bmatrix}$$

$$\mathbf{U}=\begin{bmatrix} 2 & -3 & 1 \\ 0 & 0.5 & -2.5 \\ 0 & 0 & 25 \end{bmatrix}$$

Forward-substitution:

$\mathbf{Ly}=\mathbf{b}$

$$\begin{bmatrix} 1 & 0 & 0 \\ 0.5 & 1 & 0 \\ 1.5 & 11 & 1 \end{bmatrix}\begin{Bmatrix} y_1 \\ y_2 \\ y_3 \end{Bmatrix}=\begin{Bmatrix} 7 \\ -2 \\ 0 \end{Bmatrix}$$

$$\therefore \quad y_1=7,\ y_2=-2\ -0.5(7)=-5.5$$
$$y_3=-1.5(7)+11(5.5)=50.$$

Back-substitution:

$\mathbf{Ux}=\mathbf{y}$

$$\begin{bmatrix} 2 & -3 & 1 \\ 0 & 0.5 & -2.5 \\ 0 & 0 & 25 \end{bmatrix}\begin{Bmatrix} x_1 \\ x_2 \\ x_3 \end{Bmatrix}=\begin{Bmatrix} 7 \\ -5.5 \\ 50 \end{Bmatrix}$$

$$\therefore \quad x_3=2,\ x_2=(-5.5+2.5(2))/0.5=-1$$
$$x_1=(7-2+3(-1))/2=1$$

Program 2.2. Equation solution by **LU** factorisation
The program which implements the process is listed as Program 2.2. It involves the nomenclature:

Simple variables
N number of equations to be solved

Variable length arrays
A coefficient matrix of the equations
B right-hand side vector
UPTRI upper triangular factor of A
LOWTRI lower triangular factor of A

PARAMETER *restriction*
IN ≥ N.

```
      PROGRAM P22
C
C      PROGRAM 2.2 GAUSSIAN ELIMINATION USING L*U FACTORISATION
C
C      ALTER NEXT LINE TO CHANGE PROBLEM SIZE
C
      PARAMETER (IN=20)
C
      REAL A(IN,IN),UPTRI(IN,IN),LOWTRI(IN,IN),B(IN)
C
      READ (5,*) N
      READ (5,*) ((A(I,J),J=1,N),I=1,N)
      READ (5,*) (B(I),I=1,N)
      WRITE(6,*) ('** GAUSSIAN ELIMINATION USING LU FACTORISATION **')
      WRITE(6,*)
      WRITE(6,*) ('COEFFICIENT MATRIX')
      CALL PRINTA(A,IN,N,N,6)
      WRITE(6,*)
      WRITE(6,*) ('RIGHT HAND SIDE VECTOR')
      CALL PRINTV(B,N,6)
      WRITE(6,*)
      CALL LUFAC(A,UPTRI,LOWTRI,IN,N)
      WRITE(6,*) ('UPPER TRIANGULAR FACTORS')
      CALL PRINTA(UPTRI,IN,N,N,6)
      WRITE(6,*)
      WRITE(6,*) ('LOWER TRIANGULAR FACTORS')
      CALL PRINTA(LOWTRI,IN,N,N,6)
      WRITE(6,*)
      CALL SUBFOR(LOWTRI,IN,B,N)
      CALL SUBBAC(UPTRI,IN,B,N)
      WRITE(6,*) ('SOLUTION VECTOR')
      CALL PRINTV(B,N,6)
      STOP
      END
```

Input data are as in Fig. 2.2(a), with output in Fig. 2.2(b).

The program simply consists of reading in N, **A** and **b** followed by three subroutine calls to LUFAC, SUBFOR and SUBBAC. The lower and upper triangular factors are printed out, followed by the solution which has overwritten the original right-hand side in **b**.

Number of equations	N		
	3		
Coefficient matrix	(A(I, J), J = 1, N), I = 1, N		
	10.0	1.0	−5.0
	−20.0	3.0	20.0
	5.0	3.0	5.0
Right-hand side	B(I), I = 1, N		
	1.0	2.0	6.0

Fig. 2.2(a) Input data for Program 2.2

```
** GAUSSIAN ELIMINATION USING LU FACTORISATION **

COEFFICIENT MATRIX
   .1000E+02   .1000E+01  -.5000E+01
  -.2000E+02   .3000E+01   .2000E+02
   .5000E+01   .3000E+01   .5000E+01

RIGHT HAND SIDE VECTOR
   .1000E+01   .2000E+01   .6000E+01
```

```
UPPER TRIANGULAR FACTORS
    .1000E+02    .1000E+01   -.5000E+01
    .0000E+00    .5000E+01    .1000E+02
    .0000E+00    .0000E+00    .2500E+01

LOWER TRIANGULAR FACTORS
    .1000E+01    .0000E+00    .0000E+00
   -.2000E+01    .1000E+01    .0000E+00
    .5000E+00    .5000E+00    .1000E+01

SOLUTION VECTOR
    .1000E+01   -.2000E+01    .1400E+01
```

Fig. 2.2(b) Results from Program 2.2

2.3.1 Observations on the solution process by factorisation

Comparison of outputs in Figs 2.1(b) and 2.2(b) will show that the upper triangular factor of **A** in Fig. 2.2(b) is precisely the same as the modified upper triangular part of **A** in Fig. 2.1(b). Thus Programs 2.1 and 2.2 have much in common. However, were many **b** vectors to be processed, it would merely be necessary to call LUFAC once in Program 2.2 and to create a small loop reading in each new **b** and calling SUBFOR and SUBBAC to produce the solutions. Since the time taken in LUFAC is substantially more than that taken in SUBFOR and SUBBAC, this yields great economies as N increases.

Inspection of the arithmetic in LUFAC will show that storage could be saved by overwriting **A** by LOWTRI and UPTRI, and this will be done in subsequent programs. It has not been implemented in this first factorisation program in an attempt to make the computational process as clear as possible.

2.4 Equations with symmetrical coefficient matrix

If the coefficients of the **A** matrix satisfy the condition

$$a_{ij}=a_{ji} \tag{2.18}$$

that matrix is said to be symmetrical. For example the matrix

$$\mathbf{A}=\begin{bmatrix}16 & 4 & 8\\ 4 & 5 & -4\\ 8 & -4 & 22\end{bmatrix} \tag{2.19}$$

has symmetrical coefficients. If subroutine LUFAC is used to factorise this matrix, the result will be found to be

$$\mathbf{L}=\begin{bmatrix}1 & 0 & 0\\ 0.25 & 1 & 0\\ 0.5 & -1.5 & 1\end{bmatrix} \tag{2.20}$$

and

$$\mathbf{U}=\begin{bmatrix} 16 & 4 & 8 \\ 0 & 4 & -6 \\ 0 & 0 & 9 \end{bmatrix} \tag{2.21}$$

If the rows of $\mathbf{U}$ are then divided by u_{kk}, we have

$$\mathbf{U}^1=\begin{bmatrix} 1 & 0.25 & 0.5 \\ 0 & 1 & -1.5 \\ 0 & 0 & 1 \end{bmatrix} \tag{2.22}$$

and it can be seen that $l_{ij}=u^1_{ji}$. The scaling of $\mathbf{U}$ to $\mathbf{U}^1$ is accomplished in matrix terms by

$$\mathbf{U}=\mathbf{D}\mathbf{U}^1 \tag{2.23}$$

where $\mathbf{D}$ is the diagonal matrix:

$$\mathbf{D}=\begin{bmatrix} 16 & 0 & 0 \\ 0 & 4 & 0 \\ 0 & 0 & 9 \end{bmatrix} \tag{2.24}$$

Thus, if $\mathbf{A}$ is a symmetrical matrix, we can write

$$\mathbf{A}=\mathbf{L}\mathbf{U}=\mathbf{L}\mathbf{D}\mathbf{U}^1=\mathbf{L}\mathbf{D}\mathbf{L}^{\mathrm{T}}. \tag{2.25}$$

Since the terms in $\mathbf{L}^{\mathrm{T}}$ (the 'transpose') can be inferred from the terms in $\mathbf{L}$, it will be sufficient to compute only $\mathbf{L}$ (or $\mathbf{L}^{\mathrm{T}}$), involving approximately half the work in the $\mathbf{LU}$ factorisation of unsymmetrical matrices.

Example 2.3

Use $\mathbf{LDL}^{\mathrm{T}}$ factorisation to solve the symmetric equations

$$\begin{aligned} 3x_1 - 2x_2 + x_3 &= 3 \\ -2x_1 + 3x_2 + 2x_3 &= -3 \\ x_1 + 2x_2 + 2x_3 &= 2 \end{aligned}$$

Solution 2.3

Factorise the coefficient matrix into upper and lower triangular matrices, hence

$$\mathbf{A}=\mathbf{L}\mathbf{U}$$

$$\begin{bmatrix} 3 & -2 & 1 \\ -2 & 3 & 2 \\ 1 & 2 & 2 \end{bmatrix}=\begin{bmatrix} 1 & 0 & 0 \\ l_{21} & 1 & 0 \\ l_{31} & l_{32} & 1 \end{bmatrix}\begin{bmatrix} u_{11} & u_{12} & u_{13} \\ 0 & u_{22} & u_{23} \\ 0 & 0 & u_{33} \end{bmatrix}$$

Solving for l_{ij} and u_{ij} gives

$$\mathbf{L}=\begin{bmatrix} 1 & 0 & 0 \\ -0.667 & 1 & 0 \\ 0.333 & 1.6 & 1 \end{bmatrix}$$

$$\mathbf{U}=\begin{bmatrix} 3 & -2 & 1 \\ 0 & 1.667 & 2.667 \\ 0 & 0 & -2.6 \end{bmatrix}$$

Divide each row of **U** the diagonal term

$$\therefore \qquad \mathbf{U}^1=\begin{bmatrix} 1 & -0.667 & 0.333 \\ 0 & 1 & 1.6 \\ 0 & 0 & 1 \end{bmatrix}=\mathbf{L}^{\mathrm{T}}$$

where

$$\mathbf{D}=\begin{bmatrix} 3 & 0 & 0 \\ 0 & 1.667 & 0 \\ 0 & 0 & -2.6 \end{bmatrix}$$

Hence

$$\mathbf{U}=\mathbf{D}\mathbf{U}^1=\mathbf{D}\mathbf{L}^{\mathrm{T}}$$

and $\qquad \mathbf{L}\mathbf{D}\mathbf{L}^{\mathrm{T}}\mathbf{x}=\mathbf{b}$

Forward-substitution:

$$\mathbf{L}\mathbf{y}=\mathbf{b}$$

$$\therefore \qquad \begin{bmatrix} 1 & 0 & 0 \\ -0.667 & 1 & 0 \\ 0.333 & 1.6 & 1 \end{bmatrix}\begin{Bmatrix} y_1 \\ y_2 \\ y_3 \end{Bmatrix}=\begin{Bmatrix} 3 \\ -3 \\ 2 \end{Bmatrix}$$

$$\therefore \qquad \begin{aligned} &y_1=3,\ y_2=-3+3(0.667)=-1 \\ &y_3=2-0.333(3)+1.6=2.6 \end{aligned}$$

Back-substitution:

$$\mathbf{D}\mathbf{L}^{\mathrm{T}}\mathbf{x}=\mathbf{y}$$

$$\begin{bmatrix} 3 & -2 & 1 \\ 0 & 1.667 & 2.667 \\ 0 & 0 & -2.6 \end{bmatrix}\begin{Bmatrix} x_1 \\ x_2 \\ x_3 \end{Bmatrix}=\begin{Bmatrix} 3 \\ -1 \\ 2.6 \end{Bmatrix}$$

$$x_3 = -1,\ x_2 = (-1 + 2.667)/1.667 = 1$$
$$\therefore \quad x_1 = (3 + 1 + 2)/3 = 2$$

Program 2.3. Symmetrical equation solution by $\mathbf{LDL}^{\mathrm{T}}$ factorisation

Program 2.3 enables the solution of systems of equations with symmetrical coefficients using $\mathbf{LDL}^{\mathrm{T}}$ factorisation. It resembles Program 2.2 very closely with the new decomposition routine, LDLT replacing LUFAC. However, this time storage economies are made by overwriting the original **A** by its factor ($\mathbf{U}^{\mathrm{T}}$ in this case).

The nomenclature used in the program is as follows:

Simple variables

N Number of equations to be solved

Variable length arrays

A Symmetrical coefficient matrix (overwritten by its upper triangular factor $\mathbf{U}^{\mathrm{T}}$)
D Diagonal terms for eq. 2.24 stored as a vector
B Right-hand side vector (overwritten by the solution).

PARAMETER *restriction*

$IN \geq N$

```
C
C       PROGRAM 2.3 GAUSSIAN ELIMINATION
C       USING A=L*D*LT FACTORISATION LT OVERWRITES A
C
C       ALTER NEXT LINE TO CHANGE PROBLEM SIZE
C
      PARAMETER (IN=20)
C
      REAL A(IN,IN),D(IN),B(IN)
C
      READ (5,*) N
      READ (5,*) ((A(I,J),J=1,N),I=1,N)
      READ (5,*) (B(I),I=1,N)
      WRITE(6,*) ('** GAUSSIAN ELIMINATION USING LDLT FACTORISATION **')
      WRITE(6,*) ('************ SYMMETRICAL EQUATIONS ****************')
      WRITE(6,*)
      WRITE(6,*) ('COEFFICIENT MATRIX')
      CALL PRINTA(A,IN,N,N,6)
      WRITE(6,*)
      WRITE(6,*) ('RIGHT HAND SIDE VECTOR')
      CALL PRINTV(B,N,6)
      WRITE(6,*)
      CALL LDLT(A,IN,D,N)
      WRITE(6,*) ('UT MATRIX')
      DO 1 I=1,N
    1 WRITE(6,100) (A(I,J),J=1,I)
      WRITE(6,*)
      WRITE(6,*) ('DIAGONAL TERMS')
      CALL PRINTV(D,N,6)
      CALL LDLFOR(A,IN,B,N)
      DO 2 I = 1,N
         DO 2 J = 1,N
    2 A(I,J) = A(I,J)/D(I)
      CALL SUBBAC(A,IN,B,N)
      WRITE(6,*)
      WRITE(6,*) ('SOLUTION VECTOR')
      CALL PRINTV(B,N,6)
  100 FORMAT(5E12.4)
      STOP
      END
```

Number of equations	N		
	3		
Coefficient matrix	(A(I, J), J=1, N), I=1, N		
	16.0	4.0	8.0
	4.0	5.0	−4.0
	8.0	−4.0	22.0
Right-hand side	B(I), I=1, N		
	4.0	2.0	5.0

Fig. 2.3(a) Input data for Program 2.3

```
** GAUSSIAN ELIMINATION USING LDLT FACTORISATION **
************ SYMMETRICAL EQUATIONS ****************

COEFFICIENT MATRIX
    .1600E+02    .4000E+01    .8000E+01
    .4000E+01    .5000E+01   -.4000E+01
    .8000E+01   -.4000E+01    .2200E+02

RIGHT HAND SIDE VECTOR
    .4000E+01    .2000E+01    .5000E+01

UT MATRIX
   .1600E+02
   .4000E+01    .4000E+01
   .8000E+01   -.6000E+01    .9000E+01

DIAGONAL TERMS
    .1600E+02    .4000E+01    .9000E+01

SOLUTION VECTOR
   -.2500E+00    .1000E+01    .5000E+00
```

Fig. 2.3(b) Results from Program 2.3

The input data are in Fig. 2.3(a) and the output results in Fig. 2.3(b). The program reads in the number of equations to be solved, the full (symmetrical) coefficient matrix and the right-hand side vector.

Subroutine LDLT forms the matrix $\mathbf{U}^T$ in eq. 2.21, which overwrites **A**, and the diagonal terms of matrix **D** in eq. 2.24 which are stored in a vector **D**. Both **A** and **D** are printed out.

In the program, forward substitution is then accomplished using the special subroutine LDLFOR because $\mathbf{U}^T$ is taking the place of **L**. Matrix $\mathbf{U}^T$ can then be scaled, using **D**, to give $\mathbf{U}^1$ as in eq. 2.22. Conventional back-substitution using SUBBAC completes the process and the results are printed.

A useful by-product of this factorisation is that the determinant of the coefficient matrix **A** can be found as the product of the components of **D**, that is $16\times4\times9$ in this case, or 576.

2.4.1 Quadratic form and positive definiteness

A 'quadratic form' is a second degree expression in n variables of the form

$$
\begin{aligned}
Q(x)=a_{11}x_1^2+2a_{12}x_1x_2+\cdots+2a_{1n}x_1x_n \\
+\ a_{22}x_2^2\ +\cdots+2a_{2n}x_2x_n \\
+\cdots+ \\
+\ a_{nn}x_n^2.
\end{aligned}
$$

Usually, terms like $2a_{12}x_1x_2$ are separated into two terms

$$\begin{aligned}2a_{12}x_1x_2 &= a_{12}x_1x_2 + a_{12}x_2x_1 \\ &= a_{12}x_1x_2 + a_{21}x_2x_1\end{aligned}$$

so that a symmetrical matrix version can be written

$$\begin{aligned}Q(x) = {} & a_{11}x_1^2 + a_{12}x_1x_2 + \cdots + a_{1n}x_1x_n \\ & + a_{21}x_2x_1 + a_{22}x_2^2 + \cdots + a_{2n}x_2x_n \\ & \quad \vdots \qquad\qquad \vdots \qquad\qquad \vdots \\ & + a_{n1}x_nx_1 + a_{n2}x_nx_2 + \cdots + a_{nn}x_n^2\end{aligned}$$

This quadratic form is 'positive' if it is equal to or greater than zero for all values of its variables. A positive form which is zero only for the values $x_1 = x_2 = x_3 = \cdots = x_n = 0$ is said to be 'positive definite'. A positive quadratic form which is not positive definite is said to be 'positive semi-definite'.

With our usual definition of vectors and matrices

$$\mathbf{x} = \begin{Bmatrix} x_1 \\ x_2 \\ \\ x_n \end{Bmatrix}, \qquad \mathbf{A} = \begin{bmatrix} a_{11} & a_{12} \ldots a_{1n} \\ \vdots & \vdots \\ a_{n1} & \ldots a_{nn} \end{bmatrix}$$

the quadratic form can be written compactly as

$$Q(x) = \mathbf{x}^{\mathrm{T}}\,\mathbf{A}\mathbf{x}$$

where $\mathbf{A}$ is the 'matrix of the quadratic form $Q(x)$'. $Q(x)$ is 'singular' or 'nonsingular' if $\mathbf{A}$ is zero or nonzero respectively.

For the quadratic form $\mathbf{x}^{\mathrm{T}}\mathbf{A}\mathbf{x}$ to be positive definite, the quantities

$$a_{11}, \quad \begin{vmatrix} a_{11} & a_{12} \\ a_{21} & a_{22} \end{vmatrix}, \quad \begin{vmatrix} a_{11} & a_{12} & a_{13} \\ a_{21} & a_{22} & a_{23} \\ a_{31} & a_{32} & a_{33} \end{vmatrix}, \quad \begin{vmatrix} a_{11} & \ldots & a_{1n} \\ \vdots & & \vdots \\ a_{n1} & \ldots & a_{nn} \end{vmatrix}$$

must all be positive.

Example 2.4

Show that the quadratic form

$$\{x_1\,x_2\,x_3\} \begin{bmatrix} 1 & 2 & -2 \\ 2 & 5 & -4 \\ -2 & -4 & 5 \end{bmatrix} \begin{Bmatrix} x_1 \\ x_2 \\ x_3 \end{Bmatrix} = \begin{matrix} x_1^2 + 2x_1x_2 - 2x_1x_3 \\ 2x_2x_1 + 5x_2^2 - 4x_2x_3 \\ -2x_3x_1 - 4x_3x_2 + 5x_3^2 \end{matrix}$$

is positive definite.

Solution 2.4

The three quantities

$$1, \quad \begin{vmatrix} 1 & 2 \\ 2 & 5 \end{vmatrix} = 1 \quad \text{and} \quad \begin{vmatrix} 1 & 2 & -2 \\ 2 & 5 & -4 \\ -2 & -4 & 5 \end{vmatrix} = 1$$

all are positive. The quadratic form is

$$Q(x) = (x_1 + 2x_2 - 2x_3)^2 + x_2{}^2 + x_3{}^2$$

which can only be zero if $x_2 = x_3 = 0$ and hence $x_1 = 0$ as well.

2.4.2 Cholesky's method

When the coefficient matrix **A** is symmetric and positive definite, a slightly different factorisation can be obtained by forcing **U** to be the transpose of **L**. This factorisation can be written

$$\mathbf{A} = \mathbf{L}\mathbf{L}^{\mathrm{T}} \tag{2.26}$$

or

$$\begin{bmatrix} a_{11} & a_{12} & a_{13} \\ a_{21} & a_{22} & a_{23} \\ a_{31} & a_{32} & a_{33} \end{bmatrix} = \begin{bmatrix} l_{11} & 0 & 0 \\ l_{21} & l_{22} & 0 \\ l_{31} & l_{32} & l_{33} \end{bmatrix} \begin{bmatrix} l_{11} & l_{21} & l_{31} \\ 0 & l_{22} & l_{32} \\ 0 & 0 & l_{33} \end{bmatrix} \tag{2.27}$$

In this case $l_{11}^2 = a_{11}$ and so $l_{11} = \sqrt{a_{11}}$ and in each row a square root evaluation is necessary. For the **A** given by eq. 2.19 the Cholesky factors are

$$\mathbf{L}\mathbf{L}^{\mathrm{T}} = \begin{bmatrix} 4 & 0 & 0 \\ 1 & 2 & 0 \\ 2 & -3 & 3 \end{bmatrix} \begin{bmatrix} 4 & 1 & 2 \\ 0 & 2 & -3 \\ 0 & 0 & 3 \end{bmatrix} \tag{2.28}$$

This method is programmed in the next section, which deals with the situation when **A** has a special structure.

Example 2.5

Use Cholesky factorisation to solve the system of equations

$$\begin{aligned} 16x_1 + 4x_2 + 8x_3 &= 16 \\ 4x_1 + 5x_2 - 4x_3 &= 18 \\ 8x_1 - 4x_2 + 22x_3 &= -22 \end{aligned}$$

Solution 2.5

The coefficient matrix is positive definite, hence factorise as

$$\mathbf{A}=\mathbf{L}\mathbf{L}^{\mathrm{T}}$$

$$\begin{bmatrix} 16 & 4 & 8 \\ 4 & 5 & -4 \\ 8 & -4 & 22 \end{bmatrix} = \begin{bmatrix} l_{11} & 0 & 0 \\ l_{21} & l_{22} & 0 \\ l_{31} & l_{32} & l_{33} \end{bmatrix} \begin{bmatrix} l_{11} & l_{12} & l_{13} \\ 0 & l_{22} & l_{23} \\ 0 & 0 & l_{33} \end{bmatrix}$$

Solving for l_{ij} gives

$$\mathbf{L}=\begin{bmatrix} 4 & 0 & 0 \\ 1 & 2 & 0 \\ 2 & -3 & 3 \end{bmatrix}$$

Forward-substitution:

$$\mathbf{L}\mathbf{y}=\mathbf{b}$$

$$\begin{bmatrix} 4 & 0 & 0 \\ 1 & 2 & 0 \\ 2 & -3 & 3 \end{bmatrix} \begin{Bmatrix} y_1 \\ y_2 \\ y_3 \end{Bmatrix} = \begin{Bmatrix} 16 \\ 18 \\ -22 \end{Bmatrix}$$

$$\therefore \quad y_1=4,\ y_2=\tfrac{1}{2}(18-4)=7,\ y_3=\tfrac{1}{3}(-22+21-8)=-3$$

Back-substitution:

$$\mathbf{L}^{\mathrm{T}}\mathbf{x}=\mathbf{y}$$

$$\begin{bmatrix} 4 & 1 & 2 \\ 0 & 2 & -3 \\ 0 & 0 & 3 \end{bmatrix} \begin{Bmatrix} x_1 \\ x_2 \\ x_3 \end{Bmatrix} = \begin{Bmatrix} 4 \\ 7 \\ -3 \end{Bmatrix}$$

$$\therefore \quad x_3=-1,\ x_2=\tfrac{1}{2}(7-3)=2,\ x_1=\tfrac{1}{4}(4-2-+2)=1$$

2.5 Banded equations

In many engineering applications, equations have to be solved when the coefficients have a 'banded' structure (see, for example, Chapter 8). This means that the coefficients are clustered around the diagonal stretching from the top left-hand corner of the matrix

A to the bottom right-hand corner. A typical example is given below:

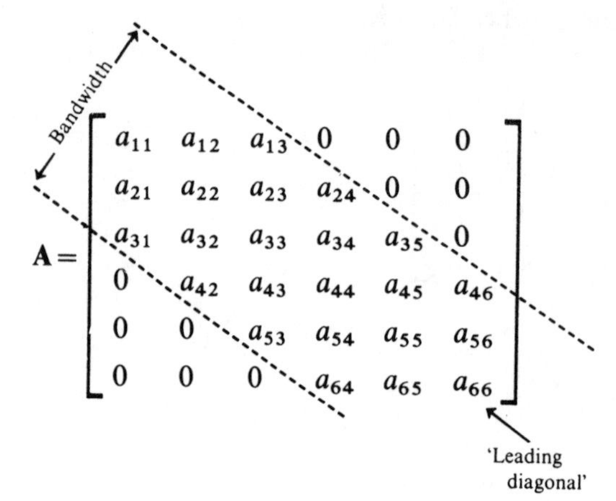

$$\mathbf{A}=\begin{bmatrix} a_{11} & a_{12} & a_{13} & 0 & 0 & 0 \\ a_{21} & a_{22} & a_{23} & a_{24} & 0 & 0 \\ a_{31} & a_{32} & a_{33} & a_{34} & a_{35} & 0 \\ 0 & a_{42} & a_{43} & a_{44} & a_{45} & a_{46} \\ 0 & 0 & a_{53} & a_{54} & a_{55} & a_{56} \\ 0 & 0 & 0 & a_{64} & a_{65} & a_{66} \end{bmatrix} \tag{2.29}$$

In this case there are never more than two non-zero coefficients to either side of the leading diagonal in any row. The 'bandwidth' of this system is said to be 5. If the coefficients are symmetrical, only the leading diagonal and two more coefficients per row need to be stored and operated on. In this case the 'half bandwidth' is said to be 2.

If we wish to store only the lower triangular factors of the symmetrical **A**, the coefficients of interest are

$$\mathbf{L}=\begin{bmatrix} l_{11} & 0 & 0 & 0 & 0 & 0 \\ l_{21} & l_{22} & 0 & 0 & 0 & 0 \\ l_{31} & l_{32} & l_{33} & 0 & 0 & 0 \\ 0 & l_{42} & l_{43} & l_{44} & 0 & 0 \\ 0 & 0 & l_{53} & l_{54} & l_{55} & 0 \\ 0 & 0 & 0 & l_{64} & l_{65} & l_{66} \end{bmatrix} \tag{2.30}$$

and it can be seen that out of 36 potential locations, only 15 are nonzero. A more economical method of storage is to keep the band in a rectangular array, by shifting the rows to obtain the structure

$$\mathbf{LB}=\begin{bmatrix} 0 & 0 & l_{11} \\ 0 & l_{21} & l_{22} \\ l_{31} & l_{32} & l_{33} \\ l_{42} & l_{43} & l_{44} \\ l_{53} & l_{54} & l_{55} \\ l_{64} & l_{65} & l_{66} \end{bmatrix} \tag{2.31}$$

This is still slightly inefficient in that zeros which are not required occur in the first two rows, but it has the advantage that there are three terms (the half bandwidth plus 1) in each row, which makes programming rather easier.

Program 2.4. Symmetrical equation solution by Cholesky's method
Program 2.4 illustrates the Cholesky solution process for symmetrical, banded equations. It makes use of two library routines, CHOFAC and CHOSUB which perform the $\mathbf{LL}^T$ factorisation and combined forward- and back-substitutions respectively. The steps followed are just eqs 2.15 to 2.17 again with $\mathbf{L}^T$ in place of $\mathbf{U}$.

```
      PROGRAM P24
C
C      PROGRAM 2.4 CHOLESKY FACTORISATION
C                  WITH BANDED STORAGE
C
C      ALTER NEXT LINE TO CHANGE PROBLEM SIZE
C
      PARAMETER (IN=20,IIW=10)
C
      REAL LB(IN,IIW),B(IN)
C
      READ (5,*) N,IW
      IWP1 = IW + 1
      READ (5,*) ((LB(I,J),J=1,IWP1),I=1,N)
      READ (5,*) (B(I),I=1,N)
      WRITE(6,*) ('*** CHOLESKY FACTORISATION WITH BANDED STORAGE ****')
      WRITE(6,*) ('************ SYMMETRICAL EQUATIONS ****************')
      WRITE(6,*)
      WRITE(6,*) ('BANDED COEFFICIENT MATRIX')
      DO 1 I=1,N
    1 WRITE(6,100) (LB(I,J),J=1,IWP1)
      WRITE(6,*)
      WRITE(6,*) ('RIGHT HAND SIDE VECTOR')
      CALL PRINTV(B,N,6)
      WRITE(6,*)
      CALL CHOFAC(LB,IN,N,IW)
      WRITE(6,*) ('LT IN BAND FORM')
      CALL PRINTA(LB,IN,N,IWP1,6)
      WRITE(6,*)
      CALL CHOSUB(LB,IN,B,N,IW)
      WRITE(6,*) ('SOLUTION VECTOR')
      CALL PRINTV(B,N,6)
  100 FORMAT(5E12.4)
      STOP
      END
```

Number of equations and half bandwidth	N 3	IW 2	
Coefficient matrix	(LB, (I, J), J=1, IW+1), I=1, N		
	0.0	0.0	16.0
	0.0	4.0	5.0
	8.0	−4.0	22.0
Right-hand side	B(I), I=1, N		
	4.0	2.0	5.0

Fig. 2.4(a) Input data for Program 2.4

```
*** CHOLESKY FACTORISATION WITH BANDED STORAGE ****
************ SYMMETRICAL EQUATIONS ****************

BANDED COEFFICIENT MATRIX
   .0000E+00    .0000E+00    .1600E+02
   .0000E+00    .4000E+01    .5000E+01
   .8000E+01   -.4000E+01    .2200E+02

RIGHT HAND SIDE VECTOR
    .4000E+01    .2000E+01    .5000E+01

LT IN BAND FORM
    .0000E+00    .0000E+00    .4000E+01
    .0000E+00    .1000E+01    .2000E+01
    .2000E+01   -.3000E+01    .3000E+01

SOLUTION VECTOR
   -.2500E+00    .1000E+01    .5000E+00
```

Fig. 2.4(b) Results from Program 2.4

Input data are listed in Fig. 2.4(a) and output results in Fig. 2.4(b). The nomenclature is as follows:

Simple variables

N Number of equations to be solved

IW Half bandwidth of symmetrical banded coefficient matrix

Variable size arrays

LB Lower triangular components of **A** stored as a rectangular band matrix with leading diagonal terms A(K, K) stored in LB (K, IW + 1).

B Right-hand side vector – overwritten by solution.

PARAMETER *restrictions*

IN ≥ N

IIW ≥ IW

The program is extremely simple given the subroutine library. The number of equations and half bandwidth are read in, followed by the lower 'half' of **A** in the appropriate form of eq. 2.31, and then the resulting factor is printed out. A call to CHOSUB completes the substitution phase and the results, held in **B**, are printed.

2.6 Compact storage for variable band widths

It is quite common to encounter symmetrical equation coefficient matrices which have the following structure:

Skyline

$$\mathbf{A} = \begin{bmatrix} a_{11} & 0 & 0 & 0 & a_{15} & 0 \\ 0 & a_{22} & a_{23} & 0 & 0 & 0 \\ 0 & a_{32} & a_{33} & 0 & a_{35} & a_{36} \\ 0 & 0 & 0 & a_{44} & a_{45} & 0 \\ a_{51} & 0 & a_{53} & a_{54} & a_{55} & a_{56} \\ 0 & 0 & a_{63} & 0 & a_{65} & a_{66} \end{bmatrix} \tag{2.32}$$

Half bandwidth

The system is banded, with a half bandwidth of 4, but also has a significant number of zeros within the band. If the progress of factorisation is monitored, it will be found that, in the lower triangle, zeros lying closer to the diagonal than a nonzero coefficient in that row will become nonzero during the calculation (will be 'filled in') whereas zeros lying further away from the diagonal than the outermost nonzero coefficient in that row will remain zero throughout the calculation and need not be stored or processed.

In the upper triangle, the same can be said of columns, and in eq. 2.32 a dashed line delineates the outermost extent of coefficients which need to be processed. Because of the appearance of this line in the upper triangle it is sometimes called the 'skyline' of the coefficients.

In the case of the coefficients in eq. 2.32 only 14 actually need be stored, compared with 30 in a fixed bandwidth method such as that used in Program 2.4. The penalty incurred by using a variable bandwidth technique is that a note must be kept of the number of coefficients to be processed in each row (column).

Program 2.5. 'Skyline' solution of symmetrical equations by Cholesky's method
Program 2.5 illustrates the procedure, using a Cholesky factorisation. The 14 components to be processed in eq. 2.32 would be stored in a vector as follows:

$$\mathbf{A}=(a_{11}\ a_{22}\ a_{32}\ a_{33}\ a_{44}\ a_{51} 0 a_{53}\ a_{54}\ a_{55}\ a_{63} 0 a_{65}\ a_{66}) \qquad (2.33)$$

```
      PROGRAM P25
C
C      PROGRAM 2.5 CHOLESKY FACTORISATION
C      SKYLINE STORAGE OF A AS A VECTOR
C
C      ALTER NEXT LINE TO CHANGE PROBLEM SIZE
C
      PARAMETER (IA=50,IB=20)
C
      REAL A(IA),B(IB)
      INTEGER KDIAG(IB)
C
      READ (5,*) N
      READ (5,*) (KDIAG(I),I=1,N)
      IR = KDIAG(N)
      READ (5,*) (A(I),I=1,IR)
      READ (5,*) (B(I),I=1,N)
      WRITE(6,*) ('** CHOLESKY FACTORISATION WITH SKYLINE STORAGE ***')
      WRITE(6,*) ('*********** SYMMETRICAL EQUATIONS ****************')
      WRITE(6,*)
      WRITE(6,*) ('COEFFICIENT VECTOR')
      CALL PRINTV(A,IR,6)
      WRITE(6,*)
      WRITE(6,*) ('DIAGONAL LOCATIONS')
      WRITE(6,100) (KDIAG(I),I=1,N)
      WRITE(6,*)
      WRITE(6,*) ('RIGHT HAND SIDE VECTOR')
      CALL PRINTV(B,N,6)
      WRITE(6,*)
      CALL SKYFAC(A,N,KDIAG)
      CALL SKYSUB(A,B,N,KDIAG)
      WRITE(6,*) ('SOLUTION VECTOR')
      CALL PRINTV(B,N,6)
  100 FORMAT(10I6)
      STOP
      END
```

The extra information about number of components per row (column) is kept in an integer vector called KDIAG. This could contain the number of components per row, or a running total, representing the 'address' of each leading diagonal component of the original array in **A**, which is the method used in this program. Thus the KDIAG associated with eqs 2.32 and 2.33 would be

$$\text{KDIAG}=\{1 \quad 2 \quad 4 \quad 5 \quad 10 \quad 14\} \tag{2.34}$$

reflecting that there are 1, 1, 2, 1, 5 and 4 coefficients to be processed in rows 1 to 6 respectively.

To test the program, we return to our familiar set of three simultaneous equations, leading to the input data of Fig. 2.5(a) and the output results of Fig. 2.5(b).

```
Number of equations    N
                       3

Skyline location       KDIAG(I), I=1, N
                        1     3     6

Coefficient vector     A(I), I=1, IR
                       16.0    4.0    5.0    8.0    -4.0    22.0

Right-hand side        B(I), I=1, N
                        4.0    2.0    5.0
```

Fig. 2.5(a) Input data for Program 2.5

```
** CHOLESKY FACTORISATION WITH SKYLINE STORAGE ***
*********** SYMMETRICAL EQUATIONS ****************

COEFFICIENT VECTOR
   .1600E+02   .4000E+01   .5000E+01   .8000E+01  -.4000E+01   .2200E+02

DIAGONAL LOCATIONS
    1    3    6

RIGHT HAND SIDE VECTOR
   .4000E+01   .2000E+01   .5000E+01

SOLUTION VECTOR
  -.2500E+00   .1000E+01   .5000E+00
```

Fig. 2.5(b) Results from Program 2.5

The nomenclature used is as follows:

Simple variables

N	The number of equations to be solved
IR	The total number of coefficients to be processed

Variable length arrays

A	Equation coefficients stored as a vector
B	Right-hand side – overwritten by the solution
KDIAG	'Address' vector holding positions of diagonal terms

PARAMETER *restrictions*

IA ≥ IR
IB ≥ N.

In the program, N is first read in, followed by the 'address' vector KDIAG as described in eq. 2.34.

Then the equation coefficients are read in, row by row, followed by the right-hand side vector. Calls to the 'skyline' factorisation and substitution routines SKYFAC and SKYSUB complete the solution process and the results, stored in B, can be printed.

2.7 Pivoting

In the solution of unsymmetrical equations using conventional Gaussian elimination (Program 2.1) or **LU** factorisation (Program 2.2) we side-stepped the problem of what to do should a leading diagonal component of the coefficient matrix be zero to start with, or become zero during the solution process. In the next program, we illustrate how to cope with this by using a row interchange technique usually called 'pivoting'. We saw that the crucial terms or pivots lay on the leading diagonal and so the technique involves searching for the largest (absolute) coefficient in the rows not so far processed, and moving it into the leading diagonal position.

For example, returning to eqs 2.3, it can be seen that the largest coefficients occur initially in row 2, and so this would be interchanged with row 1 to yield

$$\begin{aligned} 20x_3 + 3x_2 - 20x_1 &= 2 \\ -5x_3 + x_2 + 10x_1 &= 1 \\ 5x_3 + 3x_2 + 5x_1 &= 6 \end{aligned} \tag{2.35}$$

After one step of elimination (or factorisation) we would have

$$\begin{aligned} 20x_3 + 3x_2 - 20x_1 &= 2 \\ 1.75x_2 + 5x_1 &= 1.5 \\ 2.25x_2 + 10x_1 &= 5.5 \end{aligned} \tag{2.36}$$

The largest coefficient in rows 2 to N is now in row 3 and so this would be interchanged with row 2 to yield

$$\begin{aligned} 20x_3 - 20x_1 + 3x_2 &= 2 \\ 10x_1 + 2.25x_2 &= 5.5 \\ 5x_1 + 1.75x_2 &= 1.5 \end{aligned} \tag{2.37}$$

leading to the final factorisation

$$\begin{aligned} 20x_3 - 20x_1 + 3x_2 &= 2 \\ 10x_1 + 2.25x_2 &= 5.5 \\ 0.625x_2 &= -1.25 \end{aligned} \tag{2.38}$$

and the solution as before, but in the order

$(x_2, x_1, x_3) = (-2.0, 1.0, 1.4)$.

The process just described is actually called 'partial pivoting'. An extension could be made to 'total pivoting' by extending the search for the maximum modulus element to both rows and columns.

Program 2.6. Equation solution by **LU** factorisation with pivoting

In this program the interchanges are carried out in a special **LU** factorisation using subroutine LUPFAC. This stores the new row order in an integer array ROW for use in the substitution phase, carried out by subroutine LUPSOL. The factorisation arithmetic remains the same.

```
      PROGRAM P26
C
C      PROGRAM 2.6 LU FACTORISATION USING PIVOTING
C
C      ALTER NEXT LINE TO CHANGE PROBLEM SIZE
C
      PARAMETER (IN=20)
C
      REAL A(IN,IN),B(IN),X(IN)
      INTEGER ROW(IN)
C
      READ (5,*) N
      READ (5,*) ((A(I,J),J=1,N),I=1,N)
      READ (5,*) (B(I),I=1,N)
      WRITE(6,*) ('****** LU FACTORISATION WITH PIVOTING ******')
      WRITE(6,*)
      WRITE(6,*) ('COEFFICIENT MATRIX')
      CALL PRINTA(A,IN,N,N,6)
      WRITE(6,*)
      WRITE(6,*) ('RIGHT HAND SIDE VECTOR')
      CALL PRINTV(B,N,6)
      WRITE(6,*)
      CALL LUPFAC(A,IN,N,ROW)
      CALL LUPSOL(A,IN,B,X,N,ROW)
      WRITE(6,*) ('REORDERED ROW NUMBERS')
      WRITE(6,1000) (ROW(I),I=1,N)
      WRITE(6,*)
      WRITE(6,*) ('SOLUTION VECTOR')
      CALL PRINTV(X,N,6)
 1000 FORMAT (10I5)
      STOP
      END
```

Number of equations	N		
	3		
Coefficient matrix	(A(I, J), J = 1, N), I = 1, N		
	10.0	1.0	−5.0
	−20.0	3.0	20.0
	5.0	3.0	5.0
Right-hand side	B(I), I = 1, N		
	1.0	2.0	6.0

Fig. 2.6(a) Input data for Program 2.6

```
****** LU FACTORISATION WITH PIVOTING ******
COEFFICIENT MATRIX
   .1000E+02    .1000E+01   -.5000E+01
  -.2000E+02    .3000E+01    .2000E+02
   .5000E+01    .3000E+01    .5000E+01
RIGHT HAND SIDE VECTOR
   .1000E+01    .2000E+01    .6000E+01
REORDERED ROW NUMBERS
    2    3    1
SOLUTION VECTOR
   .1000E+01   -.2000E+01    .1400E+01
```

Fig. 2.6(b) Results from Program 2.6

Input data are shown in Fig. 2.6(a) and output results in Fig. 2.6(b).
The nomenclature used is as follows:

Simple variables
N The number of equations to be solved

Variable length arrays
A The matrix of coefficients (overwritten by the factors **L** and **U**).
B The right-hand side
X The solution
ROW The new row ordering

PARAMETER *restriction*
IN ≥ N.

The program reads in N, A and B, calls the factorisation and substitution routines and outputs the new row order and the solution.

For equations with unsymmetrical coefficients, pivoting is to be recommended. Even if a pivot does not become zero in a conventional elimination, a 'small' pivot is undesirable because of potential 'round-off' errors (see Chapter 1). Should the pivot become 'zero', the set of equations is singular or very nearly so. Fortunately, for equations with symmetrical coefficients which are 'positive definite' (see Section 2.3.1), pivoting is not necessary and Programs 2.3 to 2.5 can be used.

2.7.1 Ill-conditioning

Despite our best efforts to select optimum pivots, a set of linear equations may not be solvable accurately by any of the methods we have just described, especially when 'hand' calculation techniques are employed. When this is so, the set of equations is said to be 'ill-conditioned'. Fortunately, even modern electronic 'hand' calculators are capable of working to many decimal places of accuracy and so the conditioning of sets of equations is not of such great importance as it was when 'hand' calculations implied only a few decimal places of accuracy.

A very well-known example of ill-conditioned equations arises in the form of the 'Hilbert matrix' obtained from polynomial curve fitting. The set of equations so derived takes the form

$$\begin{bmatrix} \frac{1}{2} & \frac{1}{3} & \frac{1}{4} & \frac{1}{5} & \frac{1}{6} \\ \frac{1}{3} & \frac{1}{4} & \frac{1}{5} & \frac{1}{6} & \frac{1}{7} \\ \frac{1}{4} & \frac{1}{5} & \frac{1}{6} & \frac{1}{7} & \frac{1}{8} \\ \frac{1}{5} & \frac{1}{6} & \frac{1}{7} & \frac{1}{8} & \frac{1}{9} \\ \frac{1}{6} & \frac{1}{7} & \frac{1}{8} & \frac{1}{9} & \frac{1}{10} \end{bmatrix} \begin{Bmatrix} x_1 \\ x_2 \\ x_3 \\ x_4 \\ x_5 \end{Bmatrix} = \begin{Bmatrix} 1 \\ 1 \\ 1 \\ 1 \\ 1 \end{Bmatrix}$$

In Table 2.1, numerical solutions rounded off to 4, 5, 6, 7 and 8 decimal places are compared with the true solution. It can be seen that something like 8 decimal place

Table 2.1

True solution	$x_1=30$	$x_2=-420$	$x_3=1680$	$x_4=-2520$	$x_5=1260$
4 decimal places	−8.8	90.5	−196	58.9	75.6
5	55.3	−746.5	2854.9	−4105.2	1976.9
6	13.1	−229.9	1042.7	−1696.4	898.1
7	25.9	−373.7	1524.4	−2318.6	1171.4
8	29.65	−416.02	1666.62	−2502.69	1252.39

accuracy is necessary for an answer adequate for engineering purposes. When using 'mainframe' computers with 32-bit words, engineers are advised to use 'double precision', i.e. 64-bit words, which should achieve something like 15 decimal place accuracy and be sufficient for most practical purposes.

Although mathematical measures of ill-conditioning, called 'condition numbers' can be calculated, this process is expensive and not used in engineering practice. By far the best guides to solution accuracy are independent checks on the physical consistency of the results.

2.8 Equations with prescribed 'variables'

Consider a common engineering occurrence of the system of equations $\mathbf{Ax}=\mathbf{b}$ where $\mathbf{A}$ represents the stiffness of an elastic solid, $\mathbf{b}$ the forces applied to the solid and $\mathbf{x}$ the resulting displacements. In some cases we will know all the components of $\mathbf{b}$ and have to calculate all the components of $\mathbf{x}$, but in others we may be told that some of the displacements are known in advance. In this event we could eliminate the known components of $\mathbf{x}$ from the system of equations, but this involves some quite elaborate coding to reshuffle the rows and columns of $\mathbf{A}$. A simple scaling procedure can be used instead to produce the desired result without modifying the structure of $\mathbf{A}$.

Suppose that in our symmetrical system

$$\mathbf{Ax}=\begin{bmatrix} 16 & 4 & 8 \\ 4 & 5 & -4 \\ 8 & -4 & 22 \end{bmatrix}\begin{Bmatrix} x_1 \\ x_2 \\ x_3 \end{Bmatrix}=\mathbf{b} \tag{2.39}$$

We know that x_2 has the value 5, while b_1 and b_3 have the values 4 and 5 as before. We force x_2 to be 5 by scaling a_{22} up by adding a large number to it, say 10^{10}. The purpose of this is to make the diagonal term in row 2 much larger than the off-diagonal terms. That is, row 2 reads

$$4x_1+(5+10^{10})x_2-4x_3=b_2 \tag{2.40}$$

and the term in x_2 swamps the other two terms. To get the desired result of $x_2=5$, it is clear that on the right-hand side we should put the value $(5+10^{10})\times 5$. We could also use this technique to make x_2 'zero' by putting any number which is orders of magnitude less than 10^{10} in the right-hand side b_2 position.

Program 2.7. Equation solution with prescribed 'variables'
This program illustrates the process, using as a starting point Program 2.5. The input (see Fig. 2.7(a)) and program description are very similar to those for Program 2.5.

Additional data involve the number of values to be fixed (NFIX), followed by the particular number of the fixed variable (NO(I)) and the required value (VAL(I)). In the case considered here, one variable is to be fixed, namely the value x_2 which is equal to 5.

The leading diagonal terms in the compacted storage vector A can be located using KDIAG as shown and the appropriate entry in A has the number 10^{10} added to it. The appropriate term

```
      PROGRAM P27
C
C      PROGRAM 2.7 CHOLESKY FACTORISATION
C      WITH SKYLINE STORAGE
C      PRESCRIBED VARIABLES BY 'BIG SPRINGS'
C
C      ALTER NEXT LINE TO CHANGE PROBLEM SIZE
C
      PARAMETER (IA=50,IB=20,INO=5)
C
      REAL A(IA),B(IB),VAL(INO)
      INTEGER KDIAG(IB),NO(INO)
C
      READ (5,*) N
      READ (5,*) (KDIAG(I),I=1,N)
      IR = KDIAG(N)
      READ (5,*) (A(I),I=1,IR)
      READ (5,*) (B(I),I=1,N)
      READ (5,*) NFIX
      READ (5,*) (NO(I),VAL(I),I=1,NFIX)
      WRITE(6,*) ('** CHOLESKY FACTORISATION WITH SKYLINE STORAGE ***')
      WRITE(6,*) ('*********** SYMMETRICAL EQUATIONS ****************')
      WRITE(6,*) ('*********** PRESCRIBED VARIABLES *****************')
      WRITE(6,*)
      WRITE(6,*) ('COEFFICIENT VECTOR')
      CALL PRINTV(A,IR,6)
      WRITE(6,*)
      WRITE(6,*) ('DIAGONAL LOCATIONS')
      WRITE(6,100) (KDIAG(I),I=1,N)
      WRITE(6,*)
      WRITE(6,*) ('RIGHT HAND SIDE VECTOR')
      CALL PRINTV(B,N,6)
      WRITE(6,*)
      WRITE(6,*) ('FREEDOM NUMBER AND PRESCRIBED VALUE')
      DO 1 I=1,NFIX
    1 WRITE(6,200)NO(I),VAL(I)
      WRITE(6,*)
      DO 2 I=1,NFIX
      A(KDIAG(NO(I))) = A(KDIAG(NO(I))) + 1.E10
    2 B(NO(I)) = A(KDIAG(NO(I)))*VAL(I)
      CALL SKYFAC(A,N,KDIAG)
      CALL SKYSUB(A,B,N,KDIAG)
      WRITE(6,*) ('SOLUTION VECTOR')
      CALL PRINTV(B,N,6)
  100 FORMAT(10I6)
  200 FORMAT(I5,E12.4)
      STOP
      END
```

Number of equations	N 3
Skyline locations	KDIAG (I), I = 1, N 1 3 6
Coefficient vector	(A(I), I = 1, IR) 16.0 4.0 5.0 8.0 −4.0 22.0
Right-hand side	B(I), I = 1, N 4.0 2.0 5.0
Number of prescribed variables	NFIX 1
Number and magnitude of each	NO(I), VAL(I), I = 1, NFIX 2 5.0

Fig. 2.7(a) Input data for Program 2.7

```
** CHOLESKY FACTORISATION WITH SKYLINE STORAGE ***
*********** SYMMETRICAL EQUATIONS ****************
*********** PRESCRIBED VARIABLES *****************

COEFFICIENT VECTOR
   .1600E+02   .4000E+01   .5000E+01   .8000E+01  -.4000E+01   .2200E+02

DIAGONAL LOCATIONS
    1    3    6

RIGHT HAND SIDE VECTOR
   .4000E+01   .2000E+01   .5000E+01

FREEDOM NUMBER AND PRESCRIBED VALUE
   2    .5000E+01

SOLUTION VECTOR
  -.1917E+01   .5000E+01   .1833E+01
```

Fig. 2.7(b) Results from Program 2.7

in B is then set to the augmented diagonal multiplied by the desired x_2, namely 5.0. The results are listed in Fig. 2.7(b).

2.9 Iterative methods

In the previous sections of this chapter, 'direct' solution methods for systems of linear algebraic equations have been described. By 'direct' we meant that the solution method proceeded to the answer in a fixed number of arithmetic operations. Subject to rounding errors, the solutions were as 'exact' as the computer hardware permitted. There was no opportunity for intervention by the user in the solution process.

In contrast, this section deals with 'indirect' or iterative methods of solution. These proceed by the user first guessing an answer, which is then successively corrected iteration by iteration. Several methods will be described, which differ in the technique by which the corrections are made, and will be found to lead to different rates of convergence to the true solution.

The question whether convergence will be achieved at all is outside the scope of this book. In general it can be stated that equations with a coefficient matrix which is 'diagonally dominant' are likely to have convergent iterative solutions. Roughly defined, this means that the diagonal term in any row is greater than the sum of the off-diagonal terms. As we saw earlier, it is fortunate that in many engineering applications this diagonal dominance exists.

Since it is very unlikely that the user's starting guess will be accurate, the solution will normally become gradually closer and closer to the converged one, and the opportunity is afforded for user intervention in that a decision whether the solution is 'close enough' to the desired one is possible. For example, if the solution after an iteration has in some sense hardly changed from the previous iteration, the process can be terminated. Thus, in general terms, the number of arithmetic operations to reach a solution is not known in advance for these methods.

2.9.1 The iterative process

Suppose we are looking for an iterative method of solving the equations

$$\begin{aligned} 2x_1 + x_2 &= 4 \\ x_1 + 2x_2 &= 5 \end{aligned} \tag{2.41}$$

The central feature of iterative processes is to arrange to have the unknowns occurring on both sides of the equation. For example, we could write eqs 2.41 as

$$\begin{aligned} 2x_1 &= 4 - x_2 \\ 2x_2 &= 5 - x_1 \end{aligned} \tag{2.42}$$

and then make a guess at (x_1, x_2) on the right-hand side to see if it fits the left-hand side. For this purpose it would be more convenient to have the simple variables x_1 and x_2 on the left-hand side, so we can write

$$\begin{aligned} x_1 &= 2 \quad - 0.5\,x_2 \\ x_2 &= 2.5 - 0.5\,x_1 \end{aligned} \tag{2.43}$$

Now we make a guess at (x_1, x_2) on the right-hand side, say (1.0, 1.0). Equations 2.43 then yield

$$\begin{Bmatrix} x_1 \\ x_2 \end{Bmatrix} = \begin{Bmatrix} 1.5 \\ 2.0 \end{Bmatrix} \tag{2.44}$$

so the guess was wrong. However, if this simple iteration process were to converge, we could speculate that eqs 2.44 represent a better solution than the initial guess, and substitute it on the right-hand side of eqs 2.43.

This gives

$$\begin{Bmatrix} x_1 \\ x_2 \end{Bmatrix} = \begin{Bmatrix} 1.5 \\ 1.75 \end{Bmatrix} \tag{2.45}$$

and the process is continued, with the results

$$\begin{array}{c|ccccc} & \multicolumn{5}{c}{\text{Iteration number}} \\ & 0 & 1 & 2 & 3 & 4 \\ \begin{Bmatrix} x_1 \\ x_2 \end{Bmatrix} = & \begin{matrix} 1.0 \\ 1.0 \end{matrix} & \begin{matrix} 1.5 \\ 2.0 \end{matrix} & \begin{matrix} 1.0 \\ 1.75 \end{matrix} & \begin{matrix} 1.125 \\ 2.0 \end{matrix} & \begin{matrix} 1.0 \\ 1.9375 \end{matrix} \end{array} \tag{2.46}$$

In this trivial example, it is clear that the iterative process is leading to a convergence to the true solution $x_1=1.0$, $x_2=2.0$.

A general statement of eqs 2.43 for systems of equations $\mathbf{Ax}=\mathbf{b}$ is obtained by scaling the coefficients of $\mathbf{A}$ and $\mathbf{b}$ such that the leading diagonal terms of $\mathbf{A}$ are unity

$$\begin{bmatrix} 1 & a_{12} & a_{13} \\ a_{21} & 1 & a_{23} \\ a_{31} & a_{32} & 1 \end{bmatrix} \begin{Bmatrix} x_1 \\ x_2 \\ x_3 \end{Bmatrix} = \begin{Bmatrix} b_1 \\ b_2 \\ b_3 \end{Bmatrix} \tag{2.47}$$

The coefficient matrix may now be split into

$$\mathbf{A}=\mathbf{I}-\mathbf{L}-\mathbf{U} \tag{2.48}$$

in which

$$\mathbf{L}=\begin{bmatrix} 0 & 0 & 0 \\ -a_{21} & 0 & 0 \\ -a_{31} & -a_{32} & 0 \end{bmatrix} \tag{2.49}$$

and

$$\mathbf{U}=\begin{bmatrix} 0 & -a_{12} & -a_{13} \\ 0 & 0 & -a_{23} \\ 0 & 0 & 0 \end{bmatrix} \tag{2.50}$$

Note that these triangular matrices $\mathbf{L}$ and $\mathbf{U}$ have nothing to do with the '$\mathbf{LU}$' factors we met previously in direct methods. They are additive components of $\mathbf{A}$, not factors.

With these definitions, the system $\mathbf{Ax}=\mathbf{b}$ may be written

$$(\mathbf{I}-\mathbf{L}-\mathbf{U})\mathbf{x}=\mathbf{b} \tag{2.51}$$

or $$\mathbf{x}=\mathbf{b}+(\mathbf{L}+\mathbf{U})\mathbf{x} \tag{2.52}$$

in which the unknowns now appear on both sides of the equations, leading to the iterative scheme

$$\mathbf{x}^{k+1}=\mathbf{b}+(\mathbf{L}+\mathbf{U})\mathbf{x}^k \tag{2.53}$$

Example 2.6

Solve the equations

$$16x_1+4x_2+\ 8x_3=4$$
$$4x_1+5x_2-\ 4x_3=2$$
$$8x_1-4x_2+22x_3=5$$

by simple iteration.

Solution 2.6

Divide each equation by the diagonal term and rearrange as

$$\mathbf{x}^{k+1}=\mathbf{b}+(\mathbf{L}+\mathbf{U})\mathbf{x}^k$$

hence

$$\begin{Bmatrix}x_1\\x_2\\x_3\end{Bmatrix}^{k+1}=\begin{Bmatrix}0.25\\0.4\\0.2273\end{Bmatrix}+\begin{bmatrix}0 & -0.25 & -0.5\\-0.8 & 0 & 0.8\\-0.3636 & 0.1818 & 0\end{bmatrix}\begin{Bmatrix}x_1\\x_2\\x_3\end{Bmatrix}^{k}$$

let

$$\begin{Bmatrix}x_1\\x_2\\x_3\end{Bmatrix}^{0}=\begin{Bmatrix}1\\1\\1\end{Bmatrix}$$

hence

$$\begin{Bmatrix}x_1\\x_2\\x_3\end{Bmatrix}^{1}=\begin{Bmatrix}-0.5\\0.4\\0.0455\end{Bmatrix}$$

and

$$\begin{Bmatrix}x_1\\x_2\\x_3\end{Bmatrix}^{2}=\begin{Bmatrix}0.1273\\0.8364\\0.4818\end{Bmatrix}\rightarrow\begin{Bmatrix}-0.25\\1\\0.5\end{Bmatrix}\quad\text{(after many iterations)}$$

In this case, convergence is very slow (see Program 2.8). To operate this scheme on a computer, we can see that the library modules we shall need are merely a matrix-vector multiplication routine to compute $(\mathbf{L}+\mathbf{U})\mathbf{x}$ and a vector addition to add $\mathbf{b}$ to the result of the multiply. A check on convergence would also be helpful. This method is called 'Jacobi iteration'.

Program 2.8. Equation solution by Jacobi iteration
Our first program involving an iteration method uses the straightforward Jacobi technique just described. The chosen set of equations has a symmetric coefficient matrix, with a strong leading diagonal

$$\begin{aligned} 16x_1+4x_2+\ 8x_3&=4\\ 4x_1+5x_2-\ 4x_3&=2\\ 8x_1-4x_2+22x_3&=5. \end{aligned} \tag{2.54}$$

The convergence criterion adopted is a simple one, namely that the change in the largest component of the solution vector shall not be greater than a small proportion of that component, called the 'tolerance'.

```
      PROGRAM P28
C
C      PROGRAM 2.8 JACOBI ITERATION
C
C      ALTER NEXT LINE TO CHANGE PROBLEM SIZE
C
      PARAMETER (IN=20)
C
      REAL A(IN,IN),LPU(IN,IN),B(IN),X(IN),XNEW(IN),TEMP(IN)
C
      READ (5,*) N,((A(I,J),J=1,N),I=1,N)
      READ (5,*) (B(I),I=1,N)
      READ (5,*) (X(I),I=1,N)
      READ (5,*) TOL,ITS
      WRITE(6,*) ('*************** JACOBI ITERATION ****************')
      WRITE(6,*)
      WRITE(6,*) ('COEFFICIENT MATRIX')
      CALL PRINTA(A,IN,N,N,6)
      WRITE(6,*)
      WRITE(6,*) ('RIGHT HAND SIDE VECTOR')
      CALL PRINTV(B,N,6)
      WRITE(6,*)
      DO 1 I = 1,N
         DIAG = A(I,I)
         DO 2 J = 1,N
    2    A(I,J) = A(I,J)/DIAG
    1 B(I) = B(I)/DIAG
      CALL NULL(LPU,IN,N,N)
      DO 3 I = 1,N
         DO 3 J = 1,N
    3 IF (I.NE.J) LPU(I,J) = -A(I,J)
      WRITE(6,*) ('FIRST FEW ITERATIONS')
      ITERS = 0
    4 ITERS = ITERS + 1
      CALL MVMULT(LPU,IN,X,N,N,TEMP)
      CALL VECADD(B,TEMP,XNEW,N)
      CALL CHECON(XNEW,X,N,TOL,ICON)
      IF (ITERS.LE.5) CALL PRINTV(X,N,6)
      IF (ICON.EQ.0.AND.ITERS.LT.ITS) GO TO 4
      WRITE(6,*)
      WRITE(6,*) ('ITERATIONS TO CONVERGENCE')
      WRITE(6,*) ITERS
      WRITE(6,*)
      WRITE(6,*) ('SOLUTION VECTOR')
      CALL PRINTV(X,N,6)
      STOP
      END
```

Number of equation	N		
	3		
Coefficient matrix	(A(I, J), J = 1, N), I = 1, N		
	16.0	4.0	8.0
	4.0	5.0	−4.0
	8.0	−4.0	22.0
Right-hand side	B(I), I = 1, N		
	4.0	2.0	5.0
Initial guess of solution	X(I), I = 1, N		
	1.0	1.0	1.0
Tolerance	TOL		
	1.E−5		
Iteration limit	ITS		
	100		

Fig. 2.8(a) Input data for Program 2.8

```
*************** JACOBI ITERATION *****************

COEFFICIENT MATRIX
   .1600E+02   .4000E+01   .8000E+01
   .4000E+01   .5000E+01  -.4000E+01
   .8000E+01  -.4000E+01   .2200E+02

RIGHT HAND SIDE VECTOR
   .4000E+01   .2000E+01   .5000E+01

FIRST FEW ITERATIONS
  -.5000E+00   .4000E+00   .4545E-01
   .1273E+00   .8364E+00   .4818E+00
  -.2000E+00   .6836E+00   .3331E+00
  -.8744E-01   .8264E+00   .4243E+00
  -.1688E+00   .8094E+00   .4093E+00

ITERATIONS TO CONVERGENCE
51

SOLUTION VECTOR
  -.2500E+00   .1000E+01   .5000E+00
```

Fig. 2.8(b) Results from Program 2.8

The initial guessed solution is $(x_1, x_2, x_3) = (1.0, 1.0, 1.0)$. The program data are listed in Fig. 2.8(a). Apart from integer counters and temporary storage the variables used are

Simple variables

N	Number of equations to be solved
TOL	Convergence tolerance, defined above
ITERS	Number of iterations carried out
ITS	Maximum number of iterations

Variable length arrays

A	Coefficient matrix of equations
LPU	The matrix $\mathbf{L}+\mathbf{U}$ (eqs 2.53)
B	Right-hand side vector
X, XNEW	The solution to $\mathbf{Ax}=\mathbf{b}$ on successive iterations

PARAMETER *restrictions*
IN ≥ N.

The number of equations to be solved, the coefficient matrix, the right-hand side vector and the initial vector are read in, followed by the convergence tolerance and an upper limit to the number of iterations.

The program then scales the original coefficients and right-hand sides by dividing by the diagonal coefficient in each row (called DIAG in the program). The library routine NULL is used to fill the matrix LPU with zeros, and then LPU, as defined in eqs 2.53 is formed.

The guessed solution, X, is input as data and the iteration loop entered. As called for by eqs 2.53 the library routine MVMULT multiplies LPU by the current solution X and puts the temporary result in XNEW. This can then be added to right-hand side B using library routine VECADD, to give a new solution still called XNEW. This can be compared with X by using the convergence library subroutine CHECON. If XNEW agrees with X to the specified tolerance, ICON is set to 1 and the iterations terminated. Note that the CHECON subroutine also updates X to XNEW.

The program prints out the number of iterations taken and the values of the solution vector for the first 5 iterations and at convergence. This output is shown in Fig. 2.8(b), illustrating that 51 iterations are required to give a result accurate to about 4 places of decimals using this simple technique.

2.9.2 Very sparse systems

An obvious advantage of iteration methods occurs when the set of equation coefficients is very 'sparse', that is there are very few nonzero entries. In the case of direct methods, we saw that 'fill-in' occurred in the coefficient matrix during solution, except outside the 'skyline' (Section 2.4), whereas in iterative methods, the coefficient matrix retains its sparsity pattern throughout the calculation. In such cases, one would not wish to retain the matrix form for **A** in programs like Program 2.8 but rather use specialised coding involving pointers to the nonzero terms in the coefficient matrix.

2.9.3 The Gauss–Seidel method

In Jacobi's method, all of the components of $\mathbf{x}^{k+1}$ (called XNEW in Program 2.8) are evaluated using all the components of $\mathbf{x}^k$ (X in the program). Thus the new information is obtained entirely in terms of the old.

However, after the first row of eqs 2.53 has been evaluated, there is a new x_1^{k+1} available which is presumably a better approximation to the solution than x_1^k. In the Gauss–Seidel technique, the new value x_1^{k+1} is immediately substituted for x_1^k in the 'old' solution X. After evaluation of row 2, x_2^{k+1} is substituted for x_2^k and so on. The convergence of this process operating on eqs 2.41 is

	Iteration number			
	0	1	2	3
x_1 =	1.0	1.5	1.125	1.03125
x_2 =	1.0	1.75	1.9375	1.984375

(2.55)

which is clearly better than that in eqs 2.46.

The Gauss–Seidel iteration can be written

$$(\mathbf{I}-\mathbf{L})\mathbf{x}^{k+1}=\mathbf{b}+\mathbf{U}\mathbf{x}^{k} \tag{2.56}$$

which is possible because in the operation $\mathbf{U}\mathbf{x}^k$, the evaluation of row i does not depend on x_i, x_{i-1}, etc., and so they can be updated as they become available.

It will be obvious by comparing eqs 2.56 with 2.53 that for non-sparse systems the right-hand side can be computed again by using a matrix-vector multiply and a vector addition, leading to

$$(\mathbf{I}-\mathbf{L})\mathbf{x}^{k+1}=\mathbf{y} \tag{2.57}$$

These equations take the form

$$\begin{bmatrix} 1 & 0 & 0 \\ a_{21} & 1 & 0 \\ a_{31} & a_{32} & 1 \end{bmatrix} \begin{Bmatrix} x_1 \\ x_2 \\ x_3 \end{Bmatrix}^{k+1} = \begin{Bmatrix} y_1 \\ y_2 \\ y_3 \end{Bmatrix} \tag{2.58}$$

which is just one of the processes we encountered in 'direct' methods when we called it 'forward-substitution'. Library subroutine SUBFOR is available to carry out this task.

Example 2.7

Solve the equations

$$\begin{aligned} 16x_1+4x_2+8x_3&=4 \\ 4x_1+5x_2-4x_3&=2 \\ 8x_1-4x_2+22x_3&=5 \end{aligned}$$

using Gauss–Seidel iteration.

Solution 2.7

Divide each equation by the diagonal term and rearrange as

$$(\mathbf{I}-\mathbf{L})^{k+1}=\mathbf{b}+\mathbf{U}\mathbf{x}^{k}$$

Hence

$$\begin{bmatrix} 1 & 0 & 0 \\ 0.8 & 1 & 0 \\ 0.3636 & -0.1818 & 1 \end{bmatrix} \begin{Bmatrix} x_1 \\ x_2 \\ x_3 \end{Bmatrix}^{k+1} = \begin{Bmatrix} 0.25 \\ 0.4 \\ 0.2273 \end{Bmatrix} + \begin{bmatrix} 0 & -0.25 & -0.5 \\ 0 & 0 & 0.8 \\ 0 & 0 & 0 \end{bmatrix} \begin{Bmatrix} x_1 \\ x_2 \\ x_3 \end{Bmatrix}^{k}$$

Let

$$\begin{Bmatrix} x_1 \\ x_2 \\ x_3 \end{Bmatrix}^{0} = \begin{Bmatrix} 1 \\ 1 \\ 1 \end{Bmatrix}$$

$$\therefore \begin{bmatrix} 1 & 0 & 0 \\ 0.8 & 1 & 0 \\ 0.3636 & -0.1818 & 1 \end{bmatrix} \begin{Bmatrix} x_1 \\ x_2 \\ x_3 \end{Bmatrix}^1 = \begin{Bmatrix} -0.5 \\ 1.2 \\ 0.2273 \end{Bmatrix}$$

Hence by forward-substitution

$$\begin{Bmatrix} x_1 \\ x_2 \\ x_3 \end{Bmatrix}^1 = \begin{Bmatrix} -0.5 \\ 1.6 \\ 0.7 \end{Bmatrix}$$

$$\therefore \begin{bmatrix} 1 & 0 & 0 \\ 0.8 & 1 & 0 \\ 0.3636 & -0.1818 & 1 \end{bmatrix} \begin{Bmatrix} x_1 \\ x_2 \\ x_3 \end{Bmatrix}^2 = \begin{Bmatrix} -0.5 \\ 0.96 \\ 0.2273 \end{Bmatrix}$$

and also by forward-substitution

$$\begin{Bmatrix} x_1 \\ x_2 \\ x_3 \end{Bmatrix}^2 = \begin{Bmatrix} -0.5 \\ 1.36 \\ 0.6564 \end{Bmatrix} \rightarrow \begin{Bmatrix} -0.25 \\ 1 \\ 0.5 \end{Bmatrix} \quad \text{(after many iterations)}$$

Convergence is still slow (see Program 2.9), but not as slow as in Jacobi's method.

Program 2.9. Equation solution by Gauss–Seidel iteration
This program illustrates the coding of the Gauss–Seidel process. It clearly resembles Program 2.8 quite closely and the input data are the same (Fig. 2.9(a)). The same scaling of **A** is necessary but matrix **U** is required rather than **LPU**. At the same time as **U** is formed from **A**, the matrix **I** − **L** overwrites **A**. In the iteration loop, the right-hand side of eqs 2.56, called XNEW, is formed by successive calls to MVMULT and VECADD. The forward substitution then takes place using SUBFOR leaving the new solution in XNEW. As before, convergence is checked using CHECON. The results are listed in Fig. 2.9(b) and show that the number of iterations required to achieve the same tolerance has dropped to 30, compared to Jacobi's 51.

```
      PROGRAM P29
C
C      PROGRAM 2.9 GAUSS-SEIDEL ITERATION
C
C      ALTER NEXT LINE TO CHANGE PROBLEM SIZE
C
      PARAMETER (IN=20)
C
      REAL A(IN,IN),U(IN,IN),B(IN),X(IN),XNEW(IN)
C
      READ (5,*) N,((A(I,J),J=1,N),I=1,N)
      READ (5,*) (B(I),I=1,N)
```

```
      READ (5,*) (X(I),I=1,N)
      READ (5,*) TOL,ITS
      WRITE(6,*) ('*********** GAUSS-SEIDEL ITERATION **************')
      WRITE(6,*)
      WRITE(6,*) ('COEFFICIENT MATRIX')
      CALL PRINTA(A,IN,N,N,6)
      WRITE(6,*)
      WRITE(6,*) ('RIGHT HAND SIDE VECTOR')
      CALL PRINTV(B,N,6)
      WRITE(6,*)
      DO 1 I = 1,N
          DIAG = A(I,I)
          DO 2 J = 1,N
    2     A(I,J) = A(I,J)/DIAG
    1 B(I) = B(I)/DIAG
      CALL NULL(U,IN,N,N)
      DO 3 I = 1,N
          DO 3 J = I + 1,N
              U(I,J) = -A(I,J)
    3 A(I,J) = 0.0
      WRITE(6,*) ('FIRST FEW ITERATIONS')
      ITERS = 0
    4 ITERS = ITERS + 1
      CALL MVMULT(U,IN,X,N,N,XNEW)
      CALL VECADD(B,XNEW,XNEW,N)
      CALL SUBFOR(A,IN,XNEW,N)
      CALL CHECON(XNEW,X,N,TOL,ICON)
      IF (ITERS.LE.5) CALL PRINTV(X,N,6)
      IF (ICON.EQ.0.AND.ITERS.LT.ITS) GO TO 4
      WRITE(6,*)
      WRITE(6,*) ('ITERATIONS TO CONVERGENCE')
      WRITE(6,*) ITERS
      WRITE(6,*)
      WRITE(6,*) ('SOLUTION VECTOR')
      CALL PRINTV(X,N,6)
      STOP
      END
```

Number of equations	N		
	3		
Coefficient matrix	(A(I, J), J = 1, N), I = 1, N		
	16.0	4.0	8.0
	4.0	5.0	−4.0
	8.0	−4.0	22.0
Right-hand side	B(I), I = 1, N		
	4.0	2.0	5.0
Initial guess of solution	X(I), I = 1, N		
	1.0	1.0	1.0
Tolerance	TOL		
	1.E−5		
Iteration limit	ITS		
	100		

Fig. 2.9(a) Input data for Program 2.9

```
*********** GAUSS-SEIDEL ITERATION ***************

COEFFICIENT MATRIX
   .1600E+02    .4000E+01    .8000E+01
   .4000E+01    .5000E+01   -.4000E+01
   .8000E+01   -.4000E+01    .2200E+02

RIGHT HAND SIDE VECTOR
   .4000E+01    .2000E+01    .5000E+01

FIRST FEW ITERATIONS
  -.5000E+00    .1600E+01    .7000E+00
  -.5000E+00    .1360E+01    .6564E+00
  -.4182E+00    .1260E+01    .6084E+00
  -.3691E+00    .1182E+01    .5764E+00
  -.3337E+00    .1128E+01    .5537E+00

ITERATIONS TO CONVERGENCE
30

SOLUTION VECTOR
  -.2500E+00    .1000E+01    .5000E+00
```

Fig. 2.9(b) Results from Program 2.9

2.9.4 Successive overrelaxation

In this technique, the difference between successive iterations is multiplied by a scalar parameter ω called the 'overrelaxation factor':

$$\mathbf{x}^{k+1}-\mathbf{x}^{k} \Rightarrow \omega(\mathbf{x}^{k+1}-\mathbf{x}^{k}) \tag{2.59}$$

For the Gauss–Seidel method, $\omega=1$, but in general $1<\omega<2$. Thus larger than usual changes are enforced in the solution as it proceeds.

From the simple Jacobi method (eq. 2.53) we have

$$\omega\mathbf{x}^{k+1}=\omega\mathbf{b}+(\omega\mathbf{L}+\omega\mathbf{U})\mathbf{x}^{k} \tag{2.60}$$

and from eq. 2.59

$$\mathbf{x}^{k+1}=(1-\omega)\mathbf{x}^{k}+\omega\mathbf{x}^{k+1} \tag{2.61}$$

so that

$$\mathbf{x}^{k+1}-(1-\omega)\mathbf{x}^{k}=\omega\mathbf{b}+(\omega\mathbf{L}+\omega\mathbf{U})\mathbf{x}^{k} \tag{2.62}$$

or, using updating of the solution

$$(\mathbf{I}-\omega\mathbf{L})\mathbf{x}^{k+1}=\omega\mathbf{b}+[(1-\omega)\mathbf{I}+\omega\mathbf{U}]\mathbf{x}^{k} \tag{2.63}$$

Equations 2.63 have similar properties to eqs 2.56 in that the evaluation of row i on the right-hand side does not depend on x_{i-1} etc., and so updating is possible.

Program 2.10. Equation solution by successive overrelaxation
The algorithm will clearly be very similar to the Gauss–Seidel one and is illustrated in Program 2.10, with its input data in Fig. 2.10(a).

There is one extra input quantity, OMEGA, which is the overrelaxation scalar, but no other major changes. First **A** and **b** are scaled as usual and **b** given the multiple ω as required by eqs 2.63. Using library routine MSMULT, **A** is then multiplied by the scalar ω and **U** and **A** organised as before. In order to end up with $(\mathbf{I}-\omega\mathbf{L})$ and $[(1-\omega)\mathbf{I}+\omega\mathbf{U}]$ on the left- and right-hand sides of eqs 2.63 the diagonals of **A** and of **U** are replaced by 1.0 and $1.0-\omega$ respectively.

```
      PROGRAM P210
C
C      PROGRAM 2.10 SUCCESSIVE OVERRELAXATION
C
C      ALTER NEXT LINE TO CHANGE PROBLEM SIZE
C
      PARAMETER (IN=20)
C
      REAL A(IN,IN),U(IN,IN),B(IN),X(IN),XNEW(IN)
C
      READ (5,*) N,((A(I,J),J=1,N),I=1,N)
      READ (5,*) (B(I),I=1,N)
      READ (5,*) (X(I),I=1,N)
      READ (5,*) TOL,ITS,OMEGA
      WRITE(6,*) ('*********** SUCCESSIVE OVERRELAXATION ***********')
      WRITE(6,*)
      WRITE(6,*) ('COEFFICIENT MATRIX')
      CALL PRINTA(A,IN,N,N,6)
      WRITE(6,*)
      WRITE(6,*) ('RIGHT HAND SIDE VECTOR')
      CALL PRINTV(B,N,6)
      WRITE(6,*)
      WRITE(6,*) ('OVERRELAXATION SCALAR')
      WRITE(6,100)OMEGA
      WRITE(6,*)
      DO 1 I = 1,N
         DIAG = A(I,I)
         DO 2 J = 1,N
    2    A(I,J) = A(I,J)/DIAG
    1 B(I) = OMEGA*B(I)/DIAG
      CALL NULL(U,IN,N,N)
      CALL MSMULT(A,IN,OMEGA,N,N)
      DO 3 I = 1,N
         DO 3 J = I + 1,N
            U(I,J) = -A(I,J)
    3 A(I,J) = 0.0
      DO 5 I = 1,N
         A(I,I) = 1.0
    5 U(I,I) = 1.0 - OMEGA
      WRITE(6,*) ('FIRST FEW ITERATIONS')
      ITERS = 0
    4 ITERS = ITERS + 1
      CALL MVMULT(U,IN,X,N,N,XNEW)
      CALL VECADD(B,XNEW,XNEW,N)
      CALL SUBFOR(A,IN,XNEW,N)
      CALL CHECON(XNEW,X,N,TOL,ICON)
      IF (ITERS.LE.5) CALL PRINTV(X,N,6)
      IF (ICON.EQ.0.AND.ITERS.LT.ITS) GO TO 4
      WRITE(6,*)
      WRITE(6,*) ('ITERATIONS TO CONVERGENCE')
      WRITE(6,*) ITERS
      WRITE(6,*)
      WRITE(6,*) ('SOLUTION VECTOR')
      CALL PRINTV(X,N,6)
  100 FORMAT(F10.2)
      STOP
      END
```

Number of equations	N		
	3		
Coefficient matrix	(A(I, J), J = 1, N), I = 1, N		
	16.0	4.0	8.0
	4.0	5.0	−4.0
	8.0	−4.0	22.0
Right-hand side	B(I), I = 1, N		
	4.0	2.0	5.0
Initial guess of solution	X(I), I = 1, N		
	1.0	1.0	1.0
Tolerance	TOL		
	1.E−5		
Iteration limit	ITS		
	100		
Overrelaxation scalar	OMEGA		
	1.5		

Fig. 2.10(a) Input data for Program 2.10

```
*********** SUCCESSIVE OVERRELAXATION ************

COEFFICIENT MATRIX
    .1600E+02    .4000E+01    .8000E+01
    .4000E+01    .5000E+01   -.4000E+01
    .8000E+01   -.4000E+01    .2200E+02

RIGHT HAND SIDE VECTOR
    .4000E+01    .2000E+01    .5000E+01

OVERRELAXATION SCALAR
      1.50

FIRST FEW ITERATIONS
   -.1250E+01    .2800E+01    .1286E+01
   -.1015E+01    .1961E+01    .7862E+00
   -.4427E+00    .1094E+01    .4877E+00
   -.1796E+00    .8538E+00    .4279E+00
   -.1763E+00    .8981E+00    .4681E+00

ITERATIONS TO CONVERGENCE
18

SOLUTION VECTOR
   -.2500E+00    .1000E+01    .5000E+00
```

Fig. 2.10(b) Results from Program 2.10

Iteration proceeds using exactly the same loop as for Gauss–Seidel. In the case of the input shown in Fig. 2.10(a), the number of iterations to convergence has dropped to 18 (see Fig. 2.10(b)). Figure 2.11 shows the variation of iteration count with overrelaxation scalar ω. For these equations the optimum ω is about 1.4, but this will be difficult to predict.

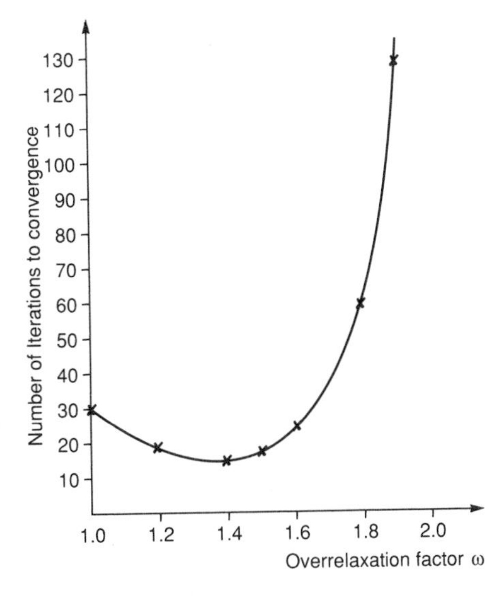

Fig. 2.11 Influence of overrelaxation factor ω on the rate of convergence of SOR

2.10 Gradient methods

The methods described in the previous sections are sometimes called 'stationary' methods because there is no attempt made in them to modify the convergence process according to a measure of the error in the trial solution. In gradient methods, by contrast, an error function is repeatedly evaluated, and used to generate new trial solutions. In the following discussion we shall confine our interest to equations with symmetric, positive definite coefficient matrices. Although it will soon be apparent that in this case the gradient methods are very simple to program, the mathematical reasoning behind them is somewhat involved. For example, the reader is referred to Jennings (1977) pp. 212–216 for a reasonably concise exposition in engineering terms.

2.10.1 The method of 'steepest descent'

For any trial solution $\mathbf{x}^k$, the error or 'residual' will clearly be expressible in terms of

$$\mathbf{r}^k = \mathbf{b} - \mathbf{A}\mathbf{x}^k \tag{2.64}$$

In the method of steepest descent, the error implicit in $\mathbf{r}^k$ is minimised according to the following algorithm

$$\begin{aligned}
&\mathbf{u}^k = \mathbf{A}\mathbf{r}^k && \text{(a)}\\
&\alpha_k = \frac{(\mathbf{r}^k)^\mathsf{T}\,\mathbf{r}^k}{(\mathbf{r}^k)^\mathsf{T}\,\mathbf{u}^k} && \text{(b)}\\
&\mathbf{x}^{k+1} = \mathbf{x}^k + \alpha_k\,\mathbf{r}^k && \text{(c)}\\
&\mathbf{r}^{k+1} = \mathbf{r}^k - \alpha_k\,\mathbf{u}^k && \text{(d)}
\end{aligned} \tag{2.65}$$

To implement this algorithm making use of a subroutine library, it will be helpful to be able to multiply a matrix by a vector in step (a), to calculate vector inner or 'dot' products in step (b), to multiply a vector by a scalar in steps (c) and (d) and to add and subtract vectors in steps (c) and (d) respectively. These five operations are catered for in the library by subroutines MVMULT, VDOTV, VSMULT, VECADD and VECSUB.

Program 2.11. Equation solution by 'steepest descent'
The program which implements eqs 2.65 is listed as Program 2.11. It uses the same data as most of the previous iterative programs (Fig. 2.12(a)) requiring as input the number of equations to be

```
      PROGRAM P211
C
C      PROGRAM 2.11 STEEPEST DESCENT
C
C      ALTER NEXT LINE TO CHANGE PROBLEM SIZE
C
      PARAMETER (IN=20)
C
      REAL A(IN,IN),B(IN),X(IN),R(IN),U(IN),TEMP(IN),XNEW(IN)
C
      READ (5,*) N,((A(I,J),J=1,N),I=1,N)
      READ (5,*) (B(I),I=1,N)
      READ (5,*) (X(I),I=1,N)
      READ (5,*) TOL,ITS
      WRITE(6,*) ('*************** STEEPEST DESCENT ****************')
      WRITE(6,*)
      WRITE(6,*) ('COEFFICIENT MATRIX')
      CALL PRINTA(A,IN,N,N,6)
      WRITE(6,*)
      WRITE(6,*) ('RIGHT HAND SIDE VECTOR')
      CALL PRINTV(B,N,6)
      WRITE(6,*)
      WRITE(6,*) ('STARTING VECTOR')
      CALL PRINTV(X,N,6)
      WRITE(6,*)
      CALL MVMULT(A,IN,X,N,N,TEMP)
      CALL VECSUB(B,TEMP,R,N)
      WRITE(6,*) ('FIRST FEW ITERATIONS')
      ITERS = 0
    1 ITERS = ITERS + 1
      CALL MVMULT(A,IN,R,N,N,U)
      CALL VDOTV(R,R,UP,N)
      CALL VDOTV(R,U,DOWN,N)
      ALPHA = UP/DOWN
      CALL VECCOP(R,TEMP,N)
      CALL VSMULT(TEMP,ALPHA,N)
      CALL VECADD(X,TEMP,XNEW,N)
      CALL VSMULT(U,ALPHA,N)
      CALL VECSUB(R,U,R,N)
      CALL CHECON(XNEW,X,N,TOL,ICON)
      IF (ITERS.LE.5) CALL PRINTV(X,N,6)
      IF (ICON.EQ.0.AND.ITERS.LT.ITS) GO TO 1
      WRITE(6,*)
      WRITE(6,*) ('ITERATIONS TO CONVERGENCE')
      WRITE(6,*)ITERS
      WRITE(6,*)
      WRITE(6,*) ('SOLUTION VECTOR')
      CALL PRINTV(X,N,6)
      STOP
      END
```

Number of equations	N		
	3		
Coefficient matrix	(A(I, J), J = 1, N), I = 1, N		
	16.0	4.0	8.0
	4.0	5.0	−4.0
	8.0	−4.0	22.0
Right-hand side	B(I), I = 1, N		
	4.0	2.0	5.0
Initial guess of solution	X(I), I = 1, N		
	1.0	1.0	1.0
Tolerance	TOL		
	1.E−5		
Iteration limit	ITS		
	100		

Fig. 2.12(a) Input data for Program 2.11

```
*************** STEEPEST DESCENT *****************

COEFFICIENT MATRIX
   .1600E+02    .4000E+01    .8000E+01
   .4000E+01    .5000E+01   -.4000E+01
   .8000E+01   -.4000E+01    .2200E+02

RIGHT HAND SIDE VECTOR
   .4000E+01    .2000E+01    .5000E+01

STARTING VECTOR
   .1000E+01    .1000E+01    .1000E+01

FIRST FEW ITERATIONS
   .9133E-01    .8864E+00    .2049E+00
  -.8584E-01    .7540E+00    .4263E+00
  -.1305E+00    .7658E+00    .3976E+00
  -.1538E+00    .8081E+00    .4512E+00
  -.1715E+00    .8257E+00    .4297E+00

ITERATIONS TO CONVERGENCE
61

SOLUTION VECTOR
  -.2500E+00    .9999E+00    .5000E+00
```

Fig. 2.12(b) Results from Program 2.11

solved (N), the equation left- and right-hand sides (A and B) and a starting guess to compute the initial value of **r** in eqs 2.64 (X). The iteration tolerance (TOL) and limit (ITS) complete the data.

The operations involved in eqs 2.64 are first performed, namely the matrix-vector multiply $\mathbf{Ax}^0$ and its subtraction from **b** to form R. The iterations in eqs 2.65 can then proceed involving MVMULT (step (a)) and two dot products giving α (ALPHA in step (b)). Vector R is temporarily copied into TEMP and step (c) completed by VSMULT and VECADD. It remains only to complete step (d) by VSMULT and VECSUB, to check the tolerance on X using CHECON, and to continue the iteration or stop if convergence has been achieved. The results are shown in Fig. 2.12(b), where it can be seen that 61 iterations are required to achieve the desired accuracy.

2.10.2 The method of 'conjugate gradients'

The results obtained by running Program 2.11 show that the method of steepest descent is not competitive with the other iteration methods tried so far. However, it can be radically improved if the descent vectors are made mutually 'conjugate' with respect to **A**. That is, descent vectors **p** satisfy

$$(\mathbf{p}^i)^\mathrm{T}\,\mathbf{A}(\mathbf{p}^j)=\mathbf{0} \quad \text{for } i\neq j \tag{2.66}$$

The equivalent algorithm to eqs 2.65 becomes

$$\mathbf{p}^0=\mathbf{r}^0=\mathbf{b}-\mathbf{A}\mathbf{x}^0 \text{ (to start the process)}$$

$$\begin{aligned}
&\mathbf{u}^k=\mathbf{A}\mathbf{p}^k && \text{(a)}\\
&\alpha_k=\frac{(\mathbf{r}^k)^\mathrm{T}\,\mathbf{r}^k}{(\mathbf{p}^k)^\mathrm{T}\,\mathbf{u}^k} && \text{(b)}\\
&\mathbf{x}^{k+1}=\mathbf{x}^k+\alpha_k\,\mathbf{p}^k && \text{(c)}\\
&\mathbf{r}^{k+1}=\mathbf{r}^k-\alpha_k\,\mathbf{u}^k && \text{(d)}\\
&\beta_k=\frac{(\mathbf{r}^{k+1})^\mathrm{T}\,\mathbf{r}^{k+1}}{(\mathbf{r}^k)^\mathrm{T}\,\mathbf{r}^k} && \text{(e)}\\
&\mathbf{p}^{k+1}=\mathbf{r}^{k+1}+\beta_k\,\mathbf{p}^k && \text{(f)}
\end{aligned} \tag{2.67}$$

Program 2.12. Equation solution by conjugate gradients
The programming of this algorithm clearly involves more steps, but uses the same library routines as the previous program. The process is listed as Program 2.12, which begins with the usual declarations of arrays following the nomenclature of eqs 2.67. The data are unchanged (Fig. 2.13(a)). The starting procedure leads to $\mathbf{r}^0$ using MVMULT and VECSUB and $\mathbf{r}^0$ is copied into $\mathbf{p}^0$ using VECCOP.

The iteration process then proceeds. Vector $\mathbf{u}^k$ is computed by MVMULT (step (a)) and ALPHA as required by step (b). Vector **p** is copied into temporary store TEMP and step (c) leads to the updated X, called XNEW as before. Routines VSMULT and VECSUB complete step (a) which gives the new **r**, and β (BETA) involves the new numerator UP2. It remains to complete step (f) by calls to VSMULT and VECADD and to check convergence using CHECON.

```
      PROGRAM P212
C
C      PROGRAM 2.12 CONJUGATE GRADIENTS
C
C      ALTER NEXT LINE TO CHANGE PROBLEM SIZE
C
      PARAMETER (IN=20)
C
      REAL A(IN,IN),B(IN),X(IN),R(IN),U(IN),TEMP(IN),P(IN),XNEW(IN)
C
      READ (5,*) N,((A(I,J),J=1,N),I=1,N)
      READ (5,*) (B(I),I=1,N)
      READ (5,*) (X(I),I=1,N)
      READ (5,*) TOL,ITS
      WRITE(6,*) ('*************** CONJUGATE GRADIENTS **************')
      WRITE(6,*)
```

```
      WRITE(6,*) ('COEFFICIENT MATRIX')
      CALL PRINTA(A,IN,N,N,6)
      WRITE(6,*)
      WRITE(6,*) ('RIGHT HAND SIDE VECTOR')
      CALL PRINTV(B,N,6)
      WRITE(6,*)
      WRITE(6,*) ('STARTING VECTOR')
      CALL PRINTV(X,N,6)
      WRITE(6,*)
      CALL MVMULT(A,IN,X,N,N,TEMP)
      CALL VECSUB(B,TEMP,R,N)
      CALL VECCOP(R,P,N)
      WRITE(6,*) ('FIRST FEW ITERATIONS')
      ITERS = 0
    1 ITERS = ITERS + 1
      CALL MVMULT(A,IN,P,N,N,U)
      CALL VDOTV(R,R,UP,N)
      CALL VDOTV(P,U,DOWN,N)
      ALPHA = UP/DOWN
      CALL VECCOP(P,TEMP,N)
      CALL VSMULT(TEMP,ALPHA,N)
      CALL VECADD(X,TEMP,XNEW,N)
      CALL VSMULT(U,ALPHA,N)
      CALL VECSUB(R,U,R,N)
      CALL VDOTV(R,R,UP2,N)
      BETA = UP2/UP
      CALL VSMULT(P,BETA,N)
      CALL VECADD(R,P,P,N)
      CALL CHECON(XNEW,X,N,TOL,ICON)
      IF (ITERS.LE.5) CALL PRINTV(X,N,6)
      IF (ICON.EQ.0.AND.ITERS.LT.ITS) GO TO 1
      WRITE(6,*)
      WRITE(6,*) ('ITERATIONS TO CONVERGENCE')
      WRITE(6,*) ITERS
      WRITE(6,*)
      WRITE(6,*) ('SOLUTION VECTOR')
      CALL PRINTV(X,N,6)
      STOP
      END
```

Number of equations	N		
	3		
Coefficient matrix	(A(I, J), J = 1, N), I = 1, N		
	16.0	4.0	8.0
	4.0	5.0	−4.0
	8.0	−4.0	22.0
Right-hand side	B(I), I = 1, N		
	4.0	2.0	5.0
Initial guess of solution	X(I), I = 1, N		
	1.0	1.0	1.0
Tolerance	TOL		
	1.E−5		
Iteration limit	ITS		
	100		

Fig. 2.13(a) Input data for Program 2.12

```
************** CONJUGATE GRADIENTS **************

COEFFICIENT MATRIX
   .1600E+02    .4000E+01    .8000E+01
   .4000E+01    .5000E+01   -.4000E+01
   .8000E+01   -.4000E+01    .2200E+02

RIGHT HAND SIDE VECTOR
   .4000E+01    .2000E+01    .5000E+01

STARTING VECTOR
   .1000E+01    .1000E+01    .1000E+01

FIRST FEW ITERATIONS
   .9133E-01    .8864E+00    .2049E+00
  -.1283E+00    .7444E+00    .4039E+00
  -.2500E+00    .1000E+01    .5000E+00
  -.2500E+00    .1000E+01    .5000E+00

ITERATIONS TO CONVERGENCE
4

SOLUTION VECTOR
  -.2500E+00    .1000E+01    .5000E+00
```

Fig. 2.13(b) Results from Program 2.12

When the program is run, the results are as shown in Fig. 2.13(b). The iteration count has dropped to 4, making this the most successful method so far. In fact theoretically, in perfect arithmetic, the method would converge in N (3 in this case) steps.

2.10.3 Convergence of iteration methods

Figure 2.14 illustrates how the five iterative methods described in this chapter converge on the solution $(x_1, x_2)=(1.0, 2.0)$ of eq. 2.41. Also shown are contours of error in the solution defined when the error in x_1 and x_2 is 10 per cent and 20 per cent respectively and the measure of total error is $\sqrt{[e(x_1)^2+e(x_2)^2]}$. The methods with poorest convergence properties (Jacobi and steepest descent) are seen to be furthest from normal to the error contours in the vicinity of the true solution.

2.10.4 Preconditioning

An increasingly popular process for large scale computation which should be mentioned is called 'preconditioning'. In the solution of the usual system

$$\mathbf{Ax}=\mathbf{b} \tag{2.68}$$

a preconditioning matrix is applied so that

$$\mathbf{PAx}=\mathbf{Pb} \tag{2.69}$$

It will be clear that if $\mathbf{P}=\mathbf{A}^{-1}$ the preconditioning process would be perfect, and no further work would be necessary. Although this level of prior knowledge can be discounted, it turns out that relatively crude approximations to $\mathbf{A}^{-1}$ can be very

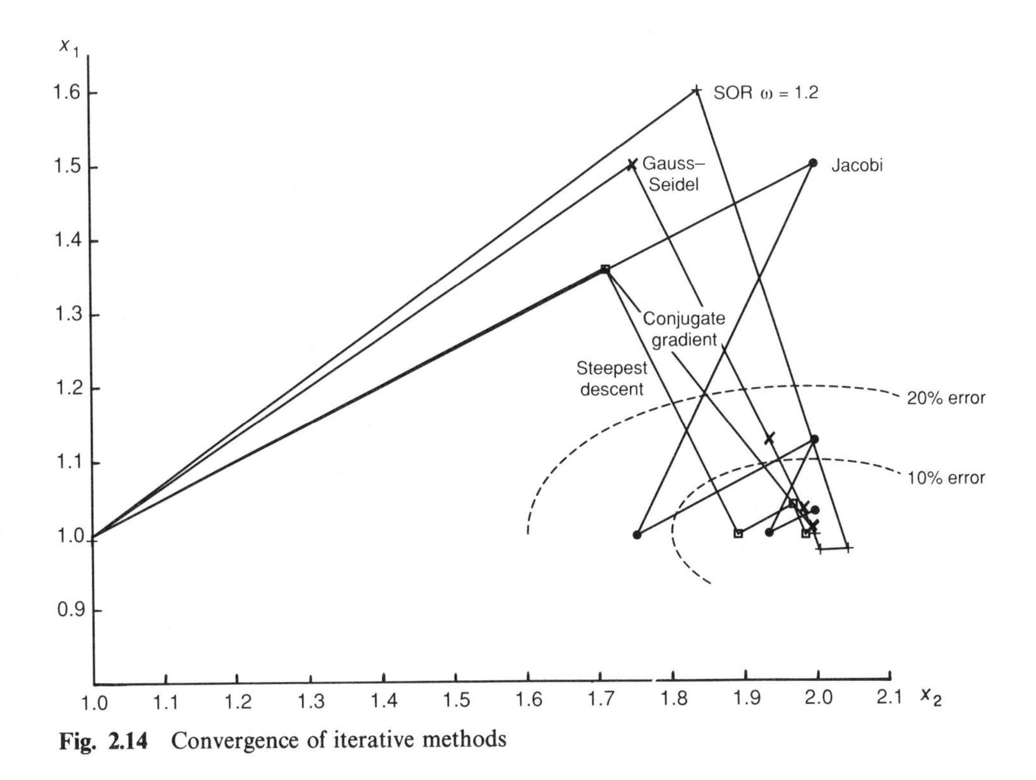

Fig. 2.14 Convergence of iterative methods

effective as preconditioners, and enable the solution of large systems of equations by conjugate gradient methods in far less than N iterations.

2.11 Comparison of direct and iterative methods

In the era of scalar digital computing, say up to 1980, it could be stated that in the majority of cases, direct solution was to be preferred to iterative solution. Exceptions were possible for very sparse systems, and some ill-conditioned systems, but the degree of security offered by direct solution was attractive. It has been shown in the examples of iterative solutions presented in this chapter that a very wide range of efficiencies (as measured by iteration count to convergence and work per iteration) is possible and so the amount of time consumed in the solution of a system of equations is rather unpredictable.

However, the widespread use of 'vector' or 'parallel' processing computers has led to a revision of previous certainties about equation solution. Comparison of Program 2.14 (conjugate gradient technique) with Program 2.4 (**LU** decomposition) will show that the former consists almost entirely of rather straightforward operations on vectors which can be processed very quickly by non-scalar machines. Even if the coefficient matrix **A** is sparse or banded, the MVMULT operation is vectorisable or parallelisable.

In contrast, the **LU** factorisation is seen to contain more complicated code involving conditional statements and variable length loops which present greater difficulties to the programmer attempting to optimise code on a non-scalar machine.

It can therefore be said that algorithm choice for linear equation solution is far from simple for large systems of equations, and depends strongly upon machine architecture. For small systems of equations, direct methods are still attractive.

2.12 Exercises

1 Solve the set of simultaneous equations

$$6x_1+3x_2+6x_3=30$$
$$2x_1+3x_2+3x_3=17$$
$$x_1+2x_2+2x_3=11$$

Answer: $x_1=1$, $x_2=2$, $x_3=3$.

2 Solve the system

$$\begin{bmatrix} 1 & 1 & 2 & -4 \\ 2 & -1 & 3 & 1 \\ 3 & 1 & -1 & 2 \\ 1 & -1 & -1 & 1 \end{bmatrix} \begin{Bmatrix} x_1 \\ x_2 \\ x_3 \\ x_4 \end{Bmatrix} = \begin{Bmatrix} 0 \\ 5 \\ 5 \\ 0 \end{Bmatrix}$$

by **LU** factorisation.
Answer:

$$\mathbf{L}=\begin{bmatrix} 1 & 0 & 0 & 0 \\ 2 & 1 & 0 & 0 \\ 3 & \frac{2}{3} & 1 & 0 \\ 1 & \frac{2}{3} & \frac{7}{19} & 1 \end{bmatrix} \quad \mathbf{U}=\begin{bmatrix} 1 & 1 & 2 & -4 \\ 0 & -3 & -1 & 9 \\ 0 & 0 & -\frac{19}{3} & 8 \\ 0 & 0 & 0 & -\frac{75}{19} \end{bmatrix}$$

and $x_1=x_2=x_3=x_4=1$.

3 Solve the symmetrical equations

$$9.3746x_1+3.0416x_2-2.4371x_3=9.2333$$
$$3.0416x_1+6.1832x_2+1.2163x_3=8.2049$$
$$-2.4371x_1+1.2163x_2+8.4429x_3=3.9339$$

by $\mathbf{LDL}^{\mathrm{T}}$ decomposition, and hence find the determinant of the coefficient matrix.
Answer: $x_1=0.8964$, $x_2=0.7651$, $x_3=0.6145$.
The terms in **D** are $D_{11}=9.3746$, $D_{22}=5.1964$, $D_{33}=7.0341$ and so the determinant of the coefficient matrix is 342.66.

4 Solve the symmetrical system

$$5x_1+6x_2-2x_3-2x_4=1$$
$$6x_1-5x_2-2x_3+2x_4=0$$
$$-2x_1-2x_2+3x_3-x_4=0$$
$$-2x_1+2x_2-x_3-3x_4=0.$$

Answer: $x_1=0.12446$, $x_2=0.07725$, $x_3=0.11159$, $x_4=-0.06867$

5 Solve the symmetrical system

$$\begin{aligned} x_1+2x_2-2x_3+x_4&=4\\ 2x_1+5x_2-2x_3+3x_4&=7\\ -2x_1-2x_2+5x_3+3x_4&=-1\\ x_1+3x_2+3x_3+2x_4&=0 \end{aligned}$$

Answer: $x_1=2$, $x_2=-1$, $x_3=-1$, $x_4=2$.

6 Attempt to solve Exercises 4 and 5 by Cholesky's method (Program 2.4).
Answer: Square roots of negative numbers will arise.

7 Solve the symmetrical banded system:

$$\begin{bmatrix} 4 & 2 & 0 & 0\\ 2 & 8 & 2 & 0\\ 0 & 2 & 8 & 2\\ 0 & 0 & 2 & 4 \end{bmatrix} \begin{Bmatrix} x_1\\ x_2\\ x_3\\ x_4 \end{Bmatrix} = \begin{Bmatrix} 4\\ 0\\ 0\\ 0 \end{Bmatrix}$$

Answer: $x_1=1.156$, $x_2=-0.311$, $x_3=0.089$, $x_4=-0.044$.

8 Solve the following system using elimination with partial pivoting

$$\begin{bmatrix} 1 & 0 & 2 & 3\\ -1 & 2 & 2 & -3\\ 0 & 1 & 1 & 4\\ 6 & 2 & 2 & 4 \end{bmatrix} \begin{Bmatrix} x_1\\ x_2\\ x_3\\ x_4 \end{Bmatrix} = \begin{Bmatrix} 1\\ -1\\ 2\\ 1 \end{Bmatrix}$$

Answer: $x_1=-\frac{13}{70}$, $x_2=\frac{8}{35}$, $x_3=-\frac{4}{35}$, $x_4=\frac{33}{70}$.
The interchanged row order is 4, 2, 1, 3.

9 Solve the following equations using elimination with partial pivoting

$$\begin{aligned} x_1+2x_2+3x_3&=2\\ 3x_1+6x_2+x_3&=14\\ x_1+x_2+x_3&=2. \end{aligned}$$

Answer: $x_1=1$, $x_2=2$, $x_3=-1$.

10 Attempt to solve Exercise 9 without pivoting.
Answer: Zero pivot found in row 2.

11 Solve the equations

$$\begin{aligned} 20x_1+2x_2-x_3&=25\\ 2x_1+13x_2-2x_3&=30\\ x_1+x_2+x_3&=2. \end{aligned}$$

using (a) Jacobi and (b) Gauss–Seidel iterations, using a starting guess $x_1=x_2=x_3=0$.
Answer: $x_1=1$, $x_2=2$, $x_3=-1$.

In case (a) the first five iterations give

x_1	x_2	x_3
0	0	0
1.25	2.308	2
1.119	2.423	−1.558
0.9298	1.8959	−1.5423
0.9833	1.9274	−0.8257
1.01598	2.02939	−0.91067

whereas in case (b) they are:

x_1	x_2	x_3
0	0	0
1.25	2.1154	−1.3654
0.97019	1.94837	−0.91857
1.00923	2.01111	−1.02034
0.99787	1.99720	−0.99507
1.00053	2.00068	−1.00120

12 Compare the iteration counts, for a solution tolerance of 1×10^{-5}, in the solution of Exercise 11 by the following methods:

Jacobi, Gauss–Seidel, SOR ($\omega = 1.2$)
steepest descent, conjugate gradients.

Answer: 16, 10, 30, does not converge, does not converge.
The last two methods are suitable only for symmetric, positive definite systems.

13 Solve Exercises 4 and 5 by the method of conjugate gradients.
Answer: For a tolerance of 1×10^{-5}, solution obtained in 5 iterations in both cases.

14 Check that the solution vectors $\{1.22 \quad -1.000 \quad 3.037\}^T$ and $\{1.222 \quad -0.75 \quad 3.33\}^T$ are both solutions to the system

$$9x_1 + 9x_2 + 8x_3 = 26$$
$$9x_1 + 8x_2 + 7x_3 = 24$$
$$8x_1 + 7x_2 + 6x_3 = 21$$

to within a tolerance of 0.1. Find the true solution. What do these results imply about the system of equations?
Answer: True solution $\{1.0 \quad 1.0 \quad 1.0\}^T$
The system is ill-conditioned.

2.13 Further reading

Bickley, W.G. and Thompson, R.S.H.G. (1964). *Matrices, Their Meaning and Manipulation*, English Universities Press, London.
Conte, S. and de Boor, C. (1980). *Elementary Numerical Analysis*, 3rd edn, McGraw-Hill, New York.
Duff, I. (1977). A survey of sparse matrix research, *Proc. IEEE*, **65**, pp. 500–535.

Evans, D. (ed.) (1985). *Sparsity and its Applications*, Cambridge University Press, Cambridge.
Fox, L. (1964). *An Introduction to Numerical Linear Algebra*, Clarendon Press, Oxford.
Frazer, R.A., Duncan, W.J. and Collar, A.R. (1938). *Elementary Matrices and some Applications to Dynamics and Differential Equations*, Cambridge University Press, Cambridge.
Froberg, C.E. (1969). *Introduction to Numerical Analysis*, 2nd edn, Addison-Wesley, Reading, Massachusetts.
Gladwell, I. and Wait, R. (eds) (1979). *A Survey of Numerical Methods for Partial Differential Equations*, Oxford University Press, Oxford.
Householder, A. (1965). *The Theory of Matrices in Numerical Analysis*, Ginn, Boston.
Jennings, A. (1977). *Matrix Computation for Engineers and Scientists*, Wiley, Chichester.
Varga, R.S. (1962). *Matrix Analysis*, Prentice-Hall, Englewood Cliffs, New Jersey.
Wilkinson, J.H. and Reinsch, C. (1971). *Linear Algebra*, Handbook for Automatic Computation, Vol. 2, Springer-Verlag, New York.

3

Nonlinear Equations

3.1 Introduction

In the previous chapter we dealt with 'linear' equations which did not involve powers of the unknowns. A common form of 'nonlinear' equations which frequently arises in practice does contain such powers, for example

$$\begin{aligned} x_1{}^3 - 2x_1x_2 + x_2{}^2 &= 4 \\ x_1 + 4x_1x_2 + 3x_2{}^2 &= 7 \end{aligned} \tag{3.1}$$

would be a pair of nonlinear equations satisfied by various combinations of x_1 and x_2.

In the simplest situation, we might have a single nonlinear equation, such as

$$y = f(x) = x^3 - x - 1 = 0 \tag{3.2}$$

A graphical interpretation helps us to understand the nature of the solutions for x which satisfy eq. 3.2. A plot of $f(x)$ versus x is shown in Fig. 3.1. Where $f(x)$ intersects the line $y=0$ is clearly a solution of the equation, often called a 'root', which in this case has the value $x \cong 1.3245$.

Note that we could also write eq. 3.2 as

$$y = g(x) = x^3 - x = 1 \tag{3.3}$$

and look for the intersection of $g(x)$ with the line $y=1$ as a solution of the equation, also shown in the figure.

Since we now know that $x \cong 1.324$ is a solution, we can factorise $f(x)$ to yield

$$f(x) \cong (x - 1.324)\,(x^2 + 1.324x + 0.755) = 0 \tag{3.4}$$

Taking the roots of the quadratic equation in eq. 3.4, we arrive at solutions

$$x \cong \frac{-1.324 \pm \sqrt{-1.267}}{2} \tag{3.5}$$

showing that the remaining two roots are imaginary ones.

It is immediately apparent that finding solutions to general sets of nonlinear equations will be quite a formidable numerical task. As in the last chapter, where we said that many physical systems produced diagonally dominant and/or symmetric systems of linear equations, it is fortunate that in many physical situations which give

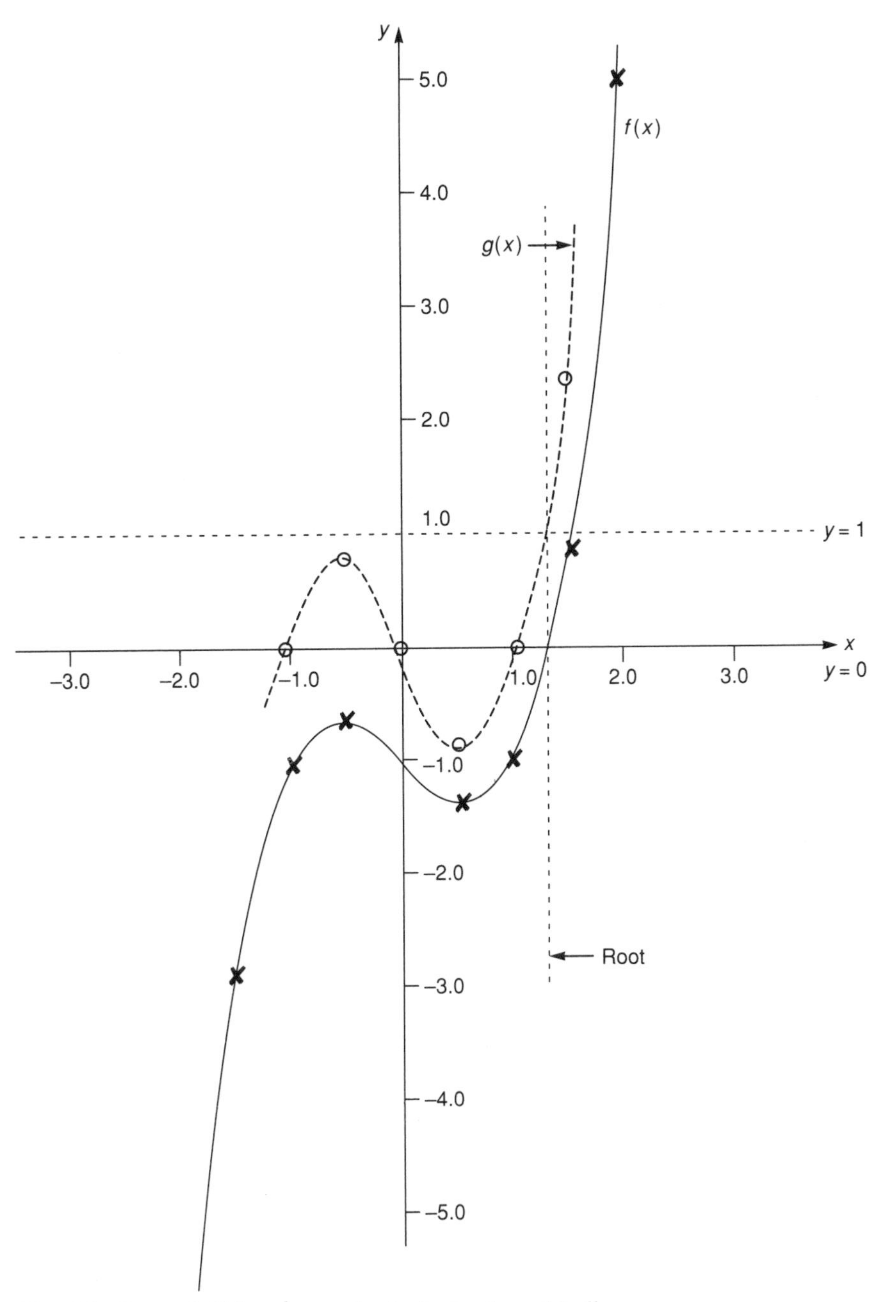

Fig. 3.1 Root of $f(x)=x^3-x-1=0$ obtained graphically

rise to nonlinear sets of equations, the nature of the problem limits the possible values that the roots may have. For example, we may know in advance that all the roots must be real, or even that they must all be real and positive. In this chapter we shall concentrate on these limited problems.

We shall also see that methods for solving nonlinear equations are intrinsically iterative in character. Referring to Fig. 3.1, it should be clear that such an iterative process might depend strongly on the quality of an initial guessed solution. For example, a guess of $x=1$ or $x=2$ in that case might have much more chance of success than a guess of $x=-100$ or $x=+100$.

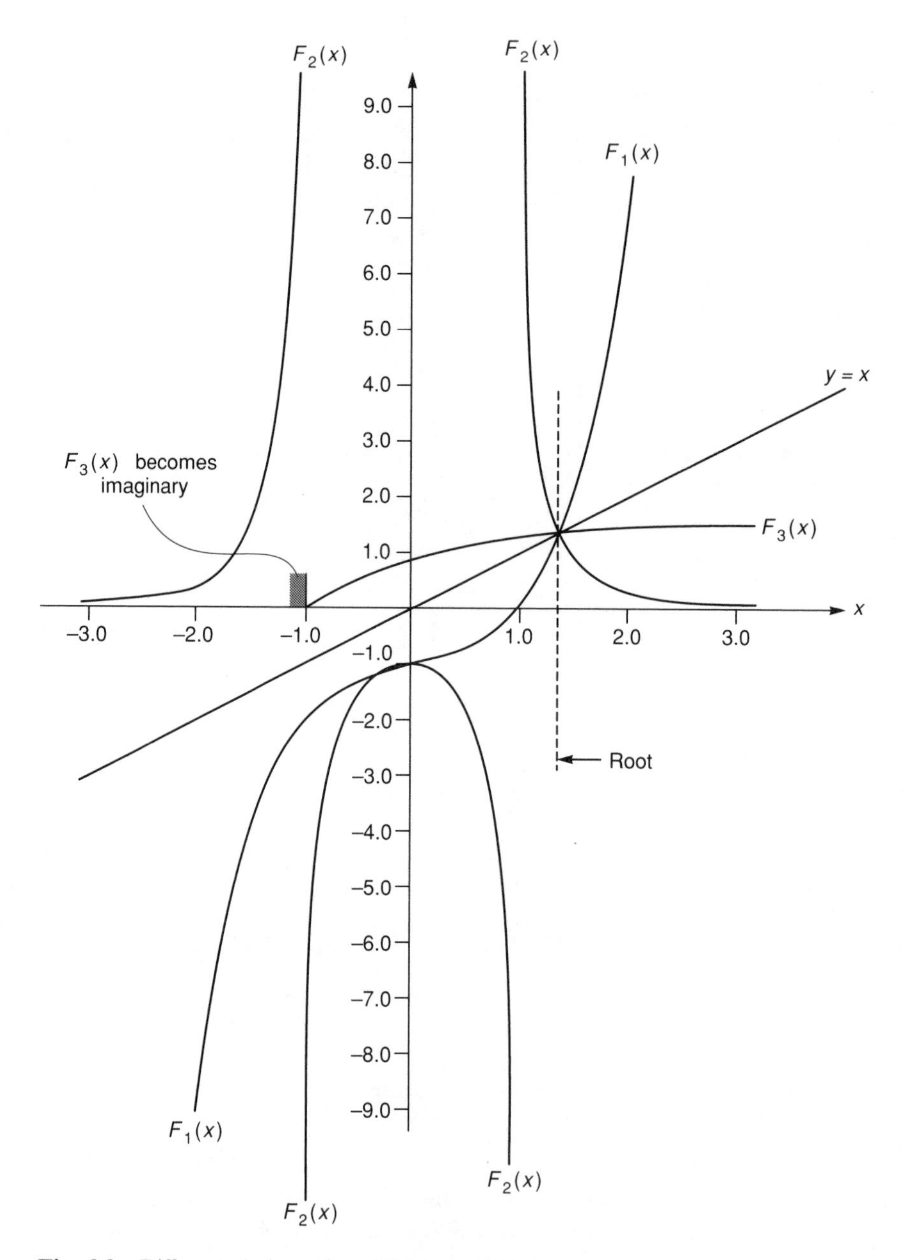

Fig. 3.2 Different choices of $x=F(x)$ in eq. 3.7

3.2 Iterative substitution

A simple iterative process is to replace an equation like eq. 3.2 which had the form

$$f(x)=x^3-x-1=0 \tag{3.2}$$

by an equivalent equation

$$x=F(x) \tag{3.6}$$

The iterative process proceeds by a guess being made for x, substitution of this guess on the right-hand side of eq. 3.6, and comparison of the result with the guessed x. In the unlikely event that equality results, the solution has been found. If not, the new $F(x)$ is assumed to be a better estimate of x and the process is repeated.

An immediate dilemma is that there is no single way of determining $F(x)$. In the case of eq. 3.2, we could write

$$\begin{aligned} x&=F_1(x)=x^3-1 && \text{(a)} \\ x&=F_2(x)=\frac{1}{x^2-1} && \text{(b)} \\ x&=F_3(x)=\sqrt[3]{x+1} && \text{(c)} \end{aligned} \tag{3.7}$$

When these are plotted, as in Fig. 3.2, further difficulties are apparent. In each case the root is correctly given by the intersection of $y=F(x)$ with $y=x$. The function $F_3(x)$, however, has no real value for $x<-1.0$ and the function $F_2(x)$ has singular points at $x=\pm 1.0$. It can also be seen that $F_1(x)$ and $F_2(x)$ are changing very rapidly in the region of the root ($x \cong 1.3245$) in contrast to $F_3(x)$ which is changing very slowly. In fact we shall see that the simple iterative substitution method just described will only yield a correct result if eq. 3.7(c) is used with a starting guess of $x>-1.0$. In order to demonstrate this, we use the simple program listed as Program 3.1.

Program 3.1. Simple iteration method for a nonlinear equation
The meanings of the quantities used in the program are as follows

X0	Starting guess for x
TOL	'Tolerance', i.e. relative change in x in consecutive iterations
ITS	Maximum number of iterations allowed
X1	New value of $x=F(x)$
ITERS	Current number of iterations
ICON	Flag which is set to 1 by subroutine CHECK if convergence achieved

The program expects $F(x)$ to be supplied by the user as a FORTRAN FUNCTION. For example, the function $F_3(x)$ is generated as shown at the end of the program in FUNCTION F using the identity

$$\sqrt[3]{x+1}=\exp[\tfrac{1}{3}\log_e(x+1)] \tag{3.8}$$

The program merely reads a starting value x_0, and calculates a new value $x_1=F(x_0)$. If x_1 is close

```
      PROGRAM P31
C
C      PROGRAM 3.1 ITERATIVE SUBSTITUTION FOR SINGLE ROOT
C
      READ (5,*) X0,TOL,ITS
      WRITE(6,*) ('**************** SIMPLE ITERATION **************')
      WRITE(6,*)
      WRITE(6,*) ('INITIAL VALUE')
      WRITE (6,100) X0
      WRITE(6,*)
      ITERS = 0
      ICON = 0
    1 X1 = F(X0)
      CALL CHECK(X1,X0,TOL,ICON)
      ITERS = ITERS + 1
      IF (ICON.EQ.0 .AND. ITERS.LT.ITS) GO TO 1
      WRITE(6,*)
      WRITE(6,*) ('SOLUTION AND ITERATIONS TO CONVERGENCE')
      WRITE (6,100) X1,ITERS
  100 FORMAT (E12.4,I10)
      STOP
      END
C
      FUNCTION F(X)
C
C      THIS FUNCTION PROVIDES THE VALUE OF F(X)
C      AND WILL VARY FROM ONE PROBLEM TO THE NEXT
C
C
      F = EXP(LOG(X+1.0)/3.0)
      RETURN
      END
```

Initial value	X0
	1.2
Tolerance	TOL
	1.E−5
Iteration limit	ITS
	100

Fig. 3.3(a) Input data for Program 3.1

```
**************** SIMPLE ITERATION **************

INITIAL VALUE
  .1200E+01


SOLUTION AND ITERATIONS TO CONVERGENCE
  .1325E+01         7
```

Fig. 3.3(b) Results from Program 3.1

enough to x_0, checked by library subroutine CHECK, the process terminates. To guard against divergent solutions, a maximum number of iterations is prescribed. Input and output are shown in Figs 3.3(a) and (b) respectively.

One might question whether the starting guess influences the computation very much (as long as, of course, $x_0 > -1.0$). Table 3.1 shows that the influence is small in this case.

Table 3.1. Influence of starting guess on iteration count

Starting value x_0	Number of iterations to convergence
0	8
1.2	7
1.3	6
10.0	9
100.0	10

If $F(x)$ is changed to $F_1(x)$, a divergent result will be obtained with $x=-20593.4$ after 7 iterations. If $F(x)$ is changed to $F_2(x)$, again divergence will be obtained with $x=0.869\times10^{-21}$ after 10 iterations. We shall merely state that convergence will only occur if $|dF/dx|<1$ in the vicinity of the root; that is, the slope of $F(x)$ is less than the slope of the line $y=x$. It is then clear from Fig. 3.2 that only the use of $F_3(x)$ will do.

Example 3.1

Use direct iteration to find a root close to $x=1.3$ of the function

$$f(x)=x^3-x-1=0$$

Solution 3.1

We could write (1) $x=x^3-1$

(2) $x=\dfrac{1}{x^2-1}$

(3) $x=\sqrt[3]{x+1}$

In this case, arrangement (3) is the only one that converges, hence

$$x_0=1.3$$

$$x_1=\sqrt[3]{1.3+1}=1.32006$$

$$x_2=\sqrt[3]{1.32006+1}=1.3238$$

$$x_3=\sqrt[3]{1.3238+1}=1.3245 \quad \text{etc.}$$

3.3 Multiple roots and other difficulties

It can readily be shown that the nonlinear equation

$$f(x)=x^4-6x^3+12x^2-10x+3=0 \tag{3.9}$$

can be factorised into

$$f(x)=(x-3)(x-1)(x-1)(x-1)=0 \tag{3.10}$$

so that of the four roots, three are coincident. This function is illustrated in Fig. 3.4. Following our experience with eq. 3.2, we may think of expressing the equation for iterative purposes as

$$x=F(x)=\sqrt[4]{6x^3-12x^2+10x-3} \tag{3.11}$$

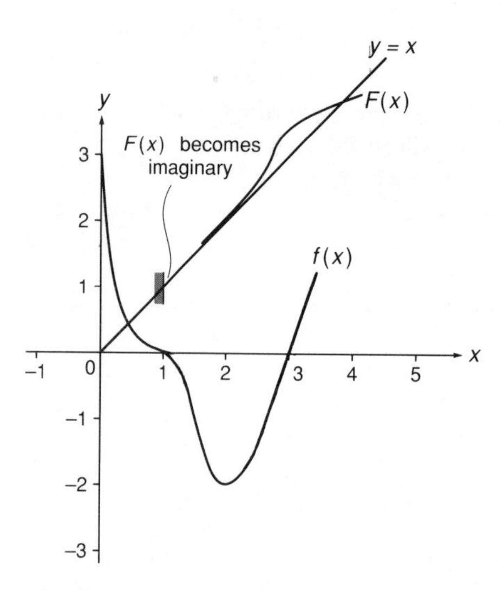

Fig. 3.4 Functions of $f(x)$ and $F(x)$ in eqs 3.10 and 3.11

This function is also shown in Fig. 3.4 where it can be seen that for $x<1.0$, $F(x)$ becomes imaginary. For x just greater than 1.0, $dF/dx>1$, although in the vicinity of the root at $x=3$, $dF/dx<1$.

When Program 3.1 is used to attempt to solve this problem, a starting guess $x_0=1.1$ converges to the root $x=3$ in 311 iterations while starting guesses of $x_0=2.0$ and $x_0=4.0$ converge to the root $x=3$ in close to 100 iterations. We can see that convergence is slow, and that it is impossible to converge on the root $x=1$ at all. We therefore turn to other methods, beginning with those based on interpolation between two estimates to find a root.

3.4 Interpolation methods

This class of methods is based on the assumption that the function changes sign in the vicinity of a root. In fact if, for the type of function shown in eq. 3.10, there had been a double root rather than a triple one, no sign change would have occurred at the double root, but this occurrence is relatively rare. Let us begin by assuming that the vicinity of a root, involving a change in sign of the function, has been located – perhaps graphically. The next two sections describe methods whereby the root can be accurately evaluated. The methods are sometimes called 'bisection' and 'false position' respectively.

3.4.1 Method of bisection

To take a typical example, we know that the function given by eq. 3.2, namely

$$f(x)=x^3-x-1=0 \tag{3.2}$$

has a root close to 1.3, and that $f(x)$ changes sign at that root (see Fig. 3.1). To carry out the bisection process, we would begin with an underestimate of the root, say $x=1.0$, and proceed to evaluate $f(x)$ at equal increments of x until a sign change occurs. This can be checked, for example, by noting that the product of $f(x_{n+1})$ and $f(x_n)$ is negative. We avoid for the moment the obvious difficulties in this process, namely that we may choose steps which are too small and involve excessive work or that for more complicated functions we may choose steps which are too big and miss two or more roots which have involved a double change of sign of the function.

Having established a change of sign of $f(x)$, we have two estimates of the root which bracket it. The value of the function half way between the estimates is then found, i.e. $f[(x_{n+1}+x_n)/2]$ or $f(x_{mid})$. If the sign of the function at this midpoint is the same as that of $f(x_n)$, the root is closer to x_{n+1} and x_{mid} replaces x_n for the next bisection. Alternatively, of course, it replaces x_{n+1}. When successive values of x_{mid} are 'close enough', i.e. within a certain tolerance, the iteration can be stopped.

Example 3.2

Use the bisection method to find a root of the function

$$f(x)=x^3-x-1=0$$

that lies in the range $1.3<x<1.4$.

Solution 3.2

A tabular approach is useful, i.e.

x	$f(x)$	
1.3	−0.1030	
1.4	0.3440	
1.35	0.1104	
1.325	0.0012	
1.3125	−0.0515	
1.31875	−0.0253	
1.32188	−0.0121	
1.32344	−0.0055	
1.32422	−0.0021	etc.

Hence at this stage the root lies in the range

$1.32422 < x < 1.325$

Program 3.2. Bisection method
The bisection method is encapsulated in Program 3.2, which employs the following nomenclature

XN	Underestimate of the root, x_n
XN1	Overestimate of the root, x_{n+1}
XMID	Average of x_n and x_{n+1}
TOL	Iterative tolerance
ITERS	Current number of iterations
ITS	Maximum number of iterations allowed
ICON	Convergence criterion in CHECK

The program's input and output are as shown in Figs 3.5(a) and (b) respectively. Numbers to be input are the first underestimate of the root, the first overestimate, the iteration tolerance and the maximum number of iterations allowed.

```
      PROGRAM P32
C
C      PROGRAM 3.2 BISECTION METHOD FOR A SINGLE ROOT
C
      READ (5,*) XN,XN1,TOL,ITS
      WRITE(6,*) ('*************** BISECTION METHOD ***************')
      WRITE(6,*)
      WRITE(6,*) ('INITIAL VALUES')
      WRITE (6,100) XN,XN1
      ITERS = 0
      ICON = 0
      XOLD = XN
    1 IF (ICON.EQ.0 .AND. ITERS.LE.ITS) THEN
          XMID = 0.5* (XN+XN1)
          PROD = F(XN)*F(XMID)
          IF (PROD.LT..0) THEN
              XN1 = XMID
          ELSE
              XN = XMID
          END IF
          CALL CHECK(XMID,XOLD,TOL,ICON)
          ITERS = ITERS + 1
          GO TO 1
      END IF
      WRITE(6,*)
      WRITE(6,*) ('SOLUTION AND ITERATIONS TO CONVERGENCE')
      WRITE (6,200) XMID,ITERS
  100 FORMAT (2E12.4)
  200 FORMAT (E12.4,I10)
      STOP
      END
C
      FUNCTION F(X)
C
C      THIS FUNCTION PROVIDES THE VALUE OF F(X)
C      AND WILL VARY FROM ONE PROBLEM TO THE NEXT
C
C
      F = X**3 - X - 1.0
      RETURN
      END
```

Initial values	XN	XN1
	1.0	2.0
Tolerance	TOL	
	1.E−5	
Iteration limit	ITS	
	100	

Fig. 3.5(a) Input data for Program 3.2

```
*************** BISECTION METHOD ***************

INITIAL VALUES
  .1000E+01   .2000E+01

SOLUTION AND ITERATIONS TO CONVERGENCE
  .1325E+01          17
```

Fig. 3.5(b) Results from Program 3.2

As long as convergence has not been achieved, or the iteration limit reached, the estimates of the root are bisected to give x_{mid}, and the lower or upper estimate are updated as required. A library routine CHECK checks if any convergence has been obtained. (This is done in the same way as CHECON uses for arrays.)

When the tolerance has been achieved, or the iterations exhausted, the current estimate of the root is printed together with the number of iterations to achieve it.

As shown in Fig. 3.5(b), for the tolerance of 1×10^{-5}, the bisection method takes 17 iterations to converge from lower and upper starting limits of 1.0 and 2.0.

3.4.2 False position method

Again this is based on finding roots of opposite sign and interpolating between them, but by a method which is generally more efficient than bisection.

Figure 3.6 shows a plot of $f_1(x)$ versus x with a root in the region of $x=2.5$. Suppose our interpolation guesses are $x_n=1.0$ and $x_{n+1}=4.0$. The false position method interpolates linearly between $[x_n, f(x_n)]$ and $[x_{n+1}, f(x_{n+1})]$, taking the intersection of the interpolating line with the x-axis as an improved guess at the root. Using the same procedure as in bisection, the improved guess replaces either the lower bound or upper bound previous guess as the case may be. The interpolation can be written

$$x_{new}=x_n-f(x_n)\left[\frac{x_{n+1}-x_n}{f(x_{n+1})-f(x_n)}\right] \tag{3.12}$$

Example 3.3

Use the false position method to find a root of the function

$f(x)=x^3-x-1=0$

in the range $1.3<x<1.4$.

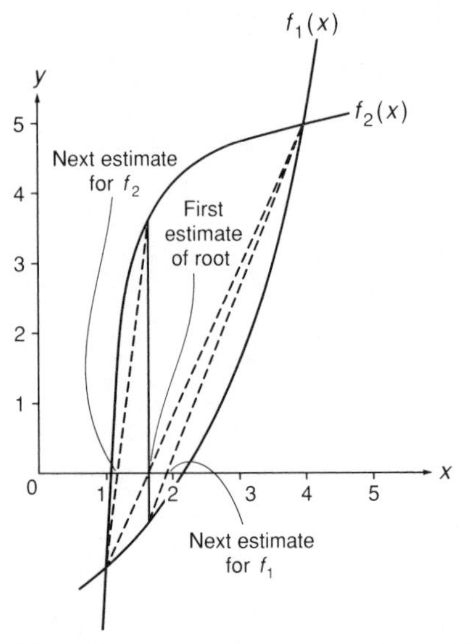

Fig. 3.6 False position method for different $f(x)$

Solution 3.3

False position formula

$$x_{new} = x_n - f(x_n)\left[\frac{x_{n+1} - x_n}{f(x_{n+1}) - (fx_n)}\right]$$

x	$f(x)$
1.3	−0.1030
1.4	0.3440
1.32304	−0.0071
1.32460	−0.0005
1.32471	−0.0000

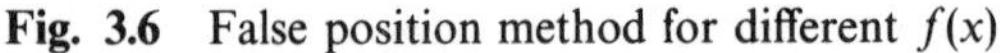

Program 3.3. False position method
This program illustrates the process described by eq. 3.12. Comparison with the previous program will show that the difference is essentially limited to one line, in which eq. 3.12 replaces the bisection statement. For clarity x_{mid}(XMID) in the bisection program is replaced by x_{new}(XNEW) in the false position program. When the program is run with the data of Fig. 3.7(a), the converged result (Fig. 3.7(b)) will be found to have been achieved in 13 iterations compared to bisection's 17.

```
      PROGRAM P33
C
C      PROGRAM 3.3 FALSE POSITION METHOD FOR A SINGLE ROOT
C
      READ (5,*) XN,XN1,TOL,ITS
      WRITE(6,*) ('********** FALSE POSITION METHOD ***************')
      WRITE(6,*)
      WRITE(6,*) ('INITIAL VALUES')
```

```
      WRITE (6,100) XN,XN1
      ITERS = 0
      ICON = 0
      XOLD = XN
    1 IF (ICON.EQ.0 .AND. ITERS.LE.ITS) THEN
          XNEW = XN - F(XN)* (XN1-XN)/ (F(XN1)-F(XN))
          PROD = F(XN)*F(XNEW)
          IF (PROD.LT..0) THEN
              XN1 = XNEW
          ELSE
              XN = XNEW
          END IF
          CALL CHECK(XNEW,XOLD,TOL,ICON)
          ITERS = ITERS + 1
          XOLD = XNEW
          GO TO 1
      END IF
      WRITE(6,*)
      WRITE(6,*) ('SOLUTION AND ITERATIONS TO CONVERGENCE')
      WRITE (6,200) XNEW,ITERS
  100 FORMAT (2E12.4)
  200 FORMAT (E12.4,I10)
      STOP
      END
C
      FUNCTION F(X)
C
C      THIS FUNCTION PROVIDES THE VALUE OF F(X)
C      AND WILL VARY FROM ONE PROBLEM TO THE NEXT
C
C
      F = X**3 - X - 1.0
      RETURN
      END
```

Initial values	XN	XN1
	1.0	2.0
Tolerance	TOL	
	1.E−5	
Iteration limit	ITS	
	100	

Fig. 3.7(a) Input data for Program 3.3

```
********** FALSE POSITION METHOD ***************

INITIAL VALUES
  .1000E+01   .2000E+01

SOLUTION AND ITERATIONS TO CONVERGENCE
  .1325E+01          13
```

Fig. 3.7(b) Results from Program 3.3

This increased efficiency will generally be found unless the function varies rather sharply close to one of the guessed roots. For example, Fig. 3.6 shows $f_2(x)$ for which linear interpolation yields a slowly converging approach to the root close to 1.0 for starting guesses of 1.0 and 4.0.

3.5 Extrapolation methods

A disadvantage of interpolation methods is the need to find sign changes in the function before carrying out the calculation. Extrapolation methods do not suffer from this problem, but this is not to say that convergence difficulties are avoided, as we shall see. Probably the most widely used of all extrapolation methods is often called the Newton–Raphson method. It bases its extrapolation procedure on the slope of the function at the guessed root.

3.5.1 Newton–Raphson method

Suppose we expand a Taylor series about a single guess at a root, x_n. For a small step h in the x direction the Taylor expansion is

$$f(x_{n+1})=f(x_n+h)=f(x_n)+hf'(x_n)+\frac{h^2}{2}f''(x_n)+\cdots \tag{3.13}$$

Consequently x_{n+1} is a root, i.e. $f(x_n+h)=0$ if, dropping terms in the expansion higher than f',

$$f(x_n)+hf'(x_n)=0$$

or (3.14)

$$f(x_n)+(x_{n+1}-x_n)f'(x_n)=0$$

Rearranging, we have the condition for a root

$$x_{n+1}=x_n-\frac{f(x_n)}{f'(x_n)} \tag{3.15}$$

Obviously this has a simple graphical interpretation as shown in Fig. 3.8. The extrapolation is just tangential to the function at the guessed point x_n. The new feature of the method is the need to calculate the derivative of the function, and this can present difficulties (see Chapter 5). However, for simple algebraic expressions like eq. 3.2 the differentiation is easily done, and the derivative returned as a FORTRAN FUNCTION.

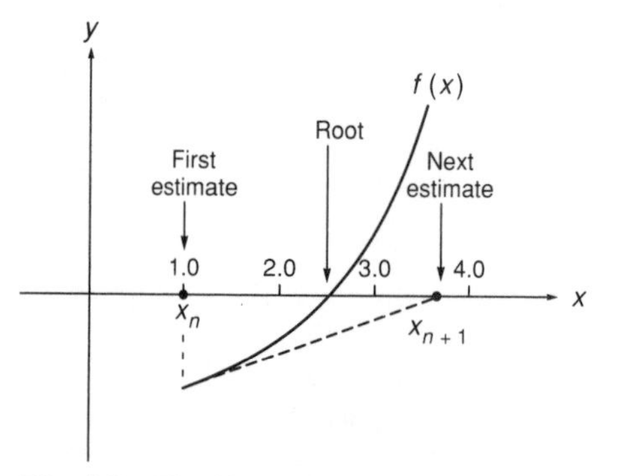

Fig. 3.8 The Newton–Raphson process

Example 3.4

Use the Newton–Raphson method to find a root close to $x=2$ of the function

$f(x)=x^3-x-1=0$

Solution 3.4

Newton–Raphson formula $x_{n+1}=x_n-\dfrac{f(x_n)}{f'(x_n)}$

hence in this case, $x_{n+1}=x_n-\dfrac{x_n{}^3-x_n-1}{3x_n{}^2-1}$

x	$f(x)$
2	5.0000
1.5455	1.1458
1.3596	0.1537
1.3258	0.0046
1.3247	0.0000

Program 3.4. Newton–Raphson method
A program for Newton–Raphson iteration is listed as Program 3.4. It is based on the simple iteration Program 3.1, and in fact differs from it only in one line, in which x_n (called X0 in the program) is updated to x_{n+1} (called X1 in the program) according to eq. 3.15. In addition to a FUNCTION F delivering $f(x)$ there is the extra FUNCTION FDASH delivering $f'(x)$.

```
      PROGRAM P34
C
C      PROGRAM 3.4 NEWTON-RAPHSON METHOD FOR SINGLE ROOT
C
      READ (5,*) X0,TOL,ITS
      WRITE(6,*) ('********** NEWTON-RAPHSON METHOD ***************')
      WRITE(6,*)
      WRITE(6,*) ('INITIAL VALUE')
      WRITE (6,100) X0
      ITERS = 0
      ICON = 0
    1 X1 = X0 - F(X0)/FDASH(X0)
      CALL CHECK(X1,X0,TOL,ICON)
      ITERS = ITERS + 1
      IF (ICON.EQ.0 .AND. ITERS.LT.ITS) GO TO 1
      WRITE(6,*)
      WRITE(6,*) ('SOLUTION AND ITERATIONS TO CONVERGENCE')
      WRITE (6,100) X1,ITERS
  100 FORMAT (E12.4I10)
      STOP
      END
C
      FUNCTION F(X)
C
```

```
C        THIS FUNCTION PROVIDES THE VALUE OF F(X)
C        AND WILL VARY FROM ONE PROBLEM TO THE NEXT
C
C
      F = X**3 - X - 1.0
      RETURN
      END
C
      FUNCTION FDASH(X)
C
C        THIS FUNCTION PROVIDES THE VALUE OF DF/DX
C        AND WILL VARY FROM ONE PROBLEM TO THE NEXT
C
C
      FDASH = 3.0*X**2 - 1.0
      RETURN
      END
```

Initial value	X0 1.2
Tolerance	TOL 1.E−5
Iteration limit	ITS 100

Fig. 3.9(a) Input data for Program 3.4

```
********** NEWTON-RAPHSON METHOD ***************

INITIAL VALUE
  .1200E+01

SOLUTION AND ITERATIONS TO CONVERGENCE
  .1325E+01          4
```

Fig. 3.9(b) Results from Program 3.4

When the program is run with the data from Fig. 3.9(a), convergence is now achieved in 4 iterations as shown in Fig. 3.9(b).

3.5.2 A modified Newton–Raphson method

For large systems of equations, the need to be forever calculating derivatives can make the iteration process expensive. A modified method can be used (there are many other 'modified' methods) in which the first evaluation of $f'(x_n)$ is used for all further extrapolations. This is shown graphically in Fig. 3.10(b), which contrasts with the basic method shown in Fig. 3.10(a).

Example 3.5

Use the modified Newton–Raphson method to find a root close to $x=2$ of the function

$f(x)=x^3-x-1=0$

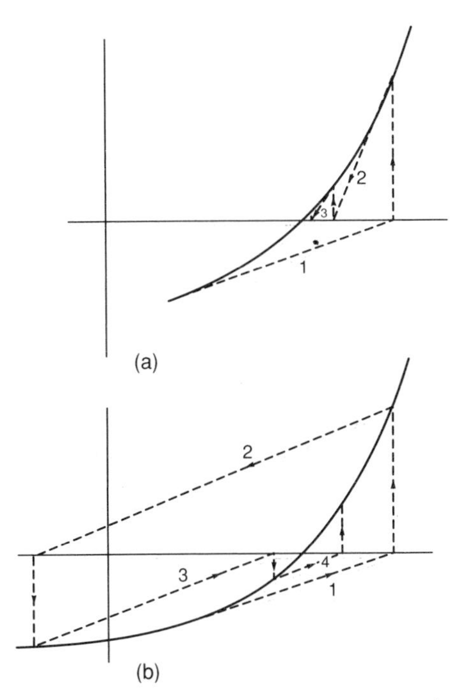

Fig. 3.10 Convergence of Newton–Raphson processes (a) Newton–Raphson, (b) modified Newton–Raphson

Solution 3.5

Modified Newton–Raphson formula $x_{n+1}=x_n-\dfrac{f(x_n)}{f'(x_0)}$

let $x_0=2$, hence in this case, $x_{n+1}=x_n-\dfrac{f(x_n)}{3(2)^2-1}$

$$\therefore \quad x_{n+1}=x_n-\frac{f(x_n)}{11}$$

x	$f(x)$	
2	5.0000	
1.5455	1.1458	
1.4413	0.5528	
1.3910	0.3006	
1.3637	0.1724	
1.3480	0.1016	
1.3388	0.0609	
1.3333	0.0368	
1.3299	0.0223	
1.3279	0.0136	
1.3267	0.0083	etc.

The convergence in this case is rather slow because the gradient at $x=2$, which remains unchanged, is a poor approximation to the 'correct' gradient at $x \simeq 1.3247$.

Program 3.5. Modified Newton–Raphson method
The program is almost identical to the previous one. The initial derivative is called FD and used thereafter as a constant. When the program is run with the data of Fig. 3.11(a) for the two different initial values, the results on each iteration are listed in Figs 3.11(b) and (c), which illustrate one of the typical problems of Newton–Raphson type methods. It can be seen that with a starting value of 1.0 the estimated root oscillates between two constant values, and will never converge to the correct solution. On the other hand, if the starting guess is changed to 1.2, convergence occurs in 8 iterations.

We can see that the process is far from automatic and that the closer to the solution the initial guess is, the better.

```
      PROGRAM P35
C
C      PROGRAM 3.5 MODIFIED NEWTON-RAPHSON FOR SINGLE ROOT
C
      READ (5,*) X0,TOL,ITS
      WRITE(6,*) ('****** MODIFIED NEWTON-RAPHSON METHOD ***********')
      WRITE(6,*)
      WRITE(6,*) ('INITIAL VALUE')
      WRITE (6,100) X0
      WRITE(6,*)
      WRITE(6,*) ('VALUE OF X AT EACH ITERATION')
      ITERS = 0
      ICON = 0
      FD = FDASH(X0)
    1 X1 = X0 - F(X0)/FD
      CALL CHECK(X1,X0,TOL,ICON)
      ITERS = ITERS + 1
      WRITE(6,100)X1,ITERS
      IF (ICON.EQ.0 .AND. ITERS.LT.ITS) GO TO 1
      WRITE(6,*)
      WRITE(6,*) ('SOLUTION AND ITERATIONS TO CONVERGENCE')
      WRITE (6,100) X1,ITERS
  100 FORMAT (E12.4,I10)
      STOP
      END
C
      FUNCTION F(X)
C
C      THIS FUNCTION PROVIDES THE VALUE OF F(X)
C      AND WILL VARY FROM ONE PROBLEM TO THE NEXT
C
C
      F = X**3 - X - 1.0
      RETURN
      END
C
      FUNCTION FDASH(X)
C
C      THIS FUNCTION PROVIDES THE VALUE OF DF/DX
C      AND WILL VARY FROM ONE PROBLEM TO THE NEXT
C
C
      FDASH = 3.0*X**2 - 1.0
      RETURN
      END
```

Initial value	X0
	1.0 (1.2)
Tolerance	TOL
	1.E−5
Iteration limit	ITS
	100

Fig. 3.11(a) Input data for Program 3.5

```
****** MODIFIED NEWTON-RAPHSON METHOD ***********

INITIAL VALUE
  .1000E+01

VALUE OF X AT EACH ITERATION
  .1500E+01           1
  .1062E+01           2
  .1494E+01           3
  .1074E+01           4
  .1492E+01           5
     .                .
     .                .
     .                .
     .                .
     .                .
  .1081E+01          96
  .1490E+01          97
  .1081E+01          98
  .1490E+01          99
  .1081E+01         100

SOLUTION AND ITERATIONS TO CONVERGENCE
  .1081E+01         100
```

Fig. 3.11(b) Results from Program 3.5 with X0 = 1.0

```
****** MODIFIED NEWTON-RAPHSON METHOD ***********

INITIAL VALUE
  .1200E+01

VALUE OF X AT EACH ITERATION
  .1342E+01           1
  .1319E+01           2
  .1326E+01           3
  .1324E+01           4
  .1325E+01           5
  .1325E+01           6
  .1325E+01           7
  .1325E+01           8

SOLUTION AND ITERATIONS TO CONVERGENCE
  .1325E+01           8
```

Fig. 3.11(c) Results from Program 3.5 with X0 = 1.2

3.6 Acceleration of convergence

For slowly convergent iterative calculations, 'Aitken's δ^2 acceleration' process is sometimes effective in extrapolating from the converging solutions to the converged result.

The method can be simply stated. If we have three estimates of a root, x_k, x_{k+1} and

x_{k+2} the δ^2 process extrapolates to a new solution

$$x_{new} = \frac{x_k x_{k+2} - x_{k+1}^2}{x_{k+2} - 2x_{k+1} + x_k} \tag{3.16}$$

The process is far from universally successful but can lead to much improved convergence rates.

3.7 Systems of nonlinear equations

For simplicity of presentation, we have so far concentrated on a single nonlinear equation. In practice we usually have many equations to solve simultaneously. For example consider the two equations

$$\begin{aligned} 2x_1^2 + x_2^2 &= 4.32 \\ x_1^2 - x_2^2 &= 0 \end{aligned} \tag{3.17}$$

By inspection, these equations have a root in the vicinity of $x_1 = 1, x_2 = 1$, and in fact in this simple case we see a solution is $x_1 = x_2 = 1.2$. Amongst the methods already described, iterative substitution and Newton–Raphson are most easily applicable and in this section we shall apply them to systems of equations.

3.7.1 Iterative substitution for systems

Equations 3.17 can be cast in the form

$$\begin{aligned} x_1 &= \sqrt{2.16 - \tfrac{1}{2}x_2^2} = F(x_1, x_2) \\ x_2 &= \sqrt{x_1^2} = x_1 \quad = G(x_1 x_2) \end{aligned} \tag{3.18}$$

and the iterative procedure followed as before. For example, a starting guess of $x_1 = x_2 = 1.0$ leads to solutions

Iteration number	0	1	2	3	4	5	6
x_1	1.0	1.29	1.29	1.16	1.16	1.22	1.19
x_2	1.0	1.0	1.29	1.29	1.16	1.16	1.19

which are clearly converging to the correct result. In the manner of Gauss–Seidel iteration, the convergence could be improved by using updated values as soon as they become available.

Program 3.6. Simple iteration for a system of equations
A program for solution of a system of equations by simple iteration now follows:

Simple variables

N	Number of equations
TOL	Iteration tolerance
ITS	Maximum number of iterations
ITERS	Current number of iterations
ICON	Convergence criterion in CHECON

Variable length arrays

X0	Values of $x_1, x_2, \ldots, x_N$ iteration
X1	Values of $x_1, x_2, \ldots, x_N$ at end of iteration

PARAMETER *restrictions*

IN≥N

```
      PROGRAM P36
C
C      PROGRAM 3.6 ITERATIVE SUBSTITUTION FOR MANY EQUATIONS
C
C      ALTER NEXT LINE TO CHANGE PROBLEM SIZE
C
      PARAMETER (IN=20)
C
      REAL X0(IN),X1(IN)
C
      READ (5,*) N,(X0(I),I=1,N)
      READ (5,*) TOL,ITS
      WRITE(6,*) ('** SIMPLE ITERATION FOR A SYSTEM OF EQUATIONS **')
      WRITE(6,*)
      WRITE(6,*) ('INITIAL VALUES')
      CALL PRINTV(X0,N,6)
      WRITE(6,*)
      ITERS = 0
      ICON = 0
    1 CALL F(X0,X1)
      CALL CHECON(X1,X0,N,TOL,ICON)
      ITERS = ITERS + 1
      IF (ICON.EQ.0 .AND. ITERS.LT.ITS) GO TO 1
      WRITE(6,*) ('SOLUTIONS AND ITERATIONS TO CONVERGENCE')
      CALL PRINTV(X1,N,6)
      WRITE(6,100) ITERS
  100 FORMAT (I10)
      STOP
      END
C
      SUBROUTINE F(X0,X1)
C
C      THIS SUBROUTINE PROVIDES THE VALUES OF THE FUNCTIONS
C      OF X AND WILL VARY FROM ONE PROBLEM TO THE NEXT
C
C
      REAL X0(*),X1(*)
C
      X1(1) = SQRT(2.16-0.5*X0(2)*X0(2))
      X1(2) = X0(1)
      RETURN
      END
```

Number of unknowns	N	
	2	
Initial values	X0(I), I = 1, N	
	1.0	1.0
Tolerance	TOL	
	1.E−5	
Iteration limit	ITS	
	100	

Fig. 3.12(a) Input data for Program 3.6

```
** SIMPLE ITERATION FOR A SYSTEM OF EQUATIONS **

INITIAL VALUES
    .1000E+01    .1000E+01

SOLUTIONS AND ITERATIONS TO CONVERGENCE
    .1200E+01    .1200E+01
        31
```

Fig. 3.12(b) Results from Program 3.6

Input data and output results are shown in Figs 3.12(a) and (b) respectively. The program is really just a copy of Program 3.1 with arrays in place of simple variables. The functions F and G in eqs 3.18 are user-supplied in the form of a subroutine called F. For these equations and the starting guess $x_1 = x_2 = 1.0$, convergence is achieved in 31 iterations.

Since, for a single equation, convergence was conditional on the magnitude of the gradient of the function, we can guess that a similar restriction will apply to systems of equations. In fact, it can be shown that equations like 3.18 will only have convergent iterative solutions of this type if

$$\left|\frac{\partial F}{\partial x_1}\right| + \left|\frac{\partial F}{\partial x_2}\right| < 1$$

and (3.19)

$$\left|\frac{\partial G}{\partial x_1}\right| + \left|\frac{\partial G}{\partial x_2}\right| < 1$$

This restriction can limit the use of the method in practice

3.7.2 Newton–Raphson for systems

This is based on Taylor expansions in several variables. For example, in the case of two equations, such as eqs 3.17, which could be written

$$\begin{aligned} f(x_1, x_2) &= 0 \\ g(x_1, x_2) &= 0 \end{aligned} \qquad (3.20)$$

suppose we make a guess $(x_1{}^k, x_2{}^k)$ which is close to a root $(x_1{}^{k+1}, x_2{}^{k+1})$. The Taylor expansions are

$$f(x_1{}^{k+1}, x_2{}^{k+1}) = f(x_1{}^k, x_2{}^k) + (x_1{}^{k+1} - x_1{}^k)\frac{\partial f(x_1{}^k, x_2{}^k)}{\partial x_1} + (x_2{}^{k+1} - x_2{}^k)\frac{\partial f(x_1{}^k, x_2{}^k)}{\partial x_2} + \cdots$$

(3.21)

$$g(x_1{}^{k+1}, x_2{}^{k+1}) = g(x_1{}^k, x_2{}^k) + (x_1{}^{k+1} - x_1{}^k)\frac{\partial g(x_1{}^k, x_2{}^k)}{\partial x_1} + (x_2{}^{k+1} - x_2{}^k)\frac{\partial g(x_1{}^k, x_2{}^k)}{\partial x_2} + \cdots$$

Again neglecting derivatives higher than the first, eqs 3.21 give an improved estimate of a root in the form

$$(x_1^{k+1}-x_1^k)\frac{\partial f(x_1^k, x_2^k)}{\partial x_1}+(x_2^{k+1}-x_2^k)\frac{\partial f(x_1^k, x_2^k)}{\partial x_2}=-f(x_1^k, x_2^k) \tag{3.22}$$

$$(x_1^{k+1}-x_1^k)\frac{\partial g(x_1^k, x_2^k)}{\partial x_1}+(x_2^{k+1}-x_2^k)\frac{\partial g(x_1^k, x_2^k)}{\partial x_2}=-g(x_1^k, x_2^k)$$

These are readily expressed in terms of changes in x_1 and x_2

$$\Delta x_1 = x_1^{k+1} - x_1^k$$

and (3.23)

$$\Delta x_2 = x_2^{k+1} - x_2^k$$

so that eqs 3.22 become

$$\Delta x_1 \frac{\partial f}{\partial x_1} + \Delta x_2 \frac{\partial f}{\partial x_2} = -f$$

$$\Delta x_1 \frac{\partial g}{\partial x_1} + \Delta x_2 \frac{\partial g}{\partial x_2} = -g \tag{3.24}$$

or

$$\begin{bmatrix} \frac{\partial f}{\partial x_1} & \frac{\partial f}{\partial x_2} \\ \frac{\partial g}{\partial x_1} & \frac{\partial g}{\partial x_2} \end{bmatrix} \begin{Bmatrix} \Delta x_1 \\ \Delta x_2 \end{Bmatrix} = \begin{Bmatrix} -f \\ -g \end{Bmatrix} \tag{3.25}$$

Thus for a system of N simultaneous nonlinear equations, we must solve N simultaneous linear equations to find the changes Δx_1 and Δx_2 in every iteration. The matrix on the left-hand side of eqs 3.25 is called the 'Jacobian' matrix **J** and its determinant simply 'the Jacobian'. Clearly, if the Jacobian is zero, the process fails, and if it is close to zero, slow convergence may be anticipated.

For a starting guess of $x_1 = x_2 = 1.0$, the iteration matrix for the first iteration on eqs 3.17 is:

$$\begin{bmatrix} 4 & 2 \\ 2 & -2 \end{bmatrix} \begin{Bmatrix} \Delta x_1 \\ \Delta x_2 \end{Bmatrix} = \begin{Bmatrix} 1.32 \\ 0 \end{Bmatrix}$$

which yields the solution $\Delta x_1 = \Delta x_2 = 0.22$ and hence an improved solution $x_1 = x_2 = 1.22$.

Example 3.6

Use the Newton–Raphson method to find a root of the equations

$$xy = 1$$
$$x^2 + y^2 = 4$$

Close to $x = 1.8$, $y = 0.5$.

Solution 3.6

Arrange equation as follows:

$$f(x, y) = xy - 1 = 0$$
$$f(x, y) = x^2 + y^2 - 4 = 0$$

$$\therefore \quad \mathbf{J} = \begin{bmatrix} y & x \\ 2x & 2y \end{bmatrix}$$

$$\mathbf{J}^{-1} = \frac{1}{2y^2 - 2x^2} \begin{bmatrix} 2y & -x \\ -2x & y \end{bmatrix}$$

1st iteration
Initial guess $x = 1.8$, $y = 0.5$

$$\mathbf{J}^{-1} = \begin{bmatrix} -0.1672 & 0.3010 \\ 0.6020 & -0.0836 \end{bmatrix}$$

$$\therefore \quad \begin{Bmatrix} \Delta x \\ \Delta y \end{Bmatrix} = \begin{bmatrix} -0.1672 & 0.3010 \\ 0.6020 & -0.0836 \end{bmatrix} \begin{Bmatrix} 0.1 \\ 0.51 \end{Bmatrix} = \begin{Bmatrix} 0.1368 \\ 0.0176 \end{Bmatrix}$$

2nd iteration

$$x = 1.8 + 0.1368 = 1.9368$$
$$y = 0.5 + 0.0176 = 0.5176$$

$$\mathbf{J}^{-1} = \begin{bmatrix} -0.1486 & 0.2780 \\ 0.5560 & -0.0743 \end{bmatrix}$$

$$\therefore \quad \begin{Bmatrix} \Delta x \\ \Delta y \end{Bmatrix} = \begin{bmatrix} -0.1486 & 0.2780 \\ 0.5560 & -0.0743 \end{bmatrix} \begin{Bmatrix} -0.0025 \\ -0.0191 \end{Bmatrix} = \begin{Bmatrix} -0.0049 \\ 0.0000 \end{Bmatrix}$$

3rd iteration

$$x = 1.9368 - 0.0049 = 1.9319$$
$$y = 0.5176 \qquad = 0.5176 \quad \text{etc.}$$

Program 3.7. Newton–Raphson method for a system of equations
A program for solution of a system of equations by the Newton–Raphson method now follows.

Simple variables

N	Number of equations
TOL	Convergence tolerance
ITS	Maximum number of iterations allowed
ITERS	Current number of iterations
ICON	Convergence criterion in CHECON

```
      PROGRAM P37
C
C      PROGRAM 3.7 NEWTON-RAPHSON FOR MANY EQUATIONS
C
C      ALTER NEXT LINE TO CHANGE PROBLEM SIZE
C
      PARAMETER (IN=20)
C
      REAL X(IN),X1(IN),DELX(IN),F(IN),DF(IN,IN)
C
      READ (5,*) N,(X(I),I=1,N)
      READ (5,*) TOL,ITS
      WRITE(6,*) ('**** NEWTON-RAPHSON FOR A SYSTEM OF EQUATIONS ****')
      WRITE(6,*)
      WRITE(6,*) ('INITIAL VALUES')
      CALL PRINTV(X,N,6)
      WRITE(6,*)
      ITERS = 0
      ICON = 0
    1 CALL FANDDF(X,F,DF,IN,N)
      CALL MATINV(DF,IN,N)
      CALL MVMULT(DF,IN,F,N,N,DELX)
      CALL VECSUB(X,DELX,X1,N)
      CALL CHECON(X1,X,N,TOL,ICON)
      ITERS = ITERS + 1
      IF (ICON.EQ.0 .AND. ITERS.LT.ITS) GO TO 1
      WRITE(6,*) ('SOLUTIONS AND ITERATIONS TO CONVERGENCE')
      CALL PRINTV(X1,N,6)
      WRITE (6,100) ITERS
  100 FORMAT (I10)
      STOP
      END
C
      SUBROUTINE FANDDF(X,F,DF,IN,N)
C
C      THIS SUBROUTINE PROVIDES THE VALUES OF THE FUNCTIONS
C      OF X AND THEIR DERIVATIVES, AND WILL VARY FROM ONE
C      PROBLEM TO THE NEXT
C
C
      REAL X(*),F(*),DF(IN,*)
C
      F(1) = 2.0*X(1)*X(1) + X(2)*X(2) - 4.32
      F(2) = X(1)*X(1) - X(2)*X(2)
      DF(1,1) = 4.0*X(1)
      DF(2,1) = 2.0*X(1)
      DF(1,2) = 2.0*X(2)
      DF(2,2) = -2.0*X(2)
      RETURN
      END
```

Variable length arrays

X	Values of $x_1, x_2, \ldots x_N$ at start of iteration
X1	Values of $x_1, x_2, \ldots x_N$ at end of iteration
DELX	Change in solution vector in an iteration
F	The function in eq. 3.19
DF	The first derivatives in eq. 3.25 arranged as a two-dimensional array

PARAMETER *restriction*

IN≥N

Number of unknowns	N	
	2	
Initial values	X0(I), I=1, N	
	1.0	1.0
Tolerance	TOL	
	1.E−5	
Iteration limit	ITS	
	100	

Fig. 3.13(a) Input data for Program 3.7

```
**** NEWTON-RAPHSON FOR A SYSTEM OF EQUATIONS ****

INITIAL VALUES
   .1000E+01    .1000E+01

SOLUTIONS AND ITERATIONS TO CONVERGENCE
   .1200E+01    .1200E+01
        4
```

Fig. 3.13(b) Results from Program 3.7

The input data and output results are given in Figs 3.13(a) and (b) respectively. The number of equations, an initial guess of the solution vector, the iteration tolerance and the number of iterations allowed must be read in by the program. The actual functions and their derivations are contained in the user-supplied subroutine FANDDF.

Library routines MATINV and MVMULT complete the equation solution required by eqs 3.25, resulting in the changes in X held in DELX. Vector X is then incremented by DELX and the CHECON subroutine called. Note that for large systems, MATINV should be replaced by the solution algorithms described in Chapter 2. Figure 3.13(b) shows that convergence was achieved in this simple case in 4 iterations.

3.7.3 Modified Newton–Raphson method for systems

The presence of subroutine MATINV inside the iteration loop will mean substantial calculations if N becomes large. We described previously a modified method which, in this case, would mean that the matrix on the left-hand side of eqs 3.25 need be inverted once only, at the first iteration. This method is programmed as Program 3.8.

Example 3.7

Use the modified Newton–Raphson method to find a root of the equations

$e^x + y = 0$

$\cosh y - x = 3.5$

Close to $x = -2.4$, $y = -0.1$

Solution 3.7

Arrange equations as follows

$f_1(x, y) = e^x + y = 0$

$f_2(x, y) = -x + \cosh y - 3.5 = 0$

$$\mathbf{J} = \begin{bmatrix} e^x & 1 \\ -1 & \sinh y \end{bmatrix}$$

$$\mathbf{J}^{-1} = \frac{1}{1 + e^x \sinh y} \begin{bmatrix} \sinh y & -1 \\ 1 & e^x \end{bmatrix}$$

Evaluate $\mathbf{J}^{-1}$ using the initial value $x = -2.4$, $y = -0.1$

$$\mathbf{J}^{-1} = \begin{bmatrix} -0.1011 & -1.0092 \\ 1.0092 & 0.0915 \end{bmatrix}$$

In the modified Newton–Raphson method, $\mathbf{J}^{-1}$ is not altered at each iteration. Hence

1st iteration

$x = -2.4$, $y = -0.1$

$$\begin{Bmatrix} \Delta x \\ \Delta y \end{Bmatrix} = \begin{bmatrix} -0.1011 & -1.0092 \\ 1.0092 & 0.0915 \end{bmatrix} \begin{Bmatrix} 0.0093 \\ 0.0950 \end{Bmatrix} = \begin{Bmatrix} -0.0968 \\ 0.0181 \end{Bmatrix}$$

2nd iteration

$x = -2.4 - 0.0968 = -2.4968$

$y = -0.1 + 0.0181 = -0.0819$

$$\begin{Bmatrix} \Delta x \\ \Delta y \end{Bmatrix} = \begin{bmatrix} -0.1011 & -1.0092 \\ 1.0092 & 0.0915 \end{bmatrix} \begin{Bmatrix} -0.0004 \\ -0.0002 \end{Bmatrix} = \begin{Bmatrix} 0.0002 \\ -0.0004 \end{Bmatrix}$$

3rd iteration

$x = -2.4968 + 0.0002 = 2.4966$

$y = -0.0819 - 0.0004 = 0.0823$ etc.

Program 3.8. Modified Newton–Raphson method for a system of equations

```
      PROGRAM P38
C
C      PROGRAM 3.8 MODIFIED  NEWTON-RAPHSON FOR MANY EQUATIONS
C
C      ALETR NEXT LINE TO CHANGE PROBLEM SIZE
C
      PARAMETER (IN=20)
C
      REAL X(IN),X1(IN),DELX(IN),F(IN),DF(IN,IN)
C
      READ (5,*) N,(X(I),I=1,N)
      READ (5,*) TOL,ITS
      WRITE(6,*) ('************ MODIFIED NEWTON-RAPHSON *************')
      WRITE(6,*) ('*********** FOR A SYSTEM OF EQUATIONS ************')
      WRITE(6,*)
      WRITE(6,*) ('INITIAL VALUES')
      CALL PRINTV(X,N,6)
      WRITE(6,*)
      ITERS = 0
      ICON = 0
      CALL FANDDF(X,F,DF,IN,N)
      CALL MATINV(DF,IN,N)
    1 CALL FUN(X,F)
      CALL MVMULT(DF,IN,F,N,N,DELX)
      CALL VECSUB(X,DELX,X1,N)
      CALL CHECON(X1,X,N,TOL,ICON)
      ITERS = ITERS + 1
      IF (ICON.EQ.0 .AND. ITERS.LT.ITS) GO TO 1
      WRITE(6,*) ('SOLUTIONS AND ITERATIONS TO CONVERGENCE')
      CALL PRINTV(X1,N,6)
      WRITE (6,100) ITERS
  100 FORMAT (I10)
      STOP
      END
C
      SUBROUTINE FANDDF(X,F,DF,IN,N)
C
C      THIS SUBROUTINE PROVIDES THE VALUES OF THE FUNCTIONS
C      OF X AND THEIR DERIVATIVES, AND WILL VARY FROM ONE
C      PROBLEM TO THE NEXT
C
C
      REAL X(*),F(*),DF(IN,*)
C
      F(1) = 2.0*X(1)*X(1) + X(2)*X(2) - 4.32
      F(2) = X(1)*X(1) - X(2)*X(2)
      DF(1,1) = 4.0*X(1)
      DF(2,1) = 2.0*X(1)
      DF(1,2) = 2.0*X(2)
      DF(2,2) = -2.0*X(2)
      RETURN
      END
C
      SUBROUTINE FUN(X,F)
C
C      THIS SUBROUTINE PROVIDES THE VALUES OF THE FUNCTIONS
```

```
C       OF X AND WILL VARY FROM ONE PROBLEM TO THE NEXT
C
C
      REAL X(*),F(*)
C
      F(1) = 2.0*X(1)*X(1) + X(2)*X(2) - 4.32
      F(2) = X(1)*X(1) - X(2)*X(2)
      RETURN
      END
```

Number of unknowns	N 2
Initial values	X0(I), I=1,N 1.0 1.0
Tolerance	TOL 1.E−5
Iteration limit	ITS 100

Fig. 3.14(a) Input data for Program 3.8

```
************ MODIFIED NEWTON-RAPHSON *************
*********** FOR A SYSTEM OF EQUATIONS ************

INITIAL VALUES
   .1000E+01    .1000E+01

SOLUTIONS AND ITERATIONS TO CONVERGENCE
   .1200E+01    .1200E+01
         7
```

Fig. 3.14(b) Results from Program 3.8

A new subroutine called FUN is used after the first iteration, and MATINV is moved outside the iteration loop. Otherwise the program is identical to Program 3.7. When run with the data given in Fig. 3.14(a) it produces the results shown in Fig. 3.14(b).

In this simple case, convergence is reached in seven iterations using the modified method and, in general, it will require more iterations than the basic Newton–Raphson procedure. It will be a question in practice of whether the increase in iterations in the modified method more than compensates for the reduction in equation solving.

3.8 Exercises

1 Find a root of the equation $x^4-8x^3+23x^2+16x-50=0$ in the vicinity of 1.0 by simple iteration.
Answer: 1.4142

2 Find the root of the equation $x^3-3x^2+2x-0.375=0$ in the vicinity of 1.0 by simple iteration in the form $x=[(x^3+2x-0.375)/3]^{1/2}$.
Answer: 0.5 in 95 iterations for a tolerance of 1×10^{-5}.

3 Find a root of the equation in Exercise 1 in the range $1.0<x<2.0$ by bisection.
Answer: 1.4142 in 17 iterations for a tolerance of 1×10^{-5}.

4 Find a root of the equation in Exercise 1 in the range $1.0<x<2.0$ by 'false position'.
Answer: 1.4142 in 3 iterations for a tolerance of 1×10^{-5}.

5 Find a root of the equation in Exercise 2 in the range $0.35<x<1.0$ by bisection.
Answer: 0.5 in 17 iterations for a tolerance of 1×10^{-5}.

6 Find a root of the equation in Exercise 2 in the range $0.4<x<0.6$ by 'false position'.
Answer: 0.5 in 12 iterations for a tolerance of 1×10^{-5}.

7 Find a root of the equation in Exercise 1 in the vicinity of 1.0 by the Newton–Raphson process.
Answer: 1.4142 in 3 iterations for a tolerance of 1×10^{-5}.

8 Find a root of the equation in Exercise 2 in the vicinity of 1.0 by the Newton–Raphson process.
Answer: 0.5 in 6 iterations for a tolerance of 1×10^{-5}.

9 Find a root of the equation in Exercise 1 in the vicinity of 1.0 by the modified Newton–Raphson process.
Answer: 1.4142 in 5 iterations for a tolerance of 1×10^{-5}.

10 Find a root of the equation in Exercise 2 in the vicinity of 1.0 by the modified Newton–Raphson process.
Answer: 0.5 in 30 iterations for a tolerance of 1×10^{-5}.

11 Solve the equations

$$x_1+x_2-x_2{}^{1/2}-0.25=0$$
$$8x_1{}^2+16x_2-8x_1x_2-5=0$$

by simple iteration from a starting guess

$$x_1=x_2=1.$$

Answer: $x_1=0.5$, $x_2=0.25$ in 11 iterations for a tolerance of 1×10^{-5}.

12 Solve the equations

$$2x_1{}^2-4x_1x_2-x_2{}^2=0$$
$$2x_2{}^2+10x_1-x_1{}^2-4x_1x_2-5=0$$

by the Newton–Raphson procedure starting from

$$x_1=x_2=1.$$

Answer: $x_1=0.58$, $x_2=0.26$ in 5 iterations for a tolerance of 1×10^{-5}.

13 Solve the equations in Exercise 12 by the modified Newton–Raphson procedure starting from

(a) $x_1=x_2=1$ and (b) $x_1=0.5$, $x_2=0.25$.

Answer: $x_1=0.58$, $x_2=0.26$ in (a) 1470 iterations and (b) 6 iterations for a tolerance of 1×10^{-5}.

14 Solve the equations in Exercise 11 by
(a) the Newton–Raphson procedure and
(b) the modified Newton–Raphson procedure starting from $x_1=1.0$, $x_2=0.1$.
Answer: $x_1=0.5$, $x_2=0.25$ in (a) 5 iterations and (b) 20 iterations for a tolerance of 1×10^{-5}.

3.9 Further reading

Baker, C. and C. Phillips (eds) (1981). *The Numerical Solution of Nonlinear Problems*, Clarendon Press, Oxford, England.
Byrne, G. and C. Hall, eds (1973). *Numerical Solution of Systems of Nonlinear Algebraic Equations*, Academic Press, New York.
Gill, P., W. Murray and M. Wright (1981). *Practical Optimisation*. Academic Press, New York.
Hiebert, K. (1982). An evaluation of mathematical software that solve systems of nonlinear equations. *ACM Trans. Math. Softw.*, **11**, 250–262.
Householder, A. (1970). *The Numerical Treatment of a Single Nonlinear Equation*, McGraw-Hill, New York.
More, J., B. Garbow and K. Hillstrom (1980). *User Guide for MINPACK-1*. Argonne Nat. Lab. Rep. ANL-80-74.
Ortega, J. and W. Rheinboldt (1970). *Iterative Solution of Nonlinear Equations in Several Variables*, Academic Press, New York.
Peters, G. and J. Wilkinson (1971). Practical problems arising in the solution of polynomial equations, *J. Inst. Math. and its Appl.* **8**, 16–35.
Rheinboldt, W. (1974). *Methods for Solving Systems of Nonlinear Equations*, Society for Industrial and Applied Mathematics, Philadelphia.
Traub, J. (1964). *Iterative Methods for the Solution of Equations*. Prentice-Hall, Englewood Cliffs, New Jersey.

4

Eigenvalue Equations

4.1 Introduction

Equations of the type

$$\mathbf{Ax} = \lambda \mathbf{x} \tag{4.1}$$

often occur in practice, for example in the analysis of oscillatory systems. We have to find a vector $\mathbf{x}$ which, when multiplied by $\mathbf{A}$ yields a scalar multiple of itself. This multiple is called an 'eigenvalue' or 'characteristic value' of $\mathbf{A}$ and we shall see there are N of these for a matrix of order N. Physically they might represent frequencies of oscillation. There are also N vectors $\mathbf{x}_i$, one associated with each of the eigenvalues λ_i. These are called 'eigenvectors' or 'characteristic vectors'. Physically they might represent the mode shapes of oscillation.

A specific example of eq. 4.1 might be

$$\begin{bmatrix} 16 & -24 & 18 \\ 3 & -2 & 0 \\ -9 & 18 & -17 \end{bmatrix} \begin{Bmatrix} x_1 \\ x_2 \\ x_3 \end{Bmatrix} = \lambda \begin{Bmatrix} x_1 \\ x_2 \\ x_3 \end{Bmatrix} \tag{4.2}$$

which can be rewritten

$$\begin{bmatrix} 16-\lambda & -24 & 18 \\ 3 & -2-\lambda & 0 \\ -9 & 18 & -17-\lambda \end{bmatrix} \begin{Bmatrix} x_1 \\ x_2 \\ x_3 \end{Bmatrix} = \begin{Bmatrix} 0 \\ 0 \\ 0 \end{Bmatrix} \tag{4.3}$$

A non-trivial solution of this set of linear simultaneous equations is only possible if the determinant of the coefficients is zero

$$\begin{vmatrix} 16-\lambda & -24 & 18 \\ 3 & -2-\lambda & 0 \\ -9 & 18 & -17-\lambda \end{vmatrix} = 0 \tag{4.4}$$

Expanding the determinant gives

$$\lambda^3 + 3\lambda^2 - 36\lambda + 32 = 0 \tag{4.5}$$

which is called the 'characteristic equation'. Clearly one way of solving eigenvalue equations would be to reduce them to an Nth degree characteristic equation and use the methods of the previous chapter to find its roots. This is sometimes done, perhaps as part of a total solution process, but on its own is not usually the best means of solving eigenvalue equations.

In the case of eq. 4.5 the characteristic equation has simple factors

$$(\lambda-4)(\lambda-1)(\lambda+8)=0 \tag{4.6}$$

and so the eigenvalues of our matrix are 4, 1 and -8.

Note that for arbitrary matrices $\mathbf{A}$ the characteristic polynomial is likely to yield imaginary as well as real roots. We shall restrict our discussion to matrices with real eigenvalues and indeed physical constraints will often mean that matrices of interest are symmetric and 'positive definite' (see Chapter 2) in which case all the eigenvalues are real and positive.

Having found an eigenvalue, its associated eigenvector can, in principle, be found by solving a set of linear simultaneous equations. For example, for the case of $\lambda=1$ in eq. 4.6

$$\begin{bmatrix} 15 & -24 & 18 \\ 3 & -3 & 0 \\ -9 & 18 & -18 \end{bmatrix} \begin{Bmatrix} x_1 \\ x_2 \\ x_3 \end{Bmatrix} = \begin{Bmatrix} 0 \\ 0 \\ 0 \end{Bmatrix} \tag{4.7}$$

Carrying out the first stage of a Gaussian elimination gives

$$\begin{bmatrix} 15 & -24 & 18 \\ & 1.8 & -3.6 \\ & 3.6 & -7.2 \end{bmatrix} \begin{Bmatrix} x_1 \\ x_2 \\ x_3 \end{Bmatrix} = \begin{Bmatrix} 0 \\ 0 \\ 0 \end{Bmatrix} \tag{4.8}$$

As we knew already from the zero determinant, the system of equations exhibits linear dependence and all we can say is that the *ratio* of x_2 to x_3 is 2:1. Similarly we find the ratio of x_1 to x_2 is 1:1 and so any vector with the ratios $x_1{:}x_2{:}x_3=2{:}2{:}1$ is an eigenvector associated with the eigenvalue $\lambda=1$.

4.2 Orthogonality and normalisation of eigenvectors

In later sections we shall calculate numerically the eigenvalues and eigenvectors of the real, symmetric matrix

$$\mathbf{A}=\begin{bmatrix} 4 & \frac{1}{2} & 0 \\ \frac{1}{2} & 4 & \frac{1}{2} \\ 0 & \frac{1}{2} & 4 \end{bmatrix} \tag{4.9}$$

and we shall see that its three eigenvalues are $4+1/\sqrt{2}$, 4 and $4-1/\sqrt{2}$ respectively.

The corresponding eigenvectors will be found to be

$$\mathbf{x}^{(1)\mathrm{T}}=\left\{\frac{1}{2} \quad \frac{1}{\sqrt{2}} \quad \frac{1}{2}\right\}$$

$$\mathbf{x}^{(2)\mathrm{T}}=\left\{\frac{1}{\sqrt{2}} \quad 0 \quad -\frac{1}{\sqrt{2}}\right\} \tag{4.10}$$

$$\mathbf{x}^{(3)\mathrm{T}}=\left\{\frac{1}{2} \quad -\frac{1}{\sqrt{2}} \quad \frac{1}{2}\right\}$$

respectively.

These vectors are said to exhibit 'orthogonality' one to the other. That is, the dot product $\mathbf{x}^{(1)\mathrm{T}}\mathbf{x}^{(1)} \neq 0$ but the dot products $\mathbf{x}^{(1)\mathrm{T}}\mathbf{x}^{(2)}$ and $\mathbf{x}^{(1)\mathrm{T}}\mathbf{x}^{(3)}$ are both zero. This property will be found to be possessed by the eigenvectors of all symmetric matrices. We can also note that the three eigenvectors in eq. 4.10 have been scaled or 'normalised' such that $\sqrt{\Sigma_{i=1}^{N} x_i^2}=1$. This is convenient because it means that the dot products like $\mathbf{x}^{(1)\mathrm{T}}\mathbf{x}^{(1)}$ are equal to one. Other normalisations are possible, for example so as to make $|x_i|_{\max}=1$.

4.3 Solution methods for eigenvalue equations

Because of the presence of the unknown vector **x** on both sides of eq. 4.1 we can see that solution methods for eigenvalue problems will be essentially iterative in character. We have already seen one such method, which involved finding the roots of the characteristic polynomial. A second class of methods comprises 'transformation methods' in which the matrix **A** in eq. 4.1 is iteratively transformed into a new matrix, say **A***, which has the same eigenvalues as **A**. However, these eigenvalues are easier to compute than the eigenvalues of the original matrix. A third class of methods comprises 'vector iterative' methods which are perhaps the most obvious of all. Just as we did in Chapter 3 for iterative substitution in the solution of nonlinear equations, a guess is made for **x** on the left-hand side of eq. 4.1, the product **Ax** is formed, and compared with the right-hand side. The guess is then iteratively adjusted until agreement is reached. In the following sections we shall deal with these classes of methods in reverse order, beginning with vector iteration.

4.4 Vector iteration

The procedure is easily described and programmed, and is sometimes called the 'power' method.

Going back to eq. 4.2 which was

$$\begin{bmatrix} 16 & -24 & 18 \\ 3 & -2 & 0 \\ -9 & 18 & -17 \end{bmatrix} \begin{Bmatrix} x_1 \\ x_2 \\ x_3 \end{Bmatrix} = \lambda \begin{Bmatrix} x_1 \\ x_2 \\ x_3 \end{Bmatrix} \tag{4.2}$$

we guess a solution for **x**, say $\{1\ \ 1\ \ 1\}^{\mathrm{T}}$. Then the matrix-by-vector multiplication on the left-hand side yields

$$\begin{bmatrix} 16 & -24 & 18 \\ 3 & -2 & 0 \\ -9 & 18 & -17 \end{bmatrix} \begin{Bmatrix} 1 \\ 1 \\ 1 \end{Bmatrix} = \begin{Bmatrix} 10 \\ 1 \\ -8 \end{Bmatrix} = 10 \begin{Bmatrix} 1.0 \\ 0.1 \\ -0.8 \end{Bmatrix} \tag{4.11}$$

where we have normalised the resulting vector by dividing by the largest absolute term in it to give $|x_i|_{\max} = 1$. The normalised **x** is then used for the next round of iteration.

This round gives

$$\begin{bmatrix} 16 & -24 & 18 \\ 3 & -2 & 0 \\ -9 & 18 & -17 \end{bmatrix} \begin{Bmatrix} 1.0 \\ 0.1 \\ -0.8 \end{Bmatrix} = \begin{Bmatrix} -0.8 \\ 2.8 \\ 6.4 \end{Bmatrix} = 6.4 \begin{Bmatrix} -0.125 \\ 0.4375 \\ 1.0 \end{Bmatrix} \tag{4.12}$$

and the next gives

$$\begin{bmatrix} 16 & -24 & 18 \\ 3 & -2 & 0 \\ -9 & 18 & -17 \end{bmatrix} \begin{Bmatrix} -0.125 \\ 0.4375 \\ 1.0 \end{Bmatrix} = \begin{Bmatrix} 5.5 \\ -1.25 \\ -8 \end{Bmatrix} = -8 \begin{Bmatrix} -0.6875 \\ 0.15625 \\ 1.0 \end{Bmatrix} \tag{4.13}$$

and finally:

$$\begin{bmatrix} 16 & -24 & 18 \\ 3 & -2 & 0 \\ -9 & 18 & -17 \end{bmatrix} \begin{Bmatrix} -0.6875 \\ 0.15625 \\ 1.0 \end{Bmatrix} = \begin{Bmatrix} 3.17 \\ -2.39 \\ -7.865 \end{Bmatrix} = -7.865 \begin{Bmatrix} -0.404 \\ 0.304 \\ 1.0 \end{Bmatrix} \tag{4.14}$$

illustrating convergence towards the eigenvalue $\lambda = -8$. Note that this is the eigenvalue of largest absolute value and the power method will usually converge to this largest eigenvalue.

Program 4.1. Vector iteration to find the 'largest' eigenvalue
Given library subroutines capable of multiplying a matrix by a vector, and checking for convergence of the iterative process, programming of the power method is extremely simple. The following nomenclature is used:

Simple variables

N	Number of equations
TOL	Convergence tolerance
ITS	Maximum number of iterations allowed
ITERS	Current number of iterations
ICON	Convergence criterion in CHECON
BIG	The eigenvalue

Variable length arrays

A	The coefficients of the matrix
X0	Eigenvector before an iteration
X1	Eigenvector after an iteration

PARAMETER *restriction*

IN ≥ N

```
      PROGRAM P41
C
C      PROGRAM 4.1 POWER METHOD FOR EIGENVALUES AND EIGENVECTORS
C
C      ALTER NEXT LINE TO CHANGE PROBLEM SIZE
C
      PARAMETER (IN=20)
C
      REAL A(IN,IN),X0(IN),X1(IN)
C
      READ (5,*) N,((A(I,J),J=1,N),I=1,N)
      READ (5,*) (X0(I),I=1,N)
      READ (5,*) TOL,ITS
      WRITE(6,*) ('** DIRECT ITERATION TO FIND LARGEST EIGENVALUE **')
      WRITE(6,*)
      WRITE(6,*) ('MATRIX A')
      CALL PRINTA(A,IN,N,N,6)
      WRITE(6,*)
      ITERS = 0
    1 CALL MVMULT(A,IN,X0,N,N,X1)
      BIG = 0.0
      DO 2 I = 1,N
    2 IF (ABS(X1(I)).GT.ABS(BIG)) BIG = X1(I)
      DO 3 I = 1,N
    3 X1(I) = X1(I)/BIG
      CALL CHECON(X1,X0,N,TOL,ICON)
      ITERS = ITERS + 1
      IF (ICON.EQ.0 .AND. ITERS.LT.ITS) GO TO 1
      SUM = 0.0
      DO 4 I = 1,N
    4 SUM = SUM + X1(I)**2
      SUM = SQRT(SUM)
      DO 5 I = 1,N
    5 X1(I) = X1(I)/SUM
      WRITE(6,*) ('LARGEST EIGENVALUE')
      WRITE (6,1000) BIG
      WRITE(6,*)
      WRITE(6,*) ('CORRESPONDING EIGENVECTOR')
      CALL PRINTV(X1,N,6)
      WRITE(6,*)
      WRITE(6,*) ('ITERATIONS TO CONVERGENCE')
      WRITE (6,2000) ITERS
 1000 FORMAT (E12.4)
 2000 FORMAT (I5)
      STOP
      END
```

Input and output for the program are shown in Figs 4.1(a) and (b) respectively. The program reads in the number of equations to be solved followed by the coefficient matrix and the initial guessed eigenvector. The tolerance required of the iteration process and the maximum number of iterations complete the data set. After a matrix-vector multiplication using library routine MVMULT, the new eigenvector is normalised so that its largest component is 1.0. The CHECON subroutine then checks if convergence has been achieved and updates the eigenvector. If more iterations are required, the program returns to the MVMULT routine and the process is repeated. After convergence, the eigenvector is normalised so that $\sqrt{\sum_{i=1}^{N} x_i^2} = 1$. Figure 4.1(b) shows that convergence to the 'largest' eigenvalue -8 is achieved in 19 iterations. Subroutines PRINTA and PRINTV output arrays and vectors respectively to the required channel (in this case ICH = 6).

Array size	N		
	3		
Array	(A(I, J), J = 1, N), I = 1, N		
	16.0	−24.0	18.0
	3.0	−2.0	0.0
	−9.0	18.0	−17.0
Initial vector	X0(I), I = 1, N		
	1.0	1.0	1.0
Tolerance	TOL		
	1.E−5		
Iteration limit	ITS		
	100		

Fig. 4.1(a) Input data for Program 4.1

```
** DIRECT ITERATION TO FIND LARGEST EIGENVALUE **

MATRIX A
    .1600E+02  -.2400E+02   .1800E+02
    .3000E+01  -.2000E+01   .0000E+00
   -.9000E+01   .1800E+02  -.1700E+02

LARGEST EIGENVALUE
 -.8000E+01

CORRESPONDING EIGENVECTOR
   -.4364E+00   .2182E+00   .8729E+00

ITERATIONS TO CONVERGENCE
  19
```

Fig. 4.1(b) Results from Program 4.1

4.4.1 Shifted iteration

The rate of convergence of the power method depends upon the nature of the eigenvalues. For closely spaced eigenvalues, convergence can be slow. For example, taking the **A** matrix of eq. 4.9 and using Program 4.1 with $\mathbf{x}_0^T = \{1\ 1\ 1\}$, convergence to $\lambda = 4.7071$ is only achieved after 26 iterations for a tolerance of 1×10^{-5}. The device of 'shifting' is based on the solution of the modified problem

$$(\mathbf{A} - p\mathbf{I})\mathbf{x} = (\lambda - p)\mathbf{x} \tag{4.15}$$

where p is a simple scalar. Thus eq. 4.1 has been changed to eq. 4.15 by simply subtracting $p\mathbf{x}$ from both sides of the equation. In the modified problem the eigenvectors **x** are the same as before but the eigenvalues of the matrix $\mathbf{A} - p\mathbf{I}$ are given by $\lambda - p$, or 'shifted' by an amount p. To recover the eigenvalues of **A** we merely have to add p to the eigenvalues of $\mathbf{A} - p\mathbf{I}$.

Program 4.2. Shifted iteration

This program only differs from the previous one in that SHIFT is read, two lines are added to perform the subtraction of $p\mathbf{I}$ from **A**, and SHIFT is added to the eigenvalue before printing. The input and output are shown in Figs 4.2(a) and (b) respectively. It can be seen that using a shift of 3.5, convergence to $\lambda_1=4.7071$ is now reached in 7 iterations for the same tolerance.

```
      PROGRAM P42
C
C         SHIFTED ITERATION USING THE POWER METHOD
C
C        ALTER NEXT LINE TO CHANGE PROBLEM SIZE
C
      PARAMETER (IN=20)
C
      REAL A(IN,IN),X0(IN),X1(IN)
C
      READ (5,*) N,((A(I,J),J=1,N),I=1,N),SHIFT
      WRITE(6,*) ('*************** SHIFTED ITERATION ***************')
      WRITE(6,*)
      WRITE(6,*) ('MATRIX A')
      CALL PRINTA(A,IN,N,N,6)
      WRITE(6,*)
      WRITE(6,*) ('SHIFT')
      WRITE(6,1000) SHIFT
      WRITE(6,*)
      DO 10 I = 1,N
   10 A(I,I) = A(I,I) - SHIFT
      READ (5,*) (X0(I),I=1,N)
      READ (5,*) TOL,ITS
      ITERS = 0
    1 CALL MVMULT(A,IN,X0,N,N,X1)
      BIG = 0.0
      DO 2 I = 1,N
    2 IF (ABS(X1(I)).GT.ABS(BIG)) BIG = X1(I)
      DO 3 I = 1,N
    3 X1(I) = X1(I)/BIG
      CALL CHECON(X1,X0,N,TOL,ICON)
      ITERS = ITERS + 1
      IF (ICON.EQ.0 .AND. ITERS.LT.ITS) GO TO 1
      SUM = 0.0
      DO 4 I = 1,N
    4 SUM = SUM + X1(I)**2
      SUM = SQRT(SUM)
      DO 5 I = 1,N
    5 X1(I) = X1(I)/SUM
      WRITE(6,*) ('EIGENVALUE')
      WRITE (6,1000) BIG + SHIFT
      WRITE(6,*)
      WRITE(6,*) ('CORRESPONDING EIGENVECTOR')
      CALL PRINTV(X1,N,6)
      WRITE(6,*)
      WRITE(6,*) ('ITERATIONS TO CONVERGENCE')
      WRITE (6,2000) ITERS
 1000 FORMAT (E12.4)
 2000 FORMAT (I5)
      STOP
      END
```

A useful feature of shifted iteration is that it enables the ready calculation of the smallest eigenvalue for systems whose eigenvalues are all positive. The power method can first be used to compute the largest eigenvalue, and this can then be used as the shift, implying that convergence will be to the unshifted eigenvalue closest to zero.

Array size	N		
	3		
Array	(A (I, J), J=1, N), I=1, N		
	4.0	0.5	0.0
	0.5	4.0	0.5
	0.0	0.5	4.0
Shift	SHIFT		
	3.5		
Initial vector	X0(I), I=1, N		
	1.0	1.0	1.0
Tolerance	TOL		
	1.E−5		
Iteration limit	ITS		
	100		

Fig. 4.2(a) Input data for Program 4.2

```
*************** SHIFTED ITERATION ***************

MATRIX A
    .4000E+01    .5000E+00    .0000E+00
    .5000E+00    .4000E+01    .5000E+00
    .0000E+00    .5000E+00    .4000E+01

SHIFT
  .3500E+01

EIGENVALUE
  .4707E+01

CORRESPONDING EIGENVECTOR
   .5000E+00    .7071E+00    .5000E+00

ITERATIONS TO CONVERGENCE
   7
```

Fig. 4.2(b) Results from Program 4.2

4.4.2 Shifted inverse iteration

A more direct way of achieving convergence of the power method on eigenvalues other than the 'largest' is to recast eq. 4.1 in the form

$$(\mathbf{A}-p\mathbf{I})^{-1}\mathbf{x}=\frac{1}{\lambda-p}\mathbf{x} \tag{4.16}$$

where p is a scalar shift and λ is an eigenvalue of $\mathbf{A}$. The eigenvectors of $(\mathbf{A}-p\mathbf{I})^{-1}$ are the same of those of $\mathbf{A}$, but it can be shown that the eigenvalues of $(\mathbf{A}-p\mathbf{I})^{-1}$ are $1/(\lambda-p)$. Hence, the largest eigenvalue of $(\mathbf{A}-p\mathbf{I})^{-1}$ leads to the eigenvalue of $\mathbf{A}$ that is *closest* to p. Thus, if the largest eigenvalue of $(\mathbf{A}-p\mathbf{I})^{-1}$ is μ, then the required eigenvalue of $\mathbf{A}$ is given by $\lambda=(1/\mu)+p$. For small matrices it would be possible just to invert $(\mathbf{A}-p\mathbf{I})$ and to use the inverse to solve eq. 4.16 iteratively in exactly the same algorithm as was used in Program 4.1. However we saw, in Chapter 2, that in general factorisation methods are the most applicable to inverse problems. Thus, whereas in the normal shifted iteration method we would have to compute in every iteration

$$(\mathbf{A}-p\mathbf{I})\mathbf{x}_0=\mathbf{x}_1, \tag{4.17}$$

in the inverse iteration we have to compute

$$(\mathbf{A}-p\mathbf{I})^{-1}\mathbf{x}_0=\mathbf{x}_1.$$

By factorising $(\mathbf{A}-p\mathbf{I})$ (see Chapter 2) using the appropriate library subroutine (LUFAC) we can write

$$(\mathbf{A}-p\mathbf{I})=\mathbf{LU} \tag{4.18}$$

and so

$$(\mathbf{LU})^{-1}\mathbf{x}_0=\mathbf{x}_1 \tag{4.19}$$

or

$$\mathbf{U}^{-1}\mathbf{L}^{-1}\mathbf{x}_0=\mathbf{x}_1 \tag{4.20}$$

If now we let

$$\mathbf{L}\mathbf{y}_0=\mathbf{x}_0 \quad \text{or} \quad \mathbf{y}_0=\mathbf{L}^{-1}\mathbf{x}_0 \tag{4.21}$$

and

$$\mathbf{U}\mathbf{x}_1=\mathbf{y}_0 \quad \text{or} \quad \mathbf{x}_1=\mathbf{U}^{-1}\mathbf{y}_0 \tag{4.22}$$

we can see that eq. 4.20 is solved for $\mathbf{x}_1$ by solving in succession $\mathbf{L}\mathbf{y}_0=\mathbf{x}_0$ for $\mathbf{y}_0$ and $\mathbf{U}\mathbf{x}_1=\mathbf{y}_0$ for $\mathbf{x}_1$. These processes are just the 'forward'- and 'back'-substitution processes we saw in Chapter 2, for which subroutines SUBFOR and SUBBAC were developed.

By altering p in a systematic way, all the eigenvalues of $\mathbf{A}$ can be found by this method.

Example 4.1

Use shifted inverse iteration to find the eigenvalue of the matrix

$$\mathbf{A}=\begin{bmatrix} 3 & 2 \\ 3 & 4 \end{bmatrix}$$

that is closest to 2.

Solution 4.1

Use simple iteration operating on the matrix

$\mathbf{B}=(\mathbf{A}-p\mathbf{I})^{-1}$ where $p=2$ in this case.

$$\therefore \quad \mathbf{B}=\begin{bmatrix}1 & 2\\3 & 2\end{bmatrix}^{-1}=-0.25\begin{bmatrix}2 & -2\\-3 & 1\end{bmatrix}$$

$$\therefore \quad \mathbf{B}=\begin{bmatrix}-0.5 & 0.5\\0.75 & -0.25\end{bmatrix}$$

Let $\mathbf{x}_0=\begin{Bmatrix}1\\1\end{Bmatrix}$ and $\mathbf{x}_1^*$ be the value of $\mathbf{x}_1$ before normalisation

$$\therefore \quad \mathbf{x}_1^*=\begin{Bmatrix}0\\0.5\end{Bmatrix},\ \mathbf{x}_1=\begin{Bmatrix}0\\1\end{Bmatrix},\ \lambda_1=0.5$$

$$\therefore \quad \mathbf{x}_2^*=\begin{Bmatrix}0.5\\-0.25\end{Bmatrix},\ \mathbf{x}_2=\begin{Bmatrix}1\\-0.5\end{Bmatrix},\ \lambda_2=0.5$$

$$\therefore \quad \mathbf{x}_3^*=\begin{Bmatrix}-0.75\\0.875\end{Bmatrix},\ \mathbf{x}_3=\begin{Bmatrix}-0.8571\\1\end{Bmatrix},\ \lambda_4=0.875$$

$$\therefore \quad \mathbf{x}_4^*=\begin{Bmatrix}0.9286\\-0.8929\end{Bmatrix},\ \mathbf{x}_4=\begin{Bmatrix}-1.04\\1\end{Bmatrix},\ \lambda_4=0.9286$$

$$\downarrow \qquad \downarrow \quad \text{after many iterations}$$

$$\begin{Bmatrix}-1\\1\end{Bmatrix} \qquad 1$$

Program 4.3. Shifted inverse iteration

The program uses exactly the same nomenclature as in Program 4.2 with the additional variable length arrays:

UPTRI	Upper triangular factor of $(\mathbf{A}-p\mathbf{I})$
LOWTRI	Lower triangular factor of $(\mathbf{A}-p\mathbf{I})$

```
      PROGRAM P43
C
C      PROGRAM 4.3 SHIFTED INVERSE ITERATION FOR EIGENVALUE
C                  CLOSEST TO A GIVEN VALUE
C
C      ALTER NEXT LINE TO CHANGE PROBLEM SIZE
C
      PARAMETER (IN=20)
C
      REAL A(IN,IN),UPTRI(IN,IN),LOWTRI(IN,IN),X0(IN),X1(IN)
C
      READ (5,*) N,((A(I,J),J=1,N),I=1,N),SHIFT
      WRITE(6,*) ('********** SHIFTED INVERSE ITERATION ************')
      WRITE(6,*)
      WRITE(6,*) ('MATRIX A')
      CALL PRINTA(A,IN,N,N,6)
      WRITE(6,*)
```

```
      WRITE(6,*) ('SHIFT')
      WRITE(6,1000) SHIFT
      WRITE(6,*)
      DO 10 I=1,N
   10 A(I,I)=A(I,I)-SHIFT
      CALL LUFAC(A,UPTRI,LOWTRI,IN,N)
      READ (5,*) (X0(I),I=1,N)
      READ (5,*) TOL,ITS
      ITERS = 0
      CALL VECCOP(X0,X1,N)
    1 CALL SUBFOR(LOWTRI,IN,X1,N)
      CALL SUBBAC(UPTRI,IN,X1,N)
      BIG = 0.0
      DO 2 I = 1,N
    2 IF (ABS(X1(I)).GT.ABS(BIG)) BIG = X1(I)
      DO 3 I = 1,N
    3 X1(I) = X1(I)/BIG
      CALL CHECON(X1,X0,N,TOL,ICON)
      ITERS = ITERS + 1
      IF (ICON.EQ.0 .AND. ITERS.LT.ITS) GO TO 1
      SUM = 0.0
      DO 4 I = 1,N
    4 SUM = SUM + X1(I)**2
      SUM = SQRT(SUM)
      DO 5 I = 1,N
    5 X1(I) = X1(I)/SUM
      WRITE(6,*) ('EIGENVALUE CLOSEST TO SHIFT')
      WRITE (6,1000) 1./BIG + SHIFT
      WRITE(6,*)
      WRITE(6,*) ('CORRESPONDING EIGENVECTOR')
      CALL PRINTV(X1,N,6)
      WRITE(6,*)
      WRITE(6,*) ('ITERATIONS TO CONVERGENCE')
      WRITE (6,2000) ITERS
 1000 FORMAT (E12.4)
 2000 FORMAT (I5)
      STOP
      END
```

Array size	N		
	3		
Array	(A (I, J), J = 1, N), I = 1, N		
	4.0	0.5	0.0
	0.5	4.0	0.5
	0.0	0.5	4.0
Shift	SHIFT		
	0.0		
Initial vector	X0(I), I = 1, N		
	1.0	1.0	1.0
Tolerance	TOL		
	1.E−5		
Iteration limit	ITS		
	100		

Fig. 4.3(a) Input data for Program 4.3

```
********** SHIFTED INVERSE ITERATION ************

MATRIX A
   .4000E+01    .5000E+00    .0000E+00
   .5000E+00    .4000E+01    .5000E+00
   .0000E+00    .5000E+00    .4000E+01

SHIFT
  .0000E+00

EIGENVALUE CLOSEST TO SHIFT
  .3293E+01

CORRESPONDING EIGENVECTOR
  -.5000E+00    .7071E+00   -.5000E+00

ITERATIONS TO CONVERGENCE
  36
```

Fig. 4.3(b) Results from Program 4.3

Typical input data and output results are shown in Figs 4.3(a) and (b) respectively. The number of equations and the coefficients of **A** are read in, followed by the scalar shift, the first guess of vector $\mathbf{x}_0$, the tolerance and the maximum number of iterations. The initial 'guess' $\mathbf{x}_0$ is copied into $\mathbf{x}_1$ for future reference, using VECCOP, and the iteration loop entered. Matrix $(\mathbf{A}-p\mathbf{I})$ is formed (still called **A**) followed by factorisation using LUFAC, and calls to SUBFOR and SUBBAC complete the determination of $\mathbf{x}_1$ following eqs 4.21 and 4.22. Vector $\mathbf{x}_1$ is then normalised to $|x_i|_{\max}=1.0$ and the convergence check invoked. When convergence is complete the converged vector $\mathbf{x}_1$ is normalised so that $\sqrt{\Sigma_{i=1}^{N} x_i^2}=1.0$, and the eigenvector, the eigenvalue of the original **A** closest to p and the number of iterations to convergence printed. For example, this process can be used, with a shift $p=0$, to find the 'smallest' eigenvalue of the matrix. For the case shown in Fig. 4.3(b), convergence to the 'smallest' eigenvalue, namely $4-1/\sqrt{2}$ is achieved in 36 iterations.

4.5 Calculation of intermediate eigenvalues – deflation

In the previous section we have shown that by using the power method, or simple variations of it, convergence to the numerically largest eigenvalue, the numerically smallest, or the eigenvalue closest to a given quantity can usually be obtained. Suppose, however, that the second largest eigenvalue of a system is to be investigated. One means of doing this is called 'deflation' and it consists essentially in removing the largest eigenvalue, once it has been computed by, for example, the power method, from the system of equations.

As an example, consider the eigenvalue problem

$$\begin{bmatrix} 2 & 1 \\ 1 & 2 \end{bmatrix} \begin{Bmatrix} x_1 \\ x_2 \end{Bmatrix} = \lambda \begin{Bmatrix} x_1 \\ x_2 \end{Bmatrix} \tag{4.23}$$

The characteristic polynomial is readily shown to be

$$(\lambda-3)(\lambda-1)=0 \tag{4.24}$$

and so the eigenvalues are $+3$ and $+1$. The eigenvalue $+3$ is associated with the normalised eigenvector $\{1/\sqrt{2} \quad 1/\sqrt{2}\}^{\mathrm{T}}$ and the eigenvalue $+1$ is associated with the eigenvector $\{1/\sqrt{2} \quad -1/\sqrt{2}\}^{\mathrm{T}}$. Since the matrix in eq. 4.23 is symmetric, these eigenvectors obey the orthogonality rules described in Section 4.1, that is,

$$\mathbf{x}_1{}^{\mathsf{T}}\mathbf{x}_1 = 1$$

and (4.25)

$$\mathbf{x}_1{}^{\mathsf{T}}\mathbf{x}_2 = 0$$

We can use 'his property to establish a modified matrix $\mathbf{A}^*$ such that

$$\mathbf{A}^* = \mathbf{A} - \lambda_1 \mathbf{x}_1 \mathbf{x}_1{}^{\mathsf{T}} \tag{4.26}$$

We now multiply this equation by any eigenvector $\mathbf{x}_i$ to give

$$\mathbf{A}^*\mathbf{x}_i = \mathbf{A}\mathbf{x}_i - \lambda_1 \mathbf{x}_1 \mathbf{x}_1{}^{\mathsf{T}}\mathbf{x}_i \tag{4.27}$$

when $i=1$ eq. 4.27 can be written as

$$\begin{aligned} \mathbf{A}^*\mathbf{x}_1 &= \mathbf{A}\mathbf{x}_1 - \lambda_1 \mathbf{x}_1 \mathbf{x}_1{}^{\mathsf{T}}\mathbf{x}_1 \\ &= \lambda_1 \mathbf{x}_1 - \lambda_1 \mathbf{x}_1 = 0 \end{aligned} \tag{4.28a}$$

and when $i > 1$, eq. 4.27 can be written as

$$\mathbf{A}^*\mathbf{x}_i = \lambda_i \mathbf{x}_i \tag{4.28b}$$

Thus the first eigenvalue of $\mathbf{A}^*$ is zero, and all other eigenvalues of $\mathbf{A}^*$ are the same as those of $\mathbf{A}$.

Having 'deflated' $\mathbf{A}$ to $\mathbf{A}^*$, the largest eigenvalue of $\mathbf{A}^*$ can be found, for example by simple iteration, and thus will equal the second largest eigenvalue of $\mathbf{A}$. Following this procedure for eq. 4.23

$$\mathbf{A} = \begin{bmatrix} 2 & 1 \\ 1 & 2 \end{bmatrix} \tag{4.29}$$

and $\lambda_1 = 3$ with $\mathbf{x}_1{}^{\mathsf{T}} = \{1/\sqrt{2} \quad 1/\sqrt{2}\}$. Thus

$$\mathbf{x}_1 \mathbf{x}_1{}^{\mathsf{T}} = \begin{Bmatrix} \dfrac{1}{\sqrt{2}} \\ \dfrac{1}{\sqrt{2}} \end{Bmatrix} \begin{Bmatrix} \dfrac{1}{\sqrt{2}} & \dfrac{1}{\sqrt{2}} \end{Bmatrix} = \begin{bmatrix} \dfrac{1}{2} & \dfrac{1}{2} \\ \dfrac{1}{2} & \dfrac{1}{2} \end{bmatrix} \tag{4.30}$$

and therefore

$$\begin{aligned} \mathbf{A}^* &= \begin{bmatrix} 2 & 1 \\ 1 & 2 \end{bmatrix} - \begin{bmatrix} \frac{3}{2} & \frac{3}{2} \\ \frac{3}{2} & \frac{3}{2} \end{bmatrix} \\ &= \begin{bmatrix} \frac{1}{2} & -\frac{1}{2} \\ -\frac{1}{2} & \frac{1}{2} \end{bmatrix} \end{aligned} \tag{4.31}$$

The eigenvalues of $\mathbf{A}^*$ are then given by the characteristic polynomial

$$\lambda(\lambda-1)=0 \tag{4.32}$$

illustrating that the remaining nonzero eigenvalue of $\mathbf{A}^*$ is the second largest eigenvalue of $\mathbf{A}$, namely $+1$.

As a further example, let us use Program 4.1 to find an eigenvalue of the deflated matrix of the $\mathbf{A}$ given by eq. 4.9. Carrying out the deflation process leads to

$$\mathbf{A}^*=\begin{bmatrix} 3-\dfrac{1}{4\sqrt{2}} & \tfrac{1}{4}-\sqrt{2} & -1-\dfrac{1}{4\sqrt{2}} \\ \tfrac{1}{4}-\sqrt{2} & 2-\dfrac{1}{2\sqrt{2}} & \tfrac{1}{4}-\sqrt{2} \\ -1-\dfrac{1}{4\sqrt{2}} & \tfrac{1}{4}-\sqrt{2} & 3-\dfrac{1}{4\sqrt{2}} \end{bmatrix} \tag{4.33}$$

For the input data of Fig. 4.1, i.e. a guessed starting eigenvector $\mathbf{x}_0{}^{\mathrm{T}}=\{1\ 1\ 1\}$, convergence occurs not to the second eigenvalue, $\lambda=4$ with associated eigenvector $\{1/\sqrt{2}\ 0\ -1/\sqrt{2}\}^{\mathrm{T}}$ but rather to the third eigenvalue $4-1/\sqrt{2}$ with associated eigenvector $\{\tfrac{1}{2}\ -1/\sqrt{2}\ \tfrac{1}{2}\}^{\mathrm{T}}$ in 2 iterations.

A change in the guessed starting eigenvector to $\mathbf{x}_0{}^{\mathrm{T}}=\{1\ 0\ -\tfrac{1}{2}\}$ leads to convergence to the second eigenvalue and eigenvector in 46 iterations for the given tolerance. This example shows that care must be taken with vector iterative methods for deflated matrices or those with closely spaced eigenvalues if one is to be sure that convergence to a desired eigenvalue has been attained.

4.6 The generalised eigenvalue problem $\mathbf{Ax}=\lambda\mathbf{Bx}$

Frequently in engineering practice there is an extra matrix on the right-hand side of the eigenvalue equation leading to the form

$$\mathbf{Ax}=\lambda\mathbf{Bx} \tag{4.34}$$

By rearrangement of eq. 4.34 we could write either of the equivalent eigenvalue equations

$$\mathbf{B}^{-1}\mathbf{Ax}=\lambda\mathbf{x} \tag{4.35a}$$

or $$\mathbf{A}^{-1}\mathbf{Bx}=\frac{1}{\lambda}\mathbf{x} \tag{4.35b}$$

The present implementation yields the largest eigenvalue $(1/\lambda)$ of eq. 4.35(b). The reciprocal of this value corresponds to the smallest eigenvalue (λ) of eq. 4.35(a).

In the case of eq. 4.34 we can first set λ to 1 and make a guess at $\mathbf{x}$ on the right-hand side. A matrix-by-vector multiplication then yields

$$\mathbf{Bx}=\mathbf{y} \tag{4.36}$$

allowing a new estimate of **x** to be established by solving the set of linear equations

$$\mathbf{Ax}=\mathbf{y} \tag{4.37}$$

by any of the techniques described in Chapter 2. For example, as was done in Program 4.3, **LU** factorisation may be employed with the factorisation phase completed outside the iteration loop.

When the new **x** has been computed, it may be normalised to give $|x_i|_{max}=1$ and the procedure can be repeated from eq. 4.36 onwards.

Program 4.4. Iterative solution of $\mathbf{Ax}=\lambda\mathbf{Bx}$

The program describing this process is easily developed from the previous programs in this chapter. Figure 4.4(a) shows data from a typical engineering situation which leads to the generalised problem of eq. 4.34.

In this case the **A** matrix represents the stiffness of a compressed strut and the **B** matrix the destabilising effect of the compressive force λ. There are four equations to be solved to a tolerance of 10^{-5} with an iteration limit of 100. The guessed starting vector $\mathbf{x}_0$ is $\{1.0\ 1.0\ 1.0\ 1.0\}^T$.

Matrices **A** and **B** are read in, and in preparation for the equation solution of eq. 4.37, **A** is factorised using subroutine LUFAC. The iteration loop begins at label 1 by the multiplication of **B** by $\mathbf{x}_0$ as required by eq. 4.36. Forward- and back-substitution complete the equation solution called for by eq. 4.37 and the resulting vector is normalised. The convergence check is invoked and control returns to label 1 if convergence is incomplete, unless the iteration limit has been reached. The final normalisation involving the sum of the squares of the components of the eigenvector is then carried out and the normalised eigenvector and number of iterations printed. In this case the reciprocal of the 'largest' eigenvalue is the 'buckling' load, which is also printed. The output is shown in Fig. 4.4(b) where the estimate of the buckling load of 9.94 after 6 iterations can be compared with the exact solution $\pi^2=9.8696$.

```
      PROGRAM P44
C
C      PROGRAM 4.4 POWER METHOD FOR A*X=LAMBDA*B*X
C
C      ALTER NEXT LINE TO CHANGE PROBLEM SIZE
C
      PARAMETER (IN=20)
C
      REAL A(IN,IN),UPTRI(IN,IN),LOWTRI(IN,IN),X0(IN),X1(IN),B(IN,IN)
C
      READ (5,*) N,((A(I,J),J=1,N),I=1,N)
      READ (5,*) ((B(I,J),J=1,N),I=1,N)
      WRITE(6,*) ('************ ITERATIVE SOLUTION OF **************')
      WRITE(6,*) ('**************   A*X=LAMBDA*B*X    ****************')
      WRITE(6,*)
      WRITE(6,*) ('MATRIX A')
      CALL PRINTA(A,IN,N,N,6)
      WRITE(6,*)
      WRITE(6,*) ('MATRIX B')
      CALL PRINTA(B,IN,N,N,6)
      WRITE(6,*)
      CALL LUFAC(A,UPTRI,LOWTRI,IN,N)
      READ (5,*) (X0(I),I=1,N)
      READ (5,*) TOL,ITS
```

```
      ITERS = 0
      CALL VECCOP(X0,X1,N)
    1 CALL MVMULT(B,IN,X0,N,N,X1)
      CALL SUBFOR(LOWTRI,IN,X1,N)
      CALL SUBBAC(UPTRI,IN,X1,N)
      BIG = 0.0
      DO 2 I = 1,N
    2 IF (ABS(X1(I)).GT.ABS(BIG)) BIG = X1(I)
      DO 3 I = 1,N
    3 X1(I) = X1(I)/BIG
      CALL CHECON(X1,X0,N,TOL,ICON)
      ITERS = ITERS + 1
      IF (ICON.EQ.0 .AND. ITERS.LT.ITS) GO TO 1
      SUM = 0.0
      DO 4 I = 1,N
    4 SUM = SUM + X1(I)**2
      SUM = SQRT(SUM)
      DO 5 I = 1,N
    5 X1(I) = X1(I)/SUM
      WRITE(6,*) ('SMALLEST EIGENVALUE (LAMBDA)')
      WRITE (6,1000) 1.0/BIG
      WRITE(6,*)
      WRITE(6,*) ('CORRESPONDING EIGENVECTOR')
      CALL PRINTV(X1,N,6)
      WRITE(6,*)
      WRITE(6,*) ('ITERATIONS TO CONVERGENCE')
      WRITE (6,2000) ITERS
 1000 FORMAT (E12.4)
 2000 FORMAT (I5)
      STOP
      END
```

Array size	N			
Array A	(A (I, J), J = 1, N), I = 1, N			
	8.0	4.0	−24.0	0.0
	4.0	16.0	0.0	4.0
	−24.0	0.0	192.0	24.0
	0.0	4.0	24.0	8.0
Array B	(B (I, J), J = 1, N), I = 1, N			
	0.06667	−0.01667	−0.1	0.0
	−0.01667	0.13333	0.0	−0.01667
	−0.1	0.0	4.8	0.1
	0.0	−0.01667	0.1	0.06667
Initial vector	X0(I), I = 1, N			
	1.0	1.0	1.0	1.0
Tolerance	TOL			
	1.E−5			
Iteration limit	ITS			
	100			

Fig. 4.4(a) Input data for Program 4.4

```
************ ITERATIVE SOLUTION OF **************
**************  A*X=LAMBDA*B*X    ****************

MATRIX A
    .8000E+01    .4000E+01   -.2400E+02    .0000E+00
    .4000E+01    .1600E+02    .0000E+00    .4000E+01
   -.2400E+02    .0000E+00    .1920E+03    .2400E+02
    .0000E+00    .4000E+01    .2400E+02    .8000E+01

MATRIX B
    .6667E-01   -.1667E-01   -.1000E+00    .0000E+00
   -.1667E-01    .1333E+00    .0000E+00   -.1667E-01
   -.1000E+00    .0000E+00    .4800E+01    .1000E+00
    .0000E+00   -.1667E-01    .1000E+00    .6667E-01

SMALLEST EIGENVALUE (LAMBDA)
  .9944E+01

CORRESPONDING EIGENVECTOR
    .6898E+00    .0000E+00    .2200E+00   -.6898E+00

ITERATIONS TO CONVERGENCE
   6
```

Fig. 4.4(b) Results from Program 4.4

4.6.1 Conversion of the generalised problem to standard form

Several solution techniques for eigenvalue problems require the equation to be in the 'standard' form of eq. 4.1, namely

$$\mathbf{Ax} = \lambda\mathbf{x} \tag{4.1}$$

If the generalised form of eq. 4.34 is encountered, it is always possible to convert it into the standard form. For example, given

$$\mathbf{Ax} = \lambda\mathbf{Bx} \tag{4.34}$$

one could think of inverting **B** (or **A**) by some procedure to yield the forms

$$\mathbf{B}^{-1}\mathbf{Ax} = \lambda\mathbf{x}$$

or (4.38)

$$\mathbf{A}^{-1}\mathbf{Bx} = \frac{1}{\lambda}\mathbf{x}$$

which are in standard form. However, even for symmetrical **B** (or **A**) the products $\mathbf{B}^{-1}\mathbf{A}$ and $\mathbf{A}^{-1}\mathbf{B}$ are not in general symmetrical. In order to preserve symmetry, the following strategy can be used.

Starting from eq. 4.34 we can factorise **B** by Cholesky's method (see Chapter 2) to give

$$\mathbf{Ax} = \lambda\mathbf{LL}^{\mathrm{T}}\mathbf{x} \tag{4.39}$$

or $$\mathbf{L}^{-1}\mathbf{Ax} = \lambda\mathbf{L}^{\mathrm{T}}\mathbf{x} \tag{4.40}$$

Now we let

$$\mathbf{L}^{-1}\mathbf{A} = \mathbf{C} \tag{4.41}$$

which can be solved for **C** by repeated forward-substitutions using the columns of **A** in the form

$$\mathbf{LC}=\mathbf{A} \tag{4.42}$$

We now have

$$\mathbf{Cx}=\lambda\mathbf{L}^{\mathrm{T}}\mathbf{x} \tag{4.43}$$

By setting

$$\mathbf{L}^{\mathrm{T}}\mathbf{x}=\mathbf{y} \tag{4.44}$$

or $\quad \mathbf{x}=\mathbf{L}^{-\mathrm{T}}\mathbf{y}$ (4.45)

eq. 4.43 becomes

$$\mathbf{CL}^{-\mathrm{T}}\mathbf{y}=\lambda\mathbf{y} \tag{4.46}$$

If we set

$$\mathbf{CL}^{-\mathrm{T}}=\mathbf{D} \tag{4.47}$$

or $\quad \mathbf{L}^{-1}\mathbf{C}^{\mathrm{T}}=\mathbf{D}^{\mathrm{T}}$

this can be solved for $\mathbf{D}^{\mathrm{T}}$ by repeated forward-substitutions using the columns of $\mathbf{C}^{\mathrm{T}}$ in the form

$$\mathbf{LD}^{\mathrm{T}}=\mathbf{C}^{\mathrm{T}} \tag{4.48}$$

It will be found that in the resulting standard form equation

$$\mathbf{D}^{\mathrm{T}}\mathbf{y}=\mathbf{Dy}=\lambda\mathbf{y} \tag{4.49}$$

the matrix **D** is symmetrical. Note that the eigenvalues of eq. 4.49 are the same as those of eq. 4.34 but the eigenvectors are **y** rather than the original **x**. These originals can be recovered quite simply from eq. 4.44 using back-substitution.

Program 4.5. Conversion of $\mathbf{Ax}=\lambda\mathbf{Bx}$ to 'standard form'
A program describing this process is now presented using as data the same problem specification as was used in the previous program. The nomenclature used is as follows:

Simple variables

N	Number of equations

Variable length arrays

A, B	Arrays **A** and **B** in $\mathbf{Ax}=\lambda\mathbf{Bx}$
D	Diagonal matrix (vector) in $\mathbf{LDL}^{\mathrm{T}}$ factorisation
C, E	Temporary storage
DD	The array $\mathbf{D}=\mathbf{D}^{\mathrm{T}}$ in eq. 4.49

PARAMETER *restriction*
IN≥N

```
      PROGRAM P45
C
C      PROGRAM 4.5 CONVERTION OF A*X=LAMBDA*B*X
C                  TO STANDARD FORM
C
C      ALTER NEXT LINE TO CHANGE PROBLEM SIZE
C
      PARAMETER (IN=20)
C
      REAL A(IN,IN),D(IN),B(IN,IN),C(IN,IN),E(IN),DD(IN,IN)
C
      READ (5,*) N,((A(I,J),J=1,N),I=1,N)
      READ (5,*) ((B(I,J),J=1,N),I=1,N)
      WRITE(6,*) ('******** CONVERSION OF  A*X=LAMBDA*B*X **********')
      WRITE(6,*) ('******** TO STANDARD SYMMETRICAL FORM ***********')
      WRITE(6,*)
      WRITE(6,*) ('MATRIX A')
      CALL PRINTA(A,IN,N,N,6)
      WRITE(6,*)
      WRITE(6,*) ('MATRIX B')
      CALL PRINTA(B,IN,N,N,6)
      WRITE(6,*)
      CALL LDLT(B,IN,D,N)
      DO 1 I = 1,N
    1 D(I) = SQRT(D(I))
      DO 2 J = 1,N
          DO 2 I = J,N
    2 B(I,J) = B(I,J)/D(J)
      DO 3 I = 1,N - 1
          DO 3 J = I + 1,N
    3 B(I,J) = B(J,I)
      DO 4 J = 1,N
          DO 5 I = 1,N
    5     E(I) = A(I,J)
          CALL SUBFOR(B,IN,E,N)
          DO 6 I = 1,N
    6     C(I,J) = E(I)
    4 CONTINUE
      WRITE(6,*) ('MATRIX C')
      CALL PRINTA(C,IN,N,N,6)
      WRITE(6,*)
      CALL MATRAN(DD,IN,C,IN,N,N)
      DO 7 J = 1,N
          DO 8 I = 1,N
    8     E(I) = DD(I,J)
          CALL SUBFOR(B,IN,E,N)
          DO 9 I = 1,N
    9     DD(I,J) = E(I)
    7 CONTINUE
      WRITE(6,*) ('FINAL SYMMETRICAL MATRIX D')
      CALL PRINTA(DD,IN,N,N,6)
      STOP
      END
```

The program begins by reading in the number of equations and the **A** and **B** matrices. Matrix **B** is factorised using library routine LDLT and the Cholesky factors are then obtained by dividing the terms of **L** by $\sqrt{d_{ii}}$.

Equation 4.42 is then solved for **C**. The columns of **A** are first copied into a temporary storage vector **E** and forward substitution using SUBFOR leads to the columns of **C**. The **C** matrix is printed out and then transposed into **DD** using library routine MATRAN. Then, in a very similar sequence of operations, eq. 4.48 is solved for **DD** ($\mathbf{DD}^T$) by first copying the columns into **E** and then calling SUBFOR. The final symmetric matrix is printed out as shown in Fig. 4.5(b). If the

Array size	N			
	3			
Array A	(A (I, J), J=1, N), I=1, N			
	8.0	4.0	−24.0	0.0
	4.0	16.0	0.0	4.0
	−24.0	0.0	192.0	24.0
	0.0	4.0	24.0	8.0
Array B	(B(I, J), J=1, N), I=1, N			
	0.06667	−0.01667	−0.1	0.0
	−0.01667	0.13333	0.0	−0.01667
	−0.1	0.0	4.8	0.1
	0.0	−0.01667	0.1	0.06667

Fig. 4.5(a) Input data for Program 4.5

```
******** CONVERSION OF  A*X=LAMBDA*B*X **********
******** TO STANDARD SYMMETRICAL FORM ***********

MATRIX A
    .8000E+01    .4000E+01   -.2400E+02    .0000E+00
    .4000E+01    .1600E+02    .0000E+00    .4000E+01
   -.2400E+02    .0000E+00    .1920E+03    .2400E+02
    .0000E+00    .4000E+01    .2400E+02    .8000E+01

MATRIX B
    .6667E-01   -.1667E-01   -.1000E+00    .0000E+00
   -.1667E-01    .1333E+00    .0000E+00   -.1667E-01
   -.1000E+00    .0000E+00    .4800E+01    .1000E+00
    .0000E+00   -.1667E-01    .1000E+00    .6667E-01

MATRIX C
    .3098E+02    .1549E+02   -.9295E+02    .0000E+00
    .1670E+02    .4730E+02   -.1670E+02    .1113E+02
   -.5029E+01    .4311E+01    .7184E+02    .1149E+02
    .4001E+01    .2400E+02    .8000E+02    .3200E+02

FINAL SYMMETRICAL MATRIX D
    .1200E+03    .6466E+02   -.1948E+02    .1549E+02
    .6466E+02    .1432E+03    .8496E+01    .6957E+02
   -.1948E+02    .8496E+01    .3011E+02    .4215E+02
    .1549E+02    .6957E+02    .4215E+02    .1333E+03
```

Fig. 4.5(b) Results from Program 4.5

'largest' eigenvalue of this matrix is calculated using Program 4.1 with a tolerance of 10^{-5}, convergence to the eigenvalue $\lambda = 240$ will be found in 15 iterations.

If **A** and **B** are switched, the eigenvalues of the resulting symmetric matrix found by this program would be the reciprocals of their previous values when **A** and **B** were not switched. For example, if **A** and **B** are switched and the resulting symmetric matrix solved by Program 4.1, the 'largest' eigenvalue is given as 0.1006. This is the reciprocal of 9.94 which was the 'smallest' eigenvalue obtained previously by Program 4.4.

4.7 Transformation methods

Returning to the basic eigenvalue equation

$$\mathbf{Ax} = \lambda \mathbf{x} \tag{4.1}$$

we can see that it can be 'transformed' into an equation with the same eigenvalues

$$\mathbf{A}^*\mathbf{x} = \lambda\mathbf{x} \tag{4.50}$$

if eq. 4.1 is 'postmultiplied' by a matrix **P** and 'premultiplied' by the inverse of **P**. This gives the equation

$$\mathbf{P}^{-1}\mathbf{A}\mathbf{P}\mathbf{x} = \lambda\mathbf{P}^{-1}\mathbf{P}\mathbf{x} \tag{4.51}$$

which is of the form of eq. 4.50 with

$$\mathbf{A}^* = \mathbf{P}^{-1}\mathbf{A}\mathbf{P} \tag{4.52}$$

The concept behind such a transformation technique is to employ it so as to make the eigenvalues of **A*** easier to find than were the eigenvalues of the original **A**.

Even if **A** were symmetrical, it is highly unlikely that **A*** in eq. 4.52 would retain symmetry. However, if **A** is symmetrical, it can be shown that the matrix

$$\mathbf{A}^* = \mathbf{P}^{\mathrm{T}}\mathbf{A}\mathbf{P} \tag{4.53}$$

will always be symmetrical. When this transformation is applied to eq. 4.1, the resulting eigenvalue problem is

$$\mathbf{A}^*\mathbf{x} = \mathbf{P}^{\mathrm{T}}\mathbf{A}\mathbf{P}\mathbf{x} = \lambda\mathbf{P}^{\mathrm{T}}\mathbf{P}\mathbf{x} \tag{4.54}$$

For the eigenvalues of **A*** to be the same as those of **A**, we have to arrange that the transformation matrix has the additional property

$$\mathbf{P}^{\mathrm{T}}\mathbf{P} = \mathbf{I} \tag{4.55}$$

Matrices of this type are said to be 'orthogonal', and a matrix which has this property is the so-called 'rotation matrix'

$$\mathbf{P} = \begin{bmatrix} \cos\alpha & -\sin\alpha \\ \sin\alpha & \cos\alpha \end{bmatrix} \tag{4.56}$$

Applying this transformation to the **A** matrix of eq. 4.23, we have

$$\begin{aligned} \mathbf{A}^* &= \begin{bmatrix} \cos\alpha & \sin\alpha \\ -\sin\alpha & \cos\alpha \end{bmatrix} \begin{bmatrix} 2 & 1 \\ 1 & 2 \end{bmatrix} \begin{bmatrix} \cos\alpha & -\sin\alpha \\ \sin\alpha & \cos\alpha \end{bmatrix} \\ &= \begin{bmatrix} 2+2\sin\alpha\cos\alpha & \cos^2\alpha-\sin^2\alpha \\ \cos^2\alpha-\sin^2\alpha & 2-2\sin\alpha\cos\alpha \end{bmatrix} \end{aligned} \tag{4.57}$$

in which **A*** is clearly symmetrical for any value of α. In this case the obvious value of α to choose to give an **A*** with known eigenvalues is such as to make **A*** a diagonal matrix. Elimination of the off-diagonal terms occurs if

$$\cos^2\alpha - \sin^2\alpha = 0 \tag{4.58}$$

in which case $\tan\alpha = 1$ and $\alpha = \pi/4$, giving $\sin\alpha = \cos\alpha = 1/\sqrt{2}$. The resulting diagonalised $\mathbf{A}^*$ then has the form

$$\mathbf{A}^*=\begin{bmatrix} 3 & 0 \\ 0 & 1 \end{bmatrix} \tag{4.59}$$

As before, this shows that the eigenvalues of $\mathbf{A}$ are $+3$ and $+1$.

For matrices $\mathbf{A}$ which are bigger than 2×2, the transformation matrix is 'padded out' by putting 1 on the leading diagonal and zeros off-diagonal in every row except the two with respect to which the rotation is to be performed. For example, if $\mathbf{A}$ is 4×4, the transformation matrix could have any of 6 forms

$$\begin{bmatrix} \cos\alpha & -\sin\alpha & 0 & 0 \\ \sin\alpha & \cos\alpha & 0 & 0 \\ 0 & 0 & 1 & 0 \\ 0 & 0 & 0 & 1 \end{bmatrix} \tag{4.60}$$

$$\begin{bmatrix} 1 & 0 & 0 & 0 \\ 0 & \cos\alpha & 0 & -\sin\alpha \\ 0 & 0 & 1 & 0 \\ 0 & \sin\alpha & 0 & \cos\alpha \end{bmatrix} \tag{4.61}$$

and so on.

Matrix 4.60 would annihilate the terms (1, 2) and (2, 1) in the original $\mathbf{A}$ matrix after the transformation via $\mathbf{P}^{\mathrm{T}}\mathbf{A}\mathbf{P}$ while matrix 4.61 would annihilate terms (2, 4) and (4, 2). The effect of the 1s and 0s is to leave the other rows and columns of $\mathbf{A}$ unchanged. This still means that off-diagonal terms which become zero during one transformation revert to nonzero values on subsequent transformations and so the method is iterative as we have come to expect.

The earliest form of this type of iteration is called 'Jacobi diagonalisation', and it proceeds by annihilating the largest off-diagonal term in each rotation.

Generalising eqs 4.57 for any symmetric matrix $\mathbf{A}$ we have

$$\mathbf{A}^*=\begin{bmatrix} \cos\alpha & \sin\alpha \\ -\sin\alpha & \cos\alpha \end{bmatrix}\begin{bmatrix} a_{ii} & a_{ij} \\ a_{ji} & a_{jj} \end{bmatrix}\begin{bmatrix} \cos\alpha & -\sin\alpha \\ \sin\alpha & \cos\alpha \end{bmatrix} \tag{4.62}$$

leading to off-diagonal terms in $\mathbf{A}^*$ of the form

$$a^*_{ij}=a^*_{ji}=(-a_{ii}+a_{jj})\cos\alpha\sin\alpha+a_{ij}(\cos^2\alpha-\sin^2\alpha) \tag{4.63}$$

which must be made zero by choosing the appropriate α.

For example, a^*_{ij} is made zero by putting

$$\tan 2\alpha=\frac{2a_{ij}}{a_{ii}-a_{jj}} \tag{4.64}$$

To make a simple program for Jacobi diagonalisation we have therefore to search for the off-diagonal term in **A** of largest modulus and find the row and column in which it lies. The rotation angle α can then be computed from eq. 4.64 and the transformation matrix **P** set up. Matrix **P** can then be transposed using a library subroutine, and the matrix products to form **A***, as required by eq. 4.53, can be carried out. The leading diagonal of **A*** will converge to the eigenvalues of **A**.

Example 4.2

Use Jacobi diagonalisation to find the eigenvalues of the symmetrical matrix

$$\mathbf{A}=\begin{bmatrix}4 & 2\\ 2 & 3\end{bmatrix}$$

Solution 4.2

As **A** is only 2×2, one iteration should be sufficient. We wish to eliminate $A_{12}=A_{21}=2$

$$\therefore \qquad \tan 2\alpha=\frac{2(2)}{(4-3)}=4$$

$$\therefore \qquad \alpha=37.98^\circ$$

Transformation matrix

$$\mathbf{P}=\begin{bmatrix}\cos\alpha & -\sin\alpha\\ \sin\alpha & \cos\alpha\end{bmatrix}=\begin{bmatrix}0.7882 & -0.6154\\ 0.6154 & 0.7882\end{bmatrix}$$

$$\therefore \qquad \mathbf{A}^*=\mathbf{P}^{\mathrm{T}}\mathbf{A}\mathbf{T}$$

$$\mathbf{P}^{\mathrm{T}}\mathbf{A}=\begin{bmatrix}0.7882 & 0.6154\\ -0.6154 & 0.7882\end{bmatrix}\begin{bmatrix}4 & 2\\ 2 & 3\end{bmatrix}=\begin{bmatrix}4.3836 & 3.4226\\ -0.8852 & 1.1338\end{bmatrix}$$

$$\therefore \qquad \mathbf{A}^*=\begin{bmatrix}4.3836 & 3.4226\\ -0.8852 & 1.1338\end{bmatrix}\begin{bmatrix}0.7882 & -0.6154\\ 0.6154 & 0.7882\end{bmatrix}=\begin{bmatrix}5.561 & 0\\ 0 & 1.438\end{bmatrix}$$

Hence the eigenvalues are 5.561 and 1.438 approximately. (The exact solution to 3 decimal places is 5.562 and 1.438.)

Program 4.6. Jacobi diagonalisation
A program that uses Jacobi diagonalisation is now illustrated which uses the following nomenclature:

Simple variables

N	Number of equations to be solved
ITS	Maximum number of iterations allowed
TOL	Iteration tolerance
ITERS	Current number of iterations
ALPHA	Rotation angle for transformations

Variable length arrays

A	The original **A** matrix
A1	Temporary storage
P	Transformation matrix
PT	Transpose of transformation matrix
ENEW	Latest vector of eigenvalues
EOLD	Previous vector of eigenvalues

PARAMETER *restriction*

$IN \geq N$

```
      PROGRAM P46
C
C      PROGRAM 4.6 JACOBI TRANSFORMATION FOR EIGENVALUES
C                  OF SYMMETRIC MATRICES
C
C      ALTER NEXT LINE TO CHANGE PROBLEM SIZE)
C
      PARAMETER (IN=10)
C
      REAL A(IN,IN),A1(IN,IN),P(IN,IN),PT(IN,IN),ENEW(IN),EOLD(IN)
C
      PI = 4.*ATAN(1.)
      READ(5,*) N,((A(I,J),J=I,N),I=1,N)
      DO 2 I = 1,N
          DO 2 J = I,N
    2 A(J,I) = A(I,J)
      READ (5,*) TOL,ITS
      WRITE(6,*) ('************ JACOBI ITERATION *******************')
      WRITE(6,*) ('******** FOR SYMMETRIC MATRICES ***************')
      WRITE(6,*)
      WRITE(6,*) ('MATRIX A')
      CALL PRINTA(A,IN,N,N,6)
      WRITE(6,*)
      CALL NULVEC(EOLD,N)
      ITERS = 0
   99 ITERS = ITERS + 1
      BIG = 0.
      DO 3 I = 1,N
          DO 3 J = I + 1,N
              IF (ABS(A(I,J)).GT.BIG) THEN
                  BIG = ABS(A(I,J))
                  HOLD = A(I,J)
                  IROW = I
                  ICOL = J
              END IF
    3 CONTINUE
      DEN = A(IROW,IROW) - A(ICOL,ICOL)
      IF (DEN.EQ.0.) THEN
          ALPHA = PI*.25
          IF (HOLD.LT.0.) ALPHA = -ALPHA
      ELSE
          ALPHA = .5*ATAN(2.*HOLD/DEN)
      END IF
      CT = COS(ALPHA)
      ST = SIN(ALPHA)
      CALL NULL(P,IN,N,N)
      DO 4 I = 1,N
    4 P(I,I) = 1.
      P(IROW,IROW) = CT
      P(ICOL,ICOL) = CT
      P(IROW,ICOL) = -ST
      P(ICOL,IROW) = ST
      ALPHA = ALPHA*180./PI
```

```
      CALL MATRAN(PT,IN,P,IN,N,N)
      CALL MATMUL(PT,IN,A,IN,A1,IN,N,N,N)
      CALL MATMUL(A1,IN,P,IN,A,IN,N,N,N)
      DO 6 I = 1,N
    6 ENEW(I) = A(I,I)
      CALL CHECON(ENEW,EOLD,N,TOL,ICON)
      IF (ITERS.LT.ITS .AND. ICON.EQ.0) GO TO 99
      WRITE(6,*) ('FINAL TRANSFORMATION MATRIX P')
      CALL PRINTA(P,IN,N,N,6)
      WRITE(6,*)
      WRITE(6,*) ('FINAL TRANSFORMED MATRIX A')
      CALL PRINTA(A,IN,N,N,6)
      WRITE(6,*)
      WRITE(6,*) ('ITERATIONS TO CONVERGENCE')
      WRITE(6,1000) ITERS
 1000 FORMAT (I5)
      STOP
      END
```

Array size	N		
	3		
Array	(A (I, J), J = I, N), I = 1, N		
(accounting for symmetry)	10.0	5.0	6.0
		20.0	4.0
			30.0
Tolerance	TOL		
	1.E−7		
Iteration limit	ITS		
	25		

Fig. 4.6(a) Input data for Program 4.6

```
************ JACOBI ITERATION *******************
******** FOR SYMMETRIC MATRICES ***************

MATRIX A
   .1000E+02   .5000E+01   .6000E+01
   .5000E+01   .2000E+02   .4000E+01
   .6000E+01   .4000E+01   .3000E+02

FINAL TRANSFORMATION MATRIX P
   .1000E+01   .0000E+00   .2089E-04
   .0000E+00   .1000E+01   .0000E+00
  -.2089E-04   .0000E+00   .1000E+01

FINAL TRANSFORMED MATRIX A
   .7142E+01   .1898E-10  -.7276E-10
   .4731E-06   .1915E+02   .9683E-06
  -.4645E-07   .9088E-06   .3371E+02

ITERATIONS TO CONVERGENCE
    7
```

Fig. 4.6(b) Results from Program 4.6

As shown in Fig. 4.6(a) the number of equations is read followed by the array **A**. Since **A** must be symmetrical, only its upper triangle is read which is then copied into the lower triangle. The convergence tolerance and iteration limit complete the data. The iteration loop begins at label 99 and takes up the rest of the program. The largest off-diagonal term is stored as HOLD with its position in row and column IROW and ICOL respectively.

Rotation angle ALPHA is then computed from eq. 4.64 and its cosine CT and sine ST. Transformation matrix **P** is then easily computed. Calls to MATRAN and MATMUL enable the product in eq. 4.53 to be calculated. The eigenvectors in ENEW can then be compared with those in EOLD using subroutine CHECON. The output in Fig. 4.6(b) shows that the eigenvalues of

$$\mathbf{A}=\begin{bmatrix}10 & 5 & 6\\ 5 & 20 & 4\\ 6 & 4 & 30\end{bmatrix} \tag{4.65}$$

are computed to the given tolerance in 7 iterations and are 7.14, 19.15 and 33.71 respectively.

4.7.1 Comments on Jacobi diagonalisation

Although Program 4.6 illustrates the transformation process well for teaching purposes, it would not be used to solve large problems. One would never, in practice, store the transformation matrices **P** and $\mathbf{P}^T$, but perhaps less obviously, the searching process itself becomes very time-consuming as N increases. Alternatives to the basic Jacobi method which have been proposed include serial elimination in which the off-diagonal elements are eliminated in a predetermined sequence, thus avoiding searching altogether, and a variation of this technique in which serial elimination is performed only on those elements whose modulus exceeds a certain value or 'threshold'. When all off-diagonal terms have been reduced to the threshold, it can be further reduced and the process continued.

Jacobi's idea can also be implemented in order to reduce **A** to a tridiagonal matrix $\mathbf{A}^*$ having the same eigenvalues (rather than a diagonal $\mathbf{A}^*$ as in the basic technique). This is called 'Givens's method', which has the advantage of being non-iterative. Of course, some method must still be found for calculating the eigenvalues of the tridiagonal $\mathbf{A}^*$. A more popular tridiagonalisation technique is called 'Householder's method' which is described in the next section.

4.7.2. Householder's transformation to tridiagonal form

Equation 4.55 gave the basic property that transformation matrices should have, namely

$$\mathbf{P}^T\mathbf{P}=\mathbf{I} \tag{4.55}$$

and the Householder technique involves choosing

$$\mathbf{P}=\mathbf{I}-2\mathbf{w}\mathbf{w}^T \tag{4.66}$$

where $\mathbf{w}$ is a column vector normalised such that

$$\mathbf{w}^{\mathsf{T}}\mathbf{w}=1. \tag{4.67}$$

For example, let

$$\mathbf{w}=\begin{Bmatrix} \frac{1}{\sqrt{2}} \\ \frac{1}{\sqrt{2}} \end{Bmatrix} \tag{4.68}$$

which has the required product. Then

$$2\mathbf{w}\mathbf{w}^{\mathsf{T}}=\begin{bmatrix} 1 & 1 \\ 1 & 1 \end{bmatrix} \tag{4.69}$$

and we see that

$$\mathbf{P}=\mathbf{I}-2\mathbf{w}\mathbf{w}^{\mathsf{T}}=\begin{bmatrix} 0 & -1 \\ -1 & 0 \end{bmatrix} \tag{4.70}$$

which has the desired property

$$\mathbf{P}^{-1}=\mathbf{P}^{\mathsf{T}}=\mathbf{P} \tag{4.71}$$

so that eq. 4.55 is automatically satisfied.

In order to eliminate terms in the first row of $\mathbf{A}$ outside the tridiagonal, the vector $\mathbf{w}$ is taken as

$$\mathbf{w}=\{0 \ w_2 \ w_3 \ w_4 \dots w_n\}^{\mathsf{T}} \tag{4.72}$$

Thus the transformation matrix for row 1 is, assuming $\mathbf{A}$ is 3×3:

$$\mathbf{P}^1=\begin{bmatrix} 1 & 0 & 0 \\ 0 & 1-2w_2{}^2 & -2w_2w_3 \\ 0 & -2w_2w_3 & 1-2w_3{}^2 \end{bmatrix} \tag{4.73}$$

When the product $\mathbf{P}^1\,\mathbf{A}\mathbf{P}^1$ is carried out, the first row of the resulting matrix contains the three terms

$$\begin{aligned} a^*{}_{11}&=a_{11} \\ a^*{}_{12}&=a_{12}-2w_2(a_{12}w_2+a_{13}w_3)=r \\ a^*{}_{13}&=a_{13}-2w_3(a_{12}w_2+a_{13}w_3)=0 \end{aligned} \tag{4.74}$$

Letting

$$h=a_{12}w_2+a_{13}w_3$$

we see that to make $a^*_{13}=0$ as required we need

$$a_{13}-2w_3h=0. \tag{4.75}$$

Equation 4.67 gives another equation in the w_i, namely

$$w_2{}^2+w_3{}^2=1 \tag{4.76}$$

allowing the w_i to be determined. The formulae derived from eq. 4.74 are

$$r^2=a_{12}{}^2+a_{13}{}^2 \tag{4.77}$$

and

$$2h^2=r^2-a_{12}r \tag{4.78}$$

Instead of computing using $\mathbf{w}$ it is convenient to use

$$\mathbf{v}=2h\mathbf{w}=\{0, a_{12}-r, a_{13}\}^{\mathrm{T}} \tag{4.79}$$

leading to the transformation matrix

$$\mathbf{P}=\mathbf{I}-\frac{1}{2h^2}\mathbf{v}\mathbf{v}^{\mathrm{T}} \tag{4.80}$$

In eq. 4.77 for the determination of r, the sign should be chosen such that r is of opposite sign to a_{12} in this case.

For a general row i, the vector $\mathbf{v}$ takes the form

$$\mathbf{v}=\{0, 0, 0, \ldots 0, a_{i,i+1}-r, a_{i,i+2}, \ldots a_{i,n}\}^{\mathrm{T}} \tag{4.81}$$

Program 4.7. Householder's reduction to tridiagonal form
The procedure is illustrated in the following program with the nomenclature as given below (note that the reduction process is not iterative in character):

Simple variables

N	Number of equations to be reduced
R	See eq. 4.74 or 4.77
H	$-\frac{1}{2h^2}$ (see eq. 4.80)

Variable length arrays

A	Original matrix to be reduced (on completion holds the tridiagonalised form)
A1	Working space
P	Transformation matrix
V	Vector $\mathbf{v}$ (see eq. 4.79)

PARAMETER *restriction*
$\mathrm{IN} \geq \mathrm{N}$

The input is shown in Fig. 4.7(a). The number of equations and upper triangle of symmetric matrix **A** are first read in. Then N−2 transformations are made for rows designated by counter

```
      PROGRAM P47
C
C      PROGRAM 4.7 HOUSEHOLDER REDUCTION OF A SYMMETRIC
C                  MATRIX TO TRIDIAGONAL FORM
C
C      ALTER NEXT LINE TO CHANGE PROBLEM SIZE
C
      PARAMETER (IN=20)
C
      REAL A(IN,IN),A1(IN,IN),P(IN,IN),V(IN)
C
      READ (5,*) N,((A(I,J),J=I,N),I=1,N)
      DO 11 I = 1,N
          DO 11 J = I,N
   11 A(J,I) = A(I,J)
      WRITE(6,*) ('** HOUSEHOLDER REDUCTION OF A SYMMETRIC MATRIX **')
      WRITE(6,*) ('************* TO TRIDIAGONAL FORM ***************')
      WRITE(6,*)
      WRITE(6,*) ('MATRIX A')
      CALL PRINTA(A,IN,N,N,6)
      WRITE(6,*)
      DO 1 K = 1,N - 2
          R = 0.0
          DO 2 L = K,N - 1
    2     R = R + A(K,L+1)*A(K,L+1)
          R = SQRT(R)
          IF (R*A(K,K+1).GT.0.0) R = -R
          H = -1.0/ (R*R-R*A(K,K+1))
          CALL NULVEC(V,N)
          V(K+1) = A(K,K+1) - R
          DO 3 L = K + 2,N
    3     V(L) = A(K,L)
          CALL VVMULT(V,V,P,IN,N,N)
          CALL MSMULT(P,IN,H,N,N)
          DO 4 L = 1,N
    4     P(L,L) = P(L,L) + 1.0
          CALL MATMUL(A,IN,P,IN,A1,IN,N,N,N)
          CALL MATMUL(P,IN,A1,IN,A,IN,N,N,N)
    1 CONTINUE
      WRITE(6,*) ('TRANSFORMED MATRIX A')
      CALL PRINTA(A,IN,N,N,6)
      STOP
      END
```

K. Values of r, h and $\mathbf{v}$ are computed and the vector product required by eq. 4.80 is carried out using library routine VVMULT. Transformation matrix $\mathbf{P}$ can then be formed and two matrix multiplications using MATMUL complete the transformation. Figure 4.7(b) shows the resulting tridiagonalised $\mathbf{A}$, whose eigenvalues would then have to be computed by some other method, perhaps by vector iteration as previously described or by a characteristic polynomial method as

Array size	N			
	4			
Array	(A(I, J), J=I, N), I=1, N			
(accounting for symmetry)	1.0	−3.0	−2.0	1.0
		10.0	−3.0	6.0
			3.0	−2.0
				1.0

Fig. 4.7(a) Input data for Program 4.7

```
** HOUSEHOLDER REDUCTION OF A SYMMETRIC MATRIX **
************* TO TRIDIAGONAL FORM ***************

MATRIX A
   .1000E+01  -.3000E+01  -.2000E+01   .1000E+01
  -.3000E+01   .1000E+02  -.3000E+01   .6000E+01
  -.2000E+01  -.3000E+01   .3000E+01  -.2000E+01
   .1000E+01   .6000E+01  -.2000E+01   .1000E+01

TRANSFORMED MATRIX A
   .1000E+01   .3742E+01   .8449E-07  -.3017E-07
   .3742E+01   .2786E+01  -.5246E+01   .0000E+00
   .8449E-07  -.5246E+01   .1020E+02  -.4480E+01
  -.3017E-07   .2384E-06  -.4480E+01   .1015E+01
```

Fig. 4.7(b) Results from Program 4.7

shown in the following section. Alternatively, another transformation method may be used, as shown in the next program.

The matrix arithmetic in this algorithm has been deliberately kept simple. In practice, more involved algorithms can greatly reduce storage and computation time in this method.

4.7.3 LR Transformation for eigenvalues of tridiagonalised matrices

A transformation method most applicable to sparsely populated (band or tridiagonalised) matrices is the so-called '**LR**' transformation. This is based on repeated factorisation using what we called in Chapter 2 '**LU**' factorisation.

Thus

$$\mathbf{A}^k = \mathbf{LU} = \mathbf{LR} \tag{4.82}$$

for any step k of the iterative transformation. The step is completed by re-multiplying the factors in reverse order, that is

$$\mathbf{A}^{k+1} = \mathbf{UL} = \mathbf{RL} \tag{4.83}$$

Since from eq. 4.82

$$\mathbf{U} = \mathbf{L}^{-1}\mathbf{A}^k \tag{4.84}$$

the multiplication in eq. 4.83 implies

$$\mathbf{A}^{k+1} = \mathbf{L}^{-1}\mathbf{A}^k\mathbf{L} \tag{4.85}$$

showing that **L** has the property required of a transformation matrix **P**. As iterations proceed, the transformed matrix $\mathbf{A}^k$ tends to an upper triangular matrix whose eigenvalues are equal to the diagonal terms.

Example 4.3

Perform '**LR**' transformation on the non-symmetric matrix

$$\begin{bmatrix} 4 & 3 \\ 2 & 1 \end{bmatrix}$$

Solution 4.3

$$\mathbf{A}^0 = \begin{bmatrix} 4 & 3 \\ 2 & 1 \end{bmatrix} = \begin{bmatrix} 1 & 0 \\ 0.5 & 1 \end{bmatrix} \begin{bmatrix} 4 & 3 \\ 0 & -0.5 \end{bmatrix}$$

$$\mathbf{A}^1 = \begin{bmatrix} 4 & 3 \\ 0 & -0.5 \end{bmatrix} \begin{bmatrix} 1 & 0 \\ 0.5 & 1 \end{bmatrix} = \begin{bmatrix} 5.5 & 3 \\ -0.25 & -0.5 \end{bmatrix}$$

$$= \begin{bmatrix} 1 & 0 \\ -0.045 & 1 \end{bmatrix} \begin{bmatrix} 5.5 & 3 \\ 0 & -0.3636 \end{bmatrix}$$

$$\mathbf{A}^2 = \begin{bmatrix} 5.5 & 3 \\ 0 & -0.3636 \end{bmatrix} \begin{bmatrix} 1 & 0 \\ -0.045 & 1 \end{bmatrix} = \begin{bmatrix} 5.365 & 3 \\ 0.0164 & -0.3636 \end{bmatrix}$$

$$= \begin{bmatrix} 1 & 0 \\ 0.0031 & 1 \end{bmatrix} \begin{bmatrix} 5.365 & 3 \\ 0.0164 & -0.3728 \end{bmatrix}$$

$$\mathbf{A}^3 = \begin{bmatrix} 5.365 & 3 \\ 0.0164 & -0.3728 \end{bmatrix} \begin{bmatrix} 1 & 0 \\ 0.0031 & 1 \end{bmatrix} = \begin{bmatrix} 5.3743 & 3 \\ -0.0012 & -0.3728 \end{bmatrix}$$

$\mathbf{A}^3$ is nearly upper triangular, hence its eigenvalues are approximately 5.37 and -0.37 which are exact to 2 decimal places.

Although the method would be implemented in practice using special storage strategies, it is illustrated in Program 4.8 for the simple case of a square matrix **A**.

Program 4.8. LR transformation

This program uses the following nomenclature:

Simple variables

N	Number of equations to be solved
ITS	Maximum number of iterations allowed
TOL	Iteration tolerance
ITERS	Current number of iterations

Variable length arrays

A	Matrix $\mathbf{A}^k$
L	Lower triangular factor of **A**
U	Upper triangular factor of **A**
EOLD	Previous estimate of eigenvalues
ENEW	New estimate of eigenvalues

PARAMETER *restriction*

IN $\geq$ N

The program begins by reading the number of equations, followed by the coefficients of **A**, the convergence tolerance and iteration limit. Input and output are shown in Figs 4.8(a) and (b) respectively. The iteration loop is then entered, and begins with a call to LUFAC which completes the factorisation of the current $\mathbf{A}^k$ into **L** and **U**. These are multiplied in reverse order using MATMUL and the new estimate of the eigenvalues is found in the diagonal terms of the new $\mathbf{A}^{k+1}$, see eq. 4.83.

```
      PROGRAM P48
C
C      PROGRAM 4.8 L-R TRANSFORMATION FOR EIGENVALUES
C
C      ALTER NEXT TO CHANGE PROBLEM SIZE
C
      PARAMETER (IN=20)
C
      REAL A(IN,IN),U(IN,IN),L(IN,IN),EOLD(IN),ENEW(IN)
C
      READ (5,*) N,((A(I,J),J=1,N),I=1,N)
      READ (5,*) TOL,ITS
      WRITE(6,*) ('************ L-R TRANSFORMATION ******************')
      WRITE(6,*)
      WRITE(6,*) ('MATRIX A')
      CALL PRINTA(A,IN,N,N,6)
      WRITE(6,*)
      CALL NULVEC(EOLD,N)
      ITERS = 0
    1 ITERS = ITERS + 1
      CALL LUFAC(A,U,L,IN,N)
      CALL MATMUL(U,IN,L,IN,A,IN,N,N,N)
      DO 2 I = 1,N
    2 ENEW(I) = A(I,I)
      CALL CHECON(ENEW,EOLD,N,TOL,ICON)
      IF (ITERS.LT.ITS .AND. ICON.EQ.0) GO TO 1
      WRITE(6,*) ('FINAL TRANSFORMED DIAGONAL OF MATRIX A')
      CALL PRINTV(ENEW,N,6)
      WRITE(6,*)
      WRITE(6,*) ('ITERATIONS TO CONVERGENCE')
      WRITE(6,1000) ITERS
 1000 FORMAT (I5)
      STOP
      END
```

Array size	N		
	3		
Array	(A (I, J), J=1, N), I=1, N		
	2.0	−1.0	0.0
	−1.0	2.0	−1.0
	0.0	−1.0	1.0
Tolerance	TOL		
	1.E−7		
Iteration limit	ITS		
	100		

Fig. 4.8(a) Input data for Program 4.8

```
************ L-R TRANSFORMATION ******************

MATRIX A
   .2000E+01  -.1000E+01   .0000E+00
  -.1000E+01   .2000E+01  -.1000E+01
   .0000E+00  -.1000E+01   .1000E+01

FINAL TRANSFORMED DIAGONAL OF MATRIX A
   .3247E+01   .1555E+01   .1981E+00

ITERATIONS TO CONVERGENCE
  22
```

Fig. 4.8(b) Results from Program 4.8

The convergence check is then called, and the converged eigenvalues printed.

In variations of this method, for example the '**QR**' technique, other matrices can be substituted for **L** but the programming follows the same lines.

4.7.4 Lanczos reduction to tridiagonal form

In Chapter 2, we saw that some iterative techniques for solving linear equations, such as the steepest descent method, could be reduced to a loop involving a single matrix by vector multiplication followed by various simple vector operations. The Lanczos method for reducing matrices to tridiagonal form, while preserving their eigenvalues, involves very similar operations, and is in fact linked to the conjugate gradient technique of Program 2.12.

The transformation matrix **P** is in this method constructed using mutually orthogonal vectors. As usual we seek an eigenvalue-preserving transformation which for symmetry was given by eq. 4.54:

$$\mathbf{P}^T\mathbf{APx}=\lambda\mathbf{P}^T\mathbf{Px} \tag{4.54}$$

A means of ensuring $\mathbf{P}^T\mathbf{P}=\mathbf{I}$ is to construct **P** from mutually orthogonal vectors, say **p**, **q**, **r** for a 3×3 matrix. Then

$$\mathbf{P}^T\mathbf{P}=\begin{bmatrix} p_1 & p_2 & p_3 \\ q_1 & q_2 & q_3 \\ r_1 & r_2 & r_3 \end{bmatrix}\begin{bmatrix} p_1 & q_1 & r_1 \\ p_2 & q_2 & r_2 \\ p_3 & q_3 & r_3 \end{bmatrix} \tag{4.86}$$

$$=\begin{bmatrix} \mathbf{p}^T\mathbf{p} & \mathbf{p}^T\mathbf{q} & \mathbf{p}^T\mathbf{r} \\ \mathbf{q}^T\mathbf{p} & \mathbf{q}^T\mathbf{q} & \mathbf{q}^T\mathbf{r} \\ \mathbf{r}^T\mathbf{p} & \mathbf{r}^T\mathbf{q} & \mathbf{r}^T\mathbf{r} \end{bmatrix}=\mathbf{I}.$$

In the Lanczos method, we require $\mathbf{P}^T\mathbf{AP}$ to be a symmetric tridiagonal matrix, say,

$$\mathbf{M}=\mathbf{P}^T\mathbf{AP}=\begin{bmatrix} \alpha_1 & \beta_1 & 0 \\ \beta_1 & \alpha_2 & \beta_2 \\ 0 & \beta_2 & \alpha_3 \end{bmatrix} \tag{4.87}$$

and so

$$\mathbf{AP}=\mathbf{PM} \tag{4.88}$$

Since **P** is made up of the orthogonal vectors

$$\mathbf{P}=[\mathbf{p}\ \ \mathbf{q}\ \ \mathbf{r}] \tag{4.89}$$

we can expand eq. 4.88 to give

$$\begin{aligned} \mathbf{Ap}&=\alpha_1\mathbf{p}+\beta_1\mathbf{q} \\ \mathbf{Aq}&=\beta_1\mathbf{p}+\alpha_2\mathbf{q}+\beta_2\mathbf{r} \\ \mathbf{Ar}&=\qquad\ \ \beta_2\mathbf{q}+\alpha_3\mathbf{r} \end{aligned} \tag{4.90}$$

To construct the 'Lanczos vectors' **p**, **q** and **r** we note from the first of eqs 4.90 that

$$\mathbf{p}^T\mathbf{A}\mathbf{p}=\alpha_1\mathbf{p}^T\mathbf{p}+\beta_1\mathbf{p}^T\mathbf{q} \tag{4.91}$$

so that if α_1 is chosen to be $\mathbf{p}^T\mathbf{A}\mathbf{p}$ then $\mathbf{p}^T\mathbf{q}=0$ for any β and so **p** and **q** are orthogonal. Knowing α_1, the first of eqs 4.90 can be solved for $\beta_1\mathbf{q}$ and since $\mathbf{q}^T\mathbf{q}=1$, normalisation of $\beta_1\mathbf{q}$ such that $|\beta_1 q_i|_{max}=1$ will yield $1/\beta_1$. We then proceed to find α_2 and β_2 in the same way and continue for all n rows of **A**. Denoting $[\mathbf{p}\ \ \mathbf{q}\ \ \mathbf{r}]$ by $[\mathbf{y}_j]$ the algorithm is as follows (assuming $\beta_0=0$):

$$\begin{aligned}
\mathbf{v}_j &=\mathbf{A}\mathbf{y}_j-\beta_{j-1}\,\mathbf{y}_{j-1}\\
\alpha_j &=\mathbf{y}_j^T\mathbf{v}_j\\
\mathbf{z}_j &=\mathbf{v}_j-\alpha_j\mathbf{y}_j\\
\beta_j &=(\mathbf{z}_j^T\mathbf{z}_j)^{1/2}\\
\mathbf{y}_{j+1} &=\frac{1}{\beta_j}\mathbf{z}_j
\end{aligned} \tag{4.92}$$

Program 4.9. Lanczos reduction to tridiagonal form
The program uses the following nomenclature:

Simple variables

N	Number of equations to be reduced

Variable length arrays

A	Matrix of (symmetrical) coefficients
ALPHA	Diagonal of tridiagonal (see eq. 4.87)
BETA	Off-diagonal of tridiagonal (see eq. 4.87)
V, Z, Y0, Y1	Temporary vectors (see eq. 4.92)

PARAMETER *restriction*
IN ≥ N

```
      PROGRAM P49
C
C      PROGRAM 4.9 LANCZOS REDUCTION OF A SYMMETRIC
C                  MATRIX TO TRIDIAGONAL FORM
C
C      ALTER NEXT LINE TO CHANGE PROBLEM SIZE
C
      PARAMETER (IN=20)
C
      REAL A(IN,IN),ALPHA(IN),BETA(0:IN-1),V(IN),Y0(IN),Y1(IN),Z(IN)
C
      READ (5,*) N,((A(I,J),J=I,N),I=1,N)
      DO 10 I = 1,N
          DO 10 J = I,N
   10 A(J,I) = A(I,J)
      READ (5,*) (Y1(I),I=1,N)
      WRITE(6,*) ('**** LANCZOS REDUCTION OF A SYMMETRIC MATRIX ****')
```

```
      WRITE(6,*) ('************* TO TRIDIAGONAL FORM ***************')
      WRITE(6,*)
      WRITE(6,*) ('MATRIX A')
      CALL PRINTA(A,IN,N,N,6)
      WRITE(6,*)
      CALL NULVEC(Y0,N)
      BETA(0) = 0.0
      DO 1 J = 1,N
          CALL MVMULT(A,IN,Y1,N,N,V)
          DO 2 I = 1,N
    2     V(I) = V(I) - BETA(J-1)*Y0(I)
          CALL VECCOP(Y1,Y0,N)
          CALL VDOTV(Y1,V,ALPHA(J),N)
          IF (J.EQ.N) GO TO 1
          DO 3 I = 1,N
    3     Z(I) = V(I) - ALPHA(J)*Y1(I)
          CALL VDOTV(Z,Z,BETA(J),N)
          BETA(J) = SQRT(BETA(J))
          DO 4 I = 1,N
    4     Y1(I) = Z(I)/BETA(J)
    1 CONTINUE
      WRITE(6,*) ('TRANSFORMED MAIN DIAGONAL OF MATRIX A')
      CALL PRINTV(ALPHA,N,6)
      WRITE(6,*)
      WRITE(6,*) ('TRANSFORMED OFF-DIAGONAL OF MATRIX A')
      WRITE(6,100) (BETA(I),I=1,N-1)
  100 FORMAT(1X,6E12.4)
      STOP
      END
```

Array size	N			
	4			
Array	(A(I, J), I=1, N), J=I, N			
(accounting for symmetry)	1.0	−3.0	−2.0	1.0
		10.0	−3.0	6.0
			3.0	−2.0
				1.0
Starting vector	Y1 (I), I=1, N			
	1.0	0.0	0.0	0.0

Fig. 4.9(a) Input data for Program 4.9

```
**** LANCZOS REDUCTION OF A SYMMETRIC MATRIX ****
************* TO TRIDIAGONAL FORM ***************

MATRIX A
   .1000E+01  -.3000E+01  -.2000E+01   .1000E+01
  -.3000E+01   .1000E+02  -.3000E+01   .6000E+01
  -.2000E+01  -.3000E+01   .3000E+01  -.2000E+01
   .1000E+01   .6000E+01  -.2000E+01   .1000E+01

TRANSFORMED MAIN DIAGONAL OF MATRIX A
   .1000E+01   .2786E+01   .1020E+02   .1015E+01

TRANSFORMED OFF-DIAGONAL OF MATRIX A
   .3742E+01   .5246E+01   .4480E+01
```

Fig. 4.9(b) Results from Program 4.9

Note that the process is not iterative. Input and output are listed in Figs 4.9(a) and (b) respectively. The number of equations, N, is first read in, followed by the upper triangle coefficients of **A** which assumes symmetry. The starting vector $\mathbf{y}_1$, which is arbitrary as long as $\mathbf{y}_1^T\mathbf{y}_1=1$, is then read in and β_0 set to 0.

The main loop carries out exactly the operations of eqs 4.92 to build up N values of α and the N$-$1 values of β which are printed at the end of the program. For the given starting vector $\mathbf{y}_1=\{1\ \ 0\ \ 0\ \ 0\}^T$, the tridiagonalisation yields the same result as Householder's, but this is not always the case.

4.8 Characteristic polynomial methods

At the beginning of this chapter we illustrated how the eigenvalues of a matrix form the roots of an nth order polynomial, called the 'characteristic polynomial'. We pointed out that the methods of Chapter 3 could, in principle, be used to evaluate these roots, but that this will rarely be an effective method of eigenvalue determination. However, there are effective methods which are based on the properties of the characteristic polynomial. These are particularly attractive when the matrix whose eigenvalues has to be found is a tridiagonal matrix, and so are especially appropriate when used in conjunction with the Householder or Lanczos transformations described in the previous section.

4.8.1 Evaluating determinants of tridiagonal matrices

In the previous section we illustrated non-iterative methods of reducing matrices to tridiagonal equivalents. The resulting eigenvalue equation becomes

$$\begin{bmatrix} \alpha_1 & \beta_1 & 0 & 0 & 0 & \cdot & & \cdot\ \cdot \\ \beta_1 & \alpha_2 & \beta_2 & 0 & 0 & & & \\ 0 & \beta_2 & \alpha_3 & \beta_3 & 0 & & & \\ 0 & 0 & \beta_3 & \alpha_4 & \beta_4 & & & \\ \cdot & & & & & \cdot & & \\ \cdot & & & & & & \cdot & \beta_{n-1} \\ \cdot & & & & & & \beta_{n-1} & \alpha_n \end{bmatrix} \begin{Bmatrix} x_1 \\ x_2 \\ \cdot \\ \cdot \\ \cdot \\ \\ x_n \end{Bmatrix} = \lambda \begin{Bmatrix} x_1 \\ x_2 \\ \cdot \\ \cdot \\ \cdot \\ \\ x_n \end{Bmatrix} \tag{4.93}$$

The problem therefore becomes one of finding the roots of the determinantal equation

$$\begin{vmatrix} x_1-\lambda & \beta_1 & 0 & 0 & 0 & \cdot & \cdot & \cdot \\ \beta_1 & \alpha_2-\lambda & \beta_2 & 0 & 0 & & & \cdot \\ 0 & \beta_2 & \alpha_3-\lambda & \beta_3 & 0 & & & \cdot \\ 0 & 0 & \beta_3 & \alpha_4-\lambda & \beta_4 & & & \cdot \\ \cdot & & & & \cdot & & & \\ \cdot & & & & & & \cdot & \beta_{n-1} \\ \cdot & & & & & & \beta_{n-1} & \alpha_n-\lambda \end{vmatrix} = 0 \tag{4.94}$$

Although we shall not find these roots directly, consider the calculation of the determinant on the left-hand side of eqs 4.94.

If $n=1$, $\det_1(\lambda)=\alpha_1-\lambda$.

For $n=2$,

$$\det_2(\lambda)=\begin{vmatrix} \alpha_1-\lambda & \beta_1 \\ \beta_1 & \alpha_2-\lambda \end{vmatrix}=(\alpha_2-\lambda)(\alpha_1-\lambda)-\beta_1{}^2 \tag{4.95}$$

If $n=3$,

$$\det_3(\lambda)=\begin{vmatrix} \alpha_1-\lambda & \beta_1 & 0 \\ \beta_1 & \alpha_2-\lambda & \beta_2 \\ 0 & \beta_2 & \alpha_3-\lambda \end{vmatrix}=(\alpha_3-\lambda)\begin{vmatrix} \alpha_1-\lambda & \beta_1 \\ \beta_1 & \alpha_2-\lambda \end{vmatrix}-\beta_2{}^2(\alpha_1-\lambda) \tag{4.96}$$

We see that a recurrence relationship builds up enabling $\det_3(\lambda)$ to be evaluated simply from a knowledge of $\det_2(\lambda)$ and $\det_1(\lambda)$. If we let $\det_0(\lambda)=1$, the general recurrence may be written

$$\det_n(\lambda)=(\alpha_n-\lambda)\ \det_{n-1}(\lambda)-\beta_{n-1}^2\ \det_{n-2}(\lambda) \tag{4.97}$$

Therefore, for any value of λ, we can quickly calculate $\det_n(\lambda)$, and if we know the range within which a root $\det_n(\lambda)=0$ must lie, its value can be computed by, for example, the bisection method of Program 3.2. The remaining difficulty is to guide the choices of λ so as to be sure of bracketing a root. This task is made much easier due to a special property possessed by the principal minors of eqs 4.94, which is called the 'Sturm sequence' property.

4.8.2 The Sturm sequence property

A specific example of the left-hand side of eqs 4.94 is shown below, for $n=5$:

$$|\mathbf{A}|=\begin{vmatrix} 2-\lambda & -1 & 0 & 0 & 0 \\ -1 & 2-\lambda & -1 & 0 & 0 \\ 0 & -1 & 2-\lambda & -1 & 0 \\ 0 & 0 & -1 & 2-\lambda & -1 \\ 0 & 0 & 0 & -1 & 2-\lambda \end{vmatrix} \tag{4.98}$$

The principal minors of $\mathbf{A}$ are the submatrices outlined by the dashed lines i.e. formed by eliminating the nth, $(n-1)$th, etc., row and column of $\mathbf{A}$. The eigenvalues of $\mathbf{A}$ and of its principal minors will be found to be given by the following table (for example, by using Program 4.8)

$\mathbf{A}_5$	$\mathbf{A}_4$	$\mathbf{A}_3$	$\mathbf{A}_2$	$\mathbf{A}_1$
3.732	3.618	3.414	3.0	2.0
3.0	2.618	2.0	1.0	
2.0	1.382	0.586		
1.0	0.382			
0.268				

The characteristic polynomials of $\mathbf{A}_n$ and their roots, the eigenvalues, are also shown in Fig. 4.10. From the tabular and graphical representations it can be seen that each succeeding set of eigenvalues n, $n-1$, $n-2$ etc., always 'separates' the preceding set, that is, the eigenvalues of $\mathbf{A}_{n-1}$ always occur in the gaps between the eigenvalues of

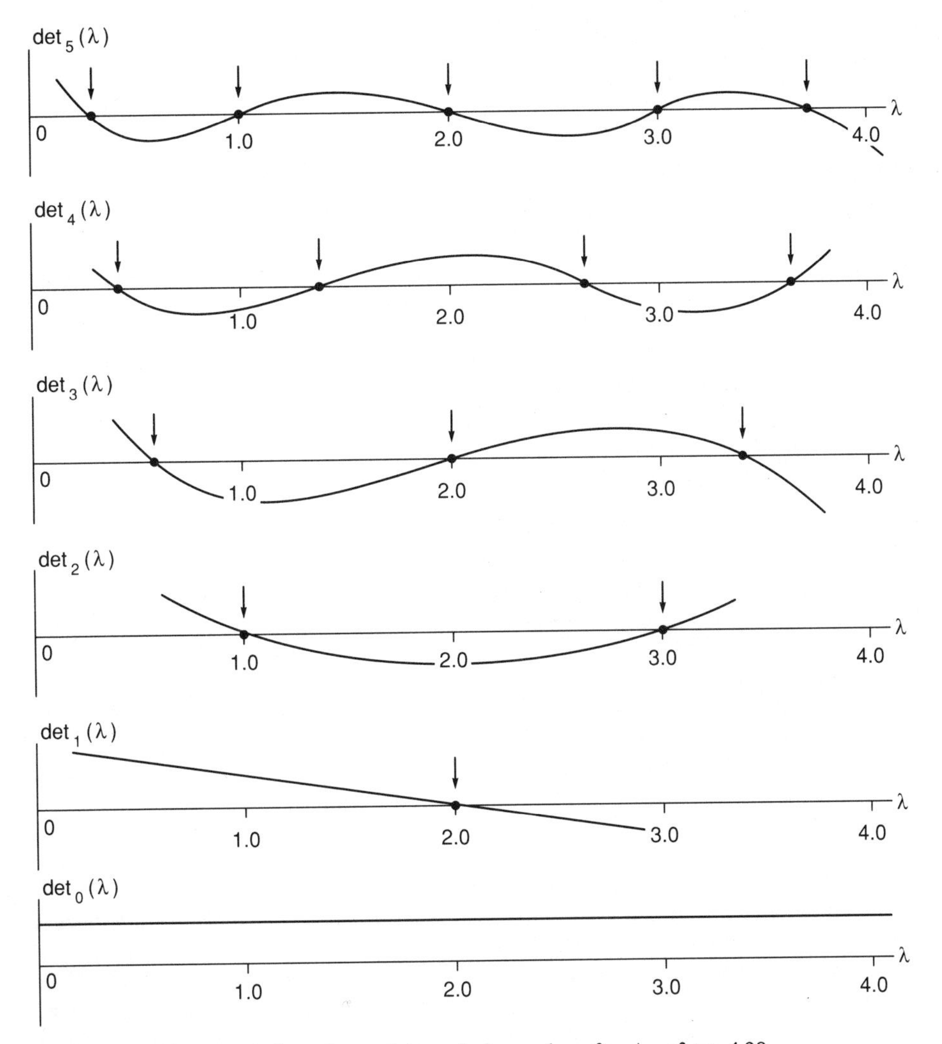

Fig. 4.10 Characteristic polynomials and eigenvalues for $\mathbf{A}_n$ of eq. 4.98

$\mathbf{A}_n$. This separation property is found for all symmetric **A** and is called the 'Strum sequence' property.

Its most useful consequence is that for any guessed λ, the number of sign changes in $\det_i(\lambda)$ for $i = 0, 1, 2, \ldots n$ is equal to the number of eigenvalues of **A** which are less than λ. Recalling that $\det_0(\lambda) = 1.0$, we can see from Fig. 4.10 the following sign change counts, noting that $\det_i(\lambda) = 0$ is *not* a change:

λ	Number of changes in sign = Number of eigenvalues less than λ
5	5
4	5
3	3
2	2
1	1

Program 4.10. Characteristic polynomial method for symmetric positive definite tridiagonal matrices

An example of a program utilising the features of the characteristic polynomial in the determination of eigenvalues is listed below. It uses the following nomenclature:

Simple variables

N	Number of tridiagonal rows
J	Eigenvalue wanted: $J = 1$ for largest, etc, $1 \leq J \leq N$
AL	Current estimate of root
ALMAX	Upper estimate of root
TOL	Iteration tolerance
ITS	Maximum number of iterations allowed
OLDL	Lower estimate of root
NUMBER	Number of sign changes of $\det_n(\lambda)$
ITERS	Current number of iterations

Variable length arrays

ALPHA	Leading diagonal entries of **A**
BETA	Off-diagonal entries of **A**
DET	$\det_n(\lambda)$ (see eq. 4.97)

PARAMETER *restriction*

IN $\geq$ N.

Input and output are shown in Figs 4.11(a) and (b) respectively. The number of equations is first read in followed by the diagonal (ALPHA) and off-diagonal (BETA) terms of the tridiagonal symmetric matrix previously obtained by methods such as Householder or Lanczos. The remaining data consist of J, the required eigenvalue where $J = 1$ is the largest etc., a starting guess of λ (AL), and upper limit to λ (ALMAX), a convergence tolerance and iteration limit. Since symmetric, positive definite **A** are implied, all of the eigenvalues will be positive.

The data relate to the example given in eqs 4.98. An upper limit of 5.0 is chosen as being bigger than the biggest eigenvalue, and the first guess for λ is chosen to be half this value, that is 2.5. The value of $\det_0(\lambda)$ is set to 1.0 and an iteration loop for the bisection process executed. The procedure continues until the iteration limit is reached or the tolerance, TOL, is satisfied by subroutine CHECK. The value of $\det_1(x)$ called DET(1) is set to $\alpha_1 - \lambda$ as required by the recursion

```
      PROGRAM P410
C
C      PROGRAM 4.10 CHARACTERISTIC POLYNOMIAL METHOD FOR
C                   EIGENVALUES OF SYMMETRIC TRIDIAGONAL
C
C      ALTER NEXT LINE TO CHANGE PROBLEM SIZE
C
      PARAMETER (IN=20)
C
      REAL ALPHA(IN),BETA(IN),DET(0:IN)
C
      READ (5,*) N,(ALPHA(I),I=1,N)
      READ (5,*) (BETA(I),I=1,N-1)
      READ (5,*) J,AL,ALMAX,TOL,ITS
      WRITE(6,*) ('**** CHARACTERISTIC POLYNOMIAL METHOD FOR *******')
      WRITE(6,*) ('* EIGENVALUES OF A SYMMETRIC TRIDIAGONAL MATRIX *')
      WRITE(6,*)
      WRITE(6,*) ('MAIN DIAGONAL')
      CALL PRINTV(ALPHA,N,6)
      WRITE(6,*)
      WRITE(6,*) ('OFF-DIAGONAL')
      CALL PRINTV(BETA,N-1,6)
      WRITE(6,*)
      WRITE(6,*) ('EIGENVALUE REQUIRED, 1=LARGEST, 2=NEXT,.. ETC.')
      WRITE(6,1000) J
      WRITE(6,*)
      DET(0) = 1.0
      AOLD = ALMAX
      WRITE(6,*) (' EIGENVALUE   DETERMINANT  NUMBER OF ROOTS SMALLER')
      WRITE(6,*) ('                                THAN CURRENT VALUE   ')
      ITERS = 0
   10 ITERS = ITERS + 1
          DET(1) = ALPHA(1) - AL
          DO 1 I = 2,N
    1     DET(I) = (ALPHA(I)-AL)*DET(I-1) - BETA(I-1)*BETA(I-1)*DET(I-2)
          NUMBER = 0
          DO 2 I = 1,N
              IF (DET(I).EQ.0.0) GO TO 2
              IF (DET(I-1).EQ.0.0) THEN
                  SIGN = DET(I)*DET(I-2)
              ELSE
                  SIGN = DET(I)*DET(I-1)
              END IF
              IF (SIGN.LT.0.0) NUMBER = NUMBER + 1
    2     CONTINUE
          IF (NUMBER.LE.N-J) THEN
              OLDL = AL
              AL = 0.5* (AL+ALMAX)
          ELSE
              ALMAX = AL
              AL = 0.5* (OLDL+AL)
          END IF
          WRITE (6,100) AL,DET(N),NUMBER
          CALL CHECK(AL,AOLD,TOL,ICON)
          IF(ITERS.NE.ITS.AND.ICON.EQ.0)GOTO 10
      WRITE(6,*)
      WRITE(6,*) ('ITERATIONS TO CONVERGENCE')
      WRITE(6,1000) ITERS
  100 FORMAT (2E12.4,I15)
 1000 FORMAT (I5)
      STOP
      END
```

Array size	N 5
Main diagonal	ALPHA (I), I = 1, N 2.0 2.0 2.0 2.0 2.0
Off-diagonal	BETA (I) = 1, N − 1 −1.0 −1.0 −1.0 −1.0
Eigenvalue required	J 1
Starting value	AL 2.5
Maximum value of root	ALMAX 5.0
Tolerance	TOL 1.E − 5
Iteration limit	ITS 100

Fig. 4.11(a) Input data for Program 4.10

```
**** CHARACTERISTIC POLYNOMIAL METHOD FOR *******
* EIGENVALUES OF A SYMMETRIC TRIDIAGONAL MATRIX *

MAIN DIAGONAL
   .2000E+01   .2000E+01   .2000E+01   .2000E+01   .2000E+01

OFF-DIAGONAL
  -.1000E+01  -.1000E+01  -.1000E+01  -.1000E+01

EIGENVALUE REQUIRED, 1=LARGEST, 2=NEXT,.. ETC.
   1

 EIGENVALUE    DETERMINANT  NUMBER OF ROOTS SMALLER
                               THAN CURRENT VALUE
  .3750E+01   -.1031E+01            3
  .3125E+01   -.2256E+00            5
  .3438E+01    .5183E+00            4
  .3594E+01    .1431E+01            4
  .3672E+01    .1129E+01            4
  .3711E+01    .6148E+00            4
  .3730E+01    .2397E+00            4
  .3740E+01    .1891E-01            4
  .3735E+01   -.1003E+00            5
  .3733E+01   -.3995E-01            5
  .3732E+01   -.1034E-01            5
  .3732E+01    .4332E-02            4
  .3732E+01   -.2990E-02            5
  .3732E+01    .6740E-03            4
  .3732E+01   -.1158E-02            5
  .3732E+01   -.2415E-03            5
  .3732E+01    .2164E-03            4

ITERATIONS TO CONVERGENCE
  17
```

Fig. 4.11(b) Results from Program 4.10

eq. 4.97 and then the other $\det_n(\lambda)$ are formed by recursion. The number of sign changes is detected by SIGN and accumulated as NUMBER.

The output in Fig. 4.11b shows that the largest eigenvalue, namely 3.732 is slightly less than the 12th estimate, since NUMBER is recorded as 5, that is, there are still 5 eigenvalues less than 3.7323. However DET(5) has been found in the bisection process to be of the order of 0.002, showing that convergence has essentially been obtained. More effective interpolation processes than bisection can of course be devised.

4.8.3 General symmetric matrices, e.g. band matrices

The principles described in the previous section can be applied to general matrices, but the simple recursion formula for finding $\det(\lambda)$ no longer applies. A way of computing $\det(\lambda)$ is to factorise $\mathbf{A}_n$, using the techniques described in Program 2.3, to yield $\mathbf{A}_n = \mathbf{LDL}^{\mathrm{T}}$. The product of the diagonal elements in $\mathbf{D}$ is the determinant of $\mathbf{A}_n$. Further useful information that can be derived from $\mathbf{D}$ is that in the factorisation of $\mathbf{A} - \lambda\mathbf{I}$, the number of negative elements in $\mathbf{D}$ is equal to the number of eigenvalues smaller than λ.

4.9 Exercises

1 Use vector iteration to find the mode corresponding to the largest eigenvalue of

$$\begin{bmatrix} 2 & 2 & 2 \\ 2 & 5 & 5 \\ 2 & 5 & 11 \end{bmatrix} \begin{Bmatrix} x_1 \\ x_2 \\ x_3 \end{Bmatrix} = \lambda \begin{Bmatrix} x_1 \\ x_2 \\ x_3 \end{Bmatrix}.$$

Answer: $\{0.2149 \quad 0.4927 \quad 0.8433\}^{\mathrm{T}}$ corresponding to the eigenvalue 14.43.

2 Use vector iteration to find the largest eigenvalue of the matrix

$$\begin{bmatrix} 3 & -1 \\ -1 & 2 \end{bmatrix}$$

and its associated eigenvector.
Answer: 3.618 associated with $\{0.8507 \quad -0.5257\}^{\mathrm{T}}$.

3 Use shifted vector iteration to find the smallest eigenvalue and eigenvector of the system given in Exercise 1.
Answer: $\{0.8360 \quad -0.5392 \quad 0.1019\}^{\mathrm{T}}$ corresponding to the eigenvalue 0.954.

4 The eigenvalues of the matrix

$$\begin{bmatrix} 5 & 1 & 0 & 0 & 0 \\ 1 & 5 & 1 & 0 & 0 \\ 0 & 1 & 5 & 1 & 0 \\ 0 & 0 & 1 & 5 & 1 \\ 0 & 0 & 0 & 1 & 5 \end{bmatrix}$$

are $5+2\cos[(i\pi)/6]$ where $i=1,2,3,4,5$. Prove this using shifted inverse iteration with shifts 6.7, 6.1, 5.3, 4.1, 3.3.
Answer: 6.732, 6.0, 5.0, 4.0, 3.268.

5 Find the eigenvalues and eigenvectors of the system

$$\begin{bmatrix} 2 & 1 \\ 1 & 1 \end{bmatrix} \begin{Bmatrix} x_1 \\ x_2 \end{Bmatrix} = \lambda \begin{bmatrix} 5 & 2 \\ 2 & 1 \end{bmatrix} \begin{Bmatrix} x_1 \\ x_2 \end{Bmatrix}$$

Answer: $\lambda_1 = 2.618$, $\mathbf{x}_1 = \{-0.3568\ 0.9342\}^T$
$\lambda_2 = 0.382$, $\mathbf{x}_2 = \{0.9342\ -0.3568\}^T$

6 Show that the system in Exercise 5 can be reduced to the 'standard form'

$$\begin{bmatrix} 0.4 & 0.2 \\ 0.2 & 2.6 \end{bmatrix} \begin{Bmatrix} x_1 \\ x_2 \end{Bmatrix} = \lambda \begin{Bmatrix} x_1 \\ x_2 \end{Bmatrix}$$

and hence find both of its eigenvalues.
How would you recover the eigenvectors of the original system?
Answer: 0.382, 2.618
See text, Section 4.6.1.

7 Use Householder's method to tridiagonalise the matrix

$$\begin{bmatrix} 1 & 1 & 1 & 1 \\ 1 & 2 & 2 & 2 \\ 1 & 2 & 3 & 3 \\ 1 & 2 & 3 & 4 \end{bmatrix}$$

Answer:

$$\begin{bmatrix} 1 & -1.732 & 0 & 0 \\ -1.732 & 7.667 & 1.247 & 0 \\ 0 & 1.247 & 0.9762 & -0.1237 \\ 0 & 0 & -0.1237 & 0.3571 \end{bmatrix}.$$

8 Use Lanczos's method to tridiagonalise the matrix in Exercise 7, using the starting vector $\{0.5\ 0.5\ 0.5\ 0.5\}^T$.
Answer:

$$\begin{bmatrix} 7.5 & 2.2913 & 0 & 0 \\ 2.2913 & 1.643 & 0.27355 & 0 \\ 0 & 0.27355 & 0.539 & 0.06943 \\ 0 & 0 & 0.06943 & 0.3182 \end{bmatrix}.$$

9 Find the eigenvalues of the tridiagonalised matrices in Exercises 7 and 8.
Answer: 8.291, 1.00, 0.4261, 0.2832 in both cases.

10 Find all the eigenvalues of the matrix

$$\begin{bmatrix} 3 & 0 & 2 \\ 0 & 5 & 0 \\ 2 & 0 & 3 \end{bmatrix}.$$

Prove that the eigenmodes associated with these eigenvalues are orthogonal.
Answer: $\lambda_1 = 1$, $\lambda_2 = 5$, $\lambda_3 = 5$, associated with the modes $\{1\ 0\ -1\}^T$, $\{1\ 0\ 1\}^T$ and $\{0\ 1\ 0\}^T$ respectively.

11 Using the characteristic polynomial method for symmetrical tridiagonal matrices, calculate all the eigenvalues of the matrix

$$\begin{bmatrix} 2 & -1 & 0 & 0 & 0 \\ -1 & 2 & -1 & 0 & 0 \\ 0 & -1 & 2 & -1 & 0 \\ 0 & 0 & -1 & 2 & -1 \\ 0 & 0 & 0 & -1 & 2 \end{bmatrix}$$

and the number of eigenvalues less than the current one.

Answer:		
	3.732299	4
	3.000004	4
	2.000004	3
	1.000004	2
	0.26795	1

4.10 Further reading

Bathe, K.J. and Wilson, E.L. (1976). *Numerical Methods in Finite Element Analysis*, Prentice-Hall, Englewood Cliffs, New Jersey.

Chatelin, F. (1987). *Eigenvalues of Matrices*, Wiley, London.

Conte, S. and de Boor, C. (1980). *Elementary Numerical Analysis*, 3rd edn, McGraw-Hill, New York.

Fox, L. (1964). *An Introduction to Numerical Linear Algebra*, Clarendon Press, Oxford.

Froberg, C.E. (1969). *Introduction to Numerical Linear Algebra*, 2nd edn, Addison-Wesley, Reading, Massachusetts.

Givens, J.W. (1954). Numerical computation of the characteristic values of a real symmetric matrix, Oak Ridge National Laboratory Report ORNL-1574.

Golub, G. and Van Loan, C. (1983). *Matrix Computations*, Johns Hopkins Press, Baltimore.

Gourlay, A.R. and Watson, G.A. (1973). *Computation Methods for Matrix Eigenproblems*, Wiley, London.

Householder, A. (1964). *The Theory of Matrices in Numerical Analysis*, Ginn, Boston.

Jennings, A. (1977). *Matrix Computation for Engineers and Scientists*, Wiley, Chichester.

Lanczos, C. (1950). An iteration method for the solution of the eigenvalue problems of linear differential and integral operators, *J. Res. Nat. Bur. Stand.*, **45**, 255–282.

Parlett, B. (1980). *The Symmetric Eigenvalue Problem*, Prentice-Hall, Englewood Cliffs, New Jersey.

Rutishauser, H. (1985). Solution of eigenvalues problems with the **LR** transformation, *Nat. Bur. Standards Appl. Math. Ser.*, **49**, 47–81.
Stewart, G.W. (1973). *Introduction to Matrix Computations*, Academic Press, New York.
Wilkinson, J.H. (1965). *The Algebraic Eigenvalue Problem*, Clarendon Press, Oxford.
Wilkinson, J.H. and Reinsch, C. (1971). *Handbook for Automatic Computation*, Vol II: Linear Algebra, Springer-Verlag, Berlin.

5

Interpolation and Curve Fitting

5.1 Introduction

This chapter is concerned with fitting mathematical functions to discrete data. Such data may come from measurements made during an experiment, or even numerical results obtained from an analysis of a large boundary or initial value problem (see Chapter 7). The functions will usually involve polynomials, which are easy to operate on, although other types of function may also be encountered.

Two general approaches will be covered. Firstly, functions which pass exactly through every point, and secondly, functions which are a 'good fit' to the points but do not necessarily pass through them. The former approach leads to 'interpolating polynomials' and the latter to 'curve fitting' or 'approximating' polynomials.

We may have several reasons for wishing to fit functions to discrete data. A common requirement is to use our derived function to interpolate between known values, or estimate derivatives and integrals. Numerical integration makes extensive use of approximating polynomials in Chapter 6, whereas estimation of derivatives from discrete data is discussed later in this chapter.

5.2 Interpolating polynomials

Firstly we consider the derivation of polynomials which pass exactly through a series of discrete data points. It can be shown that if we have $n+1$ data points, there is only one polynomial of degree n or less that satisfies this requirement. This polynomial is of degree n and is called the 'interpolating polynomial' $Q_n(x)$.

Hence, if our $n+1$ points are given as (x_i, y_i) for $i=0, 1, 2, \ldots, n$, then

$$Q_n(x_i)=y_i \quad i=0, 1, 2, \ldots, n \tag{5.1}$$

Although $Q_n(x)$ is unique for a given set of data, different methods can be employed for deriving it. In particular, if the x_i values are equally spaced, simplifications in the derivation of $Q_n(x)$ can be made.

5.2.1 Lagrangian polynomials

This approach works for any set of $n+1$ data points (x_i, y_i) for $i=0, 1, 2, \ldots, n$ leading to an interpolating polynomial given by

$$Q_n(x) = L_0(x)\,y_0 + L_1(x)\,y_1 + \cdots + L_n(x)\,y_n \tag{5.2}$$

The $L_i(x)$, $i=0, 1, 2, \ldots, n$ are 'Lagrangian' polynomials of degree n defined

$$L_i(x) = \frac{(x-x_0)(x-x_1)\ldots(x-x_{i-1})(x-x_{i+1})\ldots(x-x_{n-1})(x-x_n)}{(x_i-x_0)(x_i-x_1)\ldots(x_i-x_{i-1})(x_i-x_{i+1})\ldots(x_i-x_{n-1})(x_i-x_n)} \tag{5.3}$$

It may be noted from eq. 5.3 that the ith Lagrangian polynomial has the property

$$\left.\begin{aligned} L_i(x_j) &= 1 \text{ if } i=j \\ L_i(x_j) &= 0 \text{ if } i\neq j \end{aligned}\right\} i,\ j=0, 1, 2, \ldots, n \tag{5.4}$$

A further property of Lagrangian polynomials is that their sum equals unity, hence

$$\sum_{i=0}^{n} L_i(x) = 1 \tag{5.5}$$

Example 5.1

Use Lagrangian polynomials to derive a polynomial passing through the points

$x_0=1 \quad y_0=1$

$x_1=3 \quad y_1=5$

$x_2=6 \quad y_2=10$

and hence estimate the value of y when $x=4.5$.

Solution 5.1

There are three data points, hence there will be a second order interpolating polynomial given by

$$Q_2(x) = L_0(x)\,y_0 + L_1(x)\,y_1 + L_2(x)\,y_2$$

The three Lagrangian polynomials are derived according to eq. 5.3, hence

$$L_0(x) = \frac{(x-3)(x-6)}{(1-3)(1-6)} = \frac{1}{10}(x^2 - 9x + 18)$$

$$L_1(x) = \frac{(x-1)(x-6)}{(3-1)(3-6)} = -\frac{1}{6}(x^2 - 7x + 6)$$

$$L_2(x) = \frac{(x-1)(x-3)}{(6-1)(6-3)} = \frac{1}{15}(x^2 - 4x + 3)$$

After multiplication of each of the Lagrangian polynomials by the corresponding

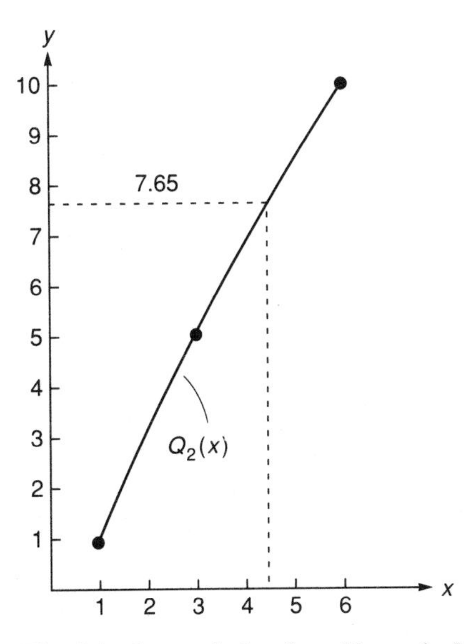

Fig. 5.1 Interpolation from Example 5.1

y-value as shown in eq. 5.2, we get the following interpolating polynomial which has been plotted in Fig. 5.1.

$$Q_2(x) = -\tfrac{1}{15}(x^2 - 34x + 18)$$

As a check, the three values of x should be substituted into the interpolating polynomial to give

$$Q_2(1) = 1, \; Q_2(3) = 5 \quad \text{and} \quad Q_2(6) = 10$$

The required interpolation is given by

$$Q_2(4.5) = 7.65$$

A disadvantage of the Lagrangian approach is the high number of arithmetic operations that must be carried out in order to compute an interpolate. Each Lagrangian polynomial is itself of order n, and must be evaluated at the required value of x.

A further problem relates to the efficiency of the Lagrangian approach if new data points are added to a set that has already been operated on. It seems that no advantage can be gained from the Lagrangian polynomials already computed, and the whole process must start again from the beginning.

Program 5.1. Interpolation using Langrangian polynomials
The first program in this chapter takes $n+1$ discrete data points of the form

$$(x_i, y_i), \; i = 0, 1, \dots n$$

and estimates the value of y at a given value of x using an interpolating polynomial. The nth order

```
      PROGRAM P51
C
C      PROGRAM 5.1  INTERPOLATION USING LAGRANGIAN POLYNOMIALS
C
C      ALTER NEXT LINE TO CHANGE PROBLEM SIZE
C
      PARAMETER (MAX=10)
C
      REAL X(MAX),Y(MAX)
C
      WRITE (6,*) ('******* LAGRANGIAN INTERPOLATION ******')
      WRITE (6,100)
      READ (5,*) NP
      DO 1 I = 1,NP
          READ (5,*) X(I),Y(I)
    1 WRITE (6,102) X(I),Y(I)
      READ (5,*) XI
      YI = 0.
      DO 2 I = 1,NP
          FAC = 1.
          DO 3 J = 1,NP
              IF (J.EQ.I) GO TO 3
              FAC = FAC* (XI-X(J))/ (X(I)-X(J))
    3     CONTINUE
    2 YI = YI + FAC*Y(I)
      WRITE (6,101) NP - 1
      WRITE (6,102) XI,YI
  100 FORMAT (/,'DATA POINTS')
  101 FORMAT (/,'INTERPOLATED POINT USING POLYNOMIAL (ORDER',I2,')')
  102 FORMAT (2E12.4)
      STOP
      END
```

interpolating polynomial that will pass exactly through the $n+1$ points is formed using Lagrangian polynomials as indicated in eqs 5.2 and 5.3. Input data consist of the number of points (NP), followed by the coordinates of the points (x, y) and the value of x at which an interpolated value of y is required.

A PARAMETER statement defines the constant MAX which must always be greater than or equal to the number of data points NP.

With the exception of simple integer counters, the variables can be summarised as follows:

NP	Number of data points
X, Y	One-dimensional arrays holding the coordinates of the data points
XI	Value of x at which interpolation is required
YI	Interpolated value of y
FAC	Running total of Lagrangian products

Number of data points	NP	
	4	
Coordinates of data points	(X(I), Y(I)), I=1, NP)	
	1.0	1.0
	3.0	5.0
	6.0	10.0
	5.0	9.0
Interpolation point	XI	
	4.5	

Fig. 5.2(a) Input data for Program 5.1

```
******* LAGRANGIAN INTERPOLATION ******

DATA POINTS
   .1000E+01    .1000E+01
   .3000E+01    .5000E+01
   .6000E+01    .1000E+02
   .5000E+01    .9000E+01

INTERPOLATED POINT USING POLYNOMIAL (ORDER 3)
   .4500E+01    .8175E+01
```

Fig. 5.2(b) Results from Program 5.1

To illustrate use of the program, the problem presented in Example 5.2 below involving interpolation between four data points is solved. The data and output from Program 5.1 are given in Figs 5.2(a) and (b) respectively. As shown in Fig. 5.2(b) the output consists of the initial data points together with the interpolated value. The cubic interpolating polynomial gives $y(4.5)=8.175$.

5.2.2 Difference methods

As alternative approach to finding the interpolating polynomials $Q_n(x)$ that will pass exactly through $n+1$ data points given as (x_i, y_i) $i=0, 1, \ldots, n$ is based on the polynomial representation given by

$$Q_n(x)=C_0+C_1(x-x_0)+C_2(x-x_0)(x-x_1)+\cdots$$
$$+C_n(x-x_0)(x-x_1)\ldots(x-x_{n-2})(x-x_{n-1}) \tag{5.6}$$

The constants $C_0, C_1, C_2, \ldots, C_n$ can be determined from the requirement that

$$Q_n(x_i)=y_i \quad i=0, 1, \ldots, n \tag{5.7}$$

which leads to

$$C_0=y_0$$

$$C_1=\frac{y_1-C_0}{(x_1-x_0)}$$

$$C_2=\frac{y_2-C_0-C_1(x_2-x_0)}{(x_2-x_0)(x_2-x_1)} \tag{5.8}$$

$$C_3=\frac{y_3-C_0-C_1(x_3-x_0)-C_2(x_3-x_0)(x_3-x_1)}{(x_3-x_0)(x_3-x_1)(x_3-x_2)}$$

etc.

The C_i's can be readily solved for and substituted into eq. 5.6. It may be noted that unlike the Lagrangian approach, if extra data are to be included, the resulting higher order interpolating polynomial is derived by the addition of terms to the lower order polynomial already found.

Example 5.2

Obtain an interpolating polynomial to pass through the following three points

$x_0=1 \qquad y_0=1$

$x_1=3 \qquad y_1=5$

$x_2=6 \qquad y_2=10$

and modify the polynomial to account for an additional point

$x_3=5 \qquad y_3=9$

Use the modified polynomial to estimate the value of y when $x=4.5$

Solution 5.2

Using the polynomial form of eq. 5.6 and the constants of eqs 5.8 we get

$$C_0=1$$

$$C_1=\frac{5-1}{3-1}=2$$

$$C_2=\frac{10-1-2(6-1)}{(6-1)(6-3)}=-\frac{1}{15}$$

Hence the required second order interpolating polynomial is given by

$$Q_2(x)=1+2(x-1)-\tfrac{1}{15}(x-1)(x-3)$$

therefore

$$Q_2(x)=-\tfrac{1}{15}(x^2-34x+18)$$

which is identical to the result of Example 5.1. The additional point (5, 9) leads to an extra constant

$$C_3=\frac{9-1-2(5-1)+\frac{1}{15}(5-1)(5-3)}{(5-1)(5-3)(5-6)}=-\tfrac{1}{15}$$

giving a cubic term which is simply added to the second order polynomial already found, hence

$$Q_3(x)=-\tfrac{1}{15}(x^2-34x+18)-\tfrac{1}{15}(x-1)(x-3)(x-6)$$

therefore

$$Q_3(x)=-\tfrac{1}{15}(x^3-9x^2-7x)$$

The cubic is shown in Fig. 5.3 and leads to the following value at $x=4.5$

$$Q_3(4.5)=8.175$$

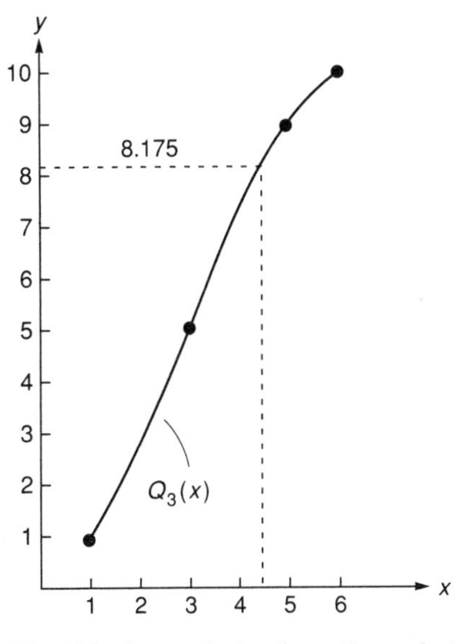

Fig. 5.3 Interpolation from Example 5.2

If the data are provided at equally spaced values of x, some simplification can be made to the derivation of the coefficients $C_0, C_1, \ldots C_n$.

Let the x-values be equidistant at h apart,

$$\text{i.e.} \quad x_{i+1} - x_i = h \tag{5.9}$$

then from eqs 5.8 we can write

$$C_0 = y_0 \tag{5.10}$$

$$\text{and} \quad C_1 = \frac{y_1 - y_0}{x_1 - x_0} = \frac{\Delta y_0}{h} \tag{5.11}$$

where a 'forward difference' notation is used,

$$\text{i.e.} \quad \Delta y_i = y_{i+1} - y_i \tag{5.12}$$

Furthermore

$$C_2 = \frac{y_2 - y_0 - \Delta y_0 (x_2 - x_0)/h}{(x_2 - x_0)(x_2 - x_1)} \tag{5.13}$$

$$= \frac{(y_2 - y_1) - (y_1 - y_0)}{2h^2}$$

$$= \frac{\Delta y_1 - \Delta y_0}{2h^2} = \frac{\Delta(\Delta y_0)}{2h^2} \tag{5.14}$$

$$\text{Hence} \quad C_2 = \frac{\Delta^2 y_0}{2h^2} \tag{5.15}$$

and it is easily shown that in general

$$C_j = \frac{\Delta^j y_0}{(j!)h^j} \tag{5.16}$$

where $\Delta^j y_i = \Delta^{j-1} y_{i+1} - \Delta^{j-1} y_i$ (5.17)

It is apparent that from eqs 5.12 and 5.17 the Δy_0 terms can be evaluated recursively. A tabular layout is useful for this purpose as shown in Table 5.1.

Table 5.1. Newton forward differences

x_0	y_0					
		Δy_0				
x_1	y_1		$\Delta^2 y_0$			
		Δy_1		$\Delta^3 y_0$		
x_2	y_2		$\Delta^2 y_1$		$\Delta^4 y_0$	
		Δy_2		$\Delta^3 y_1$		$\Delta^5 y_0$
x_3	y_3		$\Delta^2 y_2$		$\Delta^4 y_1$	
		Δy_3		$\Delta^3 y_2$		
x_4	y_4		$\Delta^2 y_3$			
		Δy_4				
x_5	y_5					

The particular scheme described in Table 5.1 is known as Newton forward differences, and is characterised by subscripts remaining the same along a downward sloping diagonal line going from top left to bottom right. Other layouts are possible, such as Newton backward differences, and Gaussian methods, but these will not be discussed here.

Example 5.3

Given the following data based on the function $y = \cos x$ where x is in degrees

x	y
20	0.93969
25	0.90631
30	0.86603
35	0.81915
40	0.76604

use a forward difference scheme to estimate cos (27°).

Solution 5.3

First we arrange the data as a table of forward differences

x	y	Δy	$\Delta^2 y$	$\Delta^3 y$	$\Delta^4 y$
20°	0.93969				
		−0.03338			
25°	0.90631		−0.00690		
		−0.04028		0.00031	
30°	0.86603		−0.00659		0.00005
		−0.04687		0.00036	
35°	0.81915		−0.00623		
		−0.05311			
40°	0.76604				

Referring to the interpolation polynomial of eq. 5.6, x_0 can be chosen to be any of the initial values of x. However, if we wish to include all five data points in the interpolating polynomial, x_0 should be put equal to the top value in the table, i.e. $x_0 = 20$. Noting that the constant interval between x-values is given by $h = 5$, the coefficients can be evaluated as follows:

j	$\Delta^j y_0$	$C_j = \Delta^j y_0/(j!)h^j$
0	0.93969	0.93969
1	−0.03338	−0.006676
2	−0.00690	−0.000138
3	0.00031	0.0000004
4	0.00005	0.000000003

Hence, written out in full

$$\begin{aligned} Q_4(x) = \ & 0.93969 - 0.006676\,(x-20) \\ & -0.000138\,(x-20)(x-25) \\ & +0.0000004\,(x-20)(x-25)(x-30) \\ & +0.000000003\,(x-20)(x-25)(x-30)(x-35) \end{aligned}$$

which is exactly the same interpolation polynomial we would have obtained using Lagrangian polynomials.

Substitution of the required value of x gives

$Q_4(27) = 0.89101$ (cf. exact solution 0.89101).

It is clear from the above expression for $Q_4(x)$ that the third and fourth order terms are contributing very little to the overall solution. If these two terms were to be discarded, the remaining second order polynomial would give

$$Q_2(x)=0.93969-0.006676\,(x-20)-0.000138\,(x-20)(x-25)$$

hence $Q_2(27)=0.89103$ which is correct to four decimal places.

The ability to truncate the interpolating polynomial if the coefficients become sufficiently small represents a potential saving on the amount of computation required. In the Lagrangian approach no such saving is possible and the full order of polynomial must be derived whether it is needed or not. The Lagrangian approach has the advantage of simplicity, however, and is relatively easy to remember for hand calculation if only a few points are provided.

5.2.3 Spline functions

The interpolating methods described so far in this chapter lead to high order polynomials, in general of order equal to one less than the number of data points. In addition to being laborious computationally, these high order polynomials can lead to undesirable maxima and minima between the given data points. If continuity of the higher derivatives of the interpolating polynomial is not essential, alternative methods can be used involving lower order polynomials.

Spline methods involve interpolation between the given data points by attaching together several low order polynomials. The best known methods involve cubic spline functions which are able to preserve continuity up to the second derivative. Physically, a spline can be visualised as a flexible elastic beam, initially straight, deformed in such a way that it passes through the required points (x_i, y_i), $i=0, 1, \ldots, n$.

Figure 5.4 shows how interpolation through four points might be achieved by the

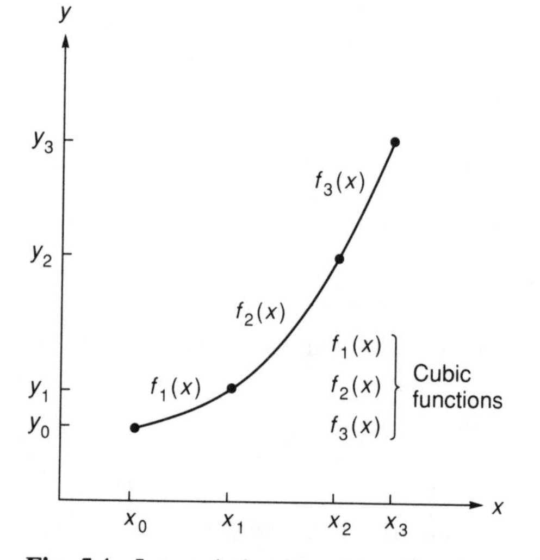

Fig. 5.4 Interpolation by cubic spline functions

use of three cubic functions $f_1(x)$, $f_2(x)$ and $f_3(x)$. In general, if we have $n+1$ points, n cubic spline functions will be required which can be written in the form

$$f_i(x)=A_{1i}+A_{2i}x+A_{3i}x^2+A_{4i}x^3, \quad i=1,2,\ldots,n \tag{5.18}$$

The $4n$ coefficients A_{ji} can be determined from the following conditions:

(a) The cubics must meet at all internal points leading to the $2n$ equations

$$\begin{aligned} f_i(x_i)&=y_i, \quad i=1,2,\ldots,n \\ f_{i+1}(x_i)&=y_i, \quad i=0,1,\ldots,n-1 \end{aligned} \tag{5.19}$$

(b) The first derivative must be continuous at all internal points leading to the $n-1$ equations

$$f_i'(x_i)=f_{i+1}'(x_i), \quad i=1,2,\ldots,n-1 \tag{5.20}$$

(c) The second derivative must also be continuous at all internal points leading to further $n-1$ equations

$$f_i''(x_i)=f_{i+1}''(x_i), \quad i=1,2,\ldots,n-1 \tag{5.21}$$

(d) The final two conditions refer to the two ends of the spline, where the second derivative is commonly put to zero, thus

$$\begin{aligned} f_1''(x_0)&=0 \\ f_n''(x_n)&=0 \end{aligned} \tag{5.22}$$

This final boundary condition preserves the 'structural' analogy, and implies a bending moment equal to zero at the ends of the 'beam'.

Although the $4n$ equations in $4n$ unknowns could be solved to obtain the required coefficients, it can be shown that a certain rearrangement of the cubic expressions between each point leads to a considerable improvement in efficiency, whereby only $n-1$ equations need to be solved.

The rearrangement involves writing the cubic expressions in terms of the second derivative at each end, thus

$$f_i(x)=\frac{f''(x_{i-1})(x_i-x)^3}{6\Delta x_i}+\frac{f''(x_i)(x-x_{i-1})^3}{6\Delta x_i}+$$
$$+\left(\frac{y_{i-1}}{\Delta x_i}-\frac{f''(x_{i-1})\Delta x_i}{6}\right)(x_i-x)+\left(\frac{y_i}{\Delta x_i}-\frac{f''(x_i)\Delta x_i}{6}\right)(x-x_{i-1}) \tag{5.23}$$

where $\Delta x_i=x_i-x_{i-1}$, $x_{i-1}\le x\le x_i$ and $i=1,2,\ldots,n$

(NB: 'Backward difference' definition of Δx_i)

It may be noted that eq. 5.23 includes only two unknown 'coefficients', namely $f''(x_{i-1})$ and $f''(x_i)$.

Differentiation of eq. 5.23 and imposition of the first derivative continuity condition leads to the following relationship

$$\Delta x_i f''(x_{i-1})+2(\Delta x_i+\Delta x_{i+1})f''(x_i)+\Delta x_{i+1} f''(x_{i+1})=6\left(\frac{\Delta y_i}{\Delta x_i}+\frac{\Delta y_{i+1}}{\Delta x_{i+1}}\right) \tag{5.24}$$

where $\Delta y_i=y_i-y_{i-1}$, $i=1,2,\ldots,n-1$

This is equivalent to a system of $n-1$ linear equations in the unknown second derivatives at the internal points, i.e.

$$\begin{bmatrix} 2(\Delta x_1+\Delta x_2) & \Delta x_2 & 0 & \ldots 0 \\ \Delta x_1 & 2(\Delta x_2+\Delta x_3) & \Delta x_3 & \ldots 0 \\ 0 & \Delta x_2 & 2(\Delta x_3+\Delta x_4) & \ldots 0 \\ 0 & 0 \ldots & \ldots & 2(\Delta x_{n-1}+\Delta x_n) \end{bmatrix} \begin{Bmatrix} f''(x_1) \\ f''(x_2) \\ \vdots \\ f''(x_{n-1}) \end{Bmatrix}$$

$$=6\begin{Bmatrix} \Delta y_1/\Delta x_1+\Delta y_2/\Delta x_2 \\ \Delta y_2/\Delta x_2+\Delta y_3/\Delta x_3 \\ \vdots \\ \Delta y_{n-1}/\Delta x_{n-1}+\Delta y_n/\Delta x_n \end{Bmatrix} \tag{5.25}$$

It may also be noted that the square matrix is tridiagonal (see Chapter 2). Once the second derivatives $f''(x_i)$, $i=1,2,\ldots,n-1$ have been obtained they are combined with the boundary conditions $f''(x_0)=f''(x_n)=0$ and the interpolating curve is fully defined.

To obtain an estimate of y corresponding to a particular value of x, it must first be ascertained which cubic function to use by observing the location of x relative to the original data points, x_i, $i=0,1,\ldots,n$. Once this is done, the appropriate values of $f''(x_i)$ and $f''(x_{i-1})$ can be substituted into eq. 5.23 and the function evaluated at the required point.

Example 5.4

Given the three data points

i	x_i	y_i
0	0.0	1.0
1	1.0	2.0
2	1.5	2.2

use cubic spline functions to estimate y when $x=1.1$.

Solution 5.4

In this example there will be two cubic splines attached at $x=1$. If we assume that the second derivative at each end equals zero, only one second derivative at $x=1$ remains to be found.

From eq. 5.25 we can write for this case

$$2(\Delta x_1+\Delta x_2)f''(x_1)=6\left(\frac{\Delta y_1}{\Delta x_1}+\frac{\Delta y_2}{\Delta x_2}\right)$$

The required backward differences can be obtained in tabular form, i.e.

i	x_i	$\Delta x_i = x_i - x_{i-1}$	y_i	$\Delta y_i = y_i - y_{i-1}$
0	0.0		1.0	
		1.0		1.0
1	1.0		2.0	
		0.5		0.2
2	1.5		2.2	

hence

$$2(1.0+0.5)f''(x_1)=6\left(\frac{1.0}{1.0}+\frac{0.2}{0.5}\right)$$

therefore

$$f''(x_1)=2.8$$

Interpolation is required at $x=1.1$ which lies in the range $1.0\leq 1.1\leq 1.5$, hence the required cubic from eq. 5.23 corresponds to $i=2$. Thus,

$$f_2(1.1)=\frac{2.8(1.5-1.1)^3}{6(0.5)}+\left(\frac{2}{0.5}-\frac{2.8(0.5)}{6}\right)(1.5-1.1)+\frac{2.2}{0.5}(1.1-1)=2.0064$$

Program 5.2. Interpolation using cubic spline functions
This program takes $n+1$ discrete data points of the form

$$(x_i, y_i), \quad i=0,1,\ldots,n$$

and estimates the value of y at a given value of x using cubic spline interpolating functions. Input data consist of the number of points (NP), followed by the coordinates of the points (x, y) and the value of x at which an interpolated value of y is required.

A PARAMETER statement defines the constant MAX, which must always be greater than or equal to the number of data points NP.

With the exception of simple integer counters, the variables can be summarised as follows

NP	Number of data points
N	NP-1
X, Y	1-d arrays holding coordinates of the data points
XI	Value of x at which interpolation is required
YI	Interpolated value of y
DX, DY	1-d arrays holding first backward differences of data points

K 2-d 'tridiagonal' square array from eq. 5.25
F 1-d array holding right-hand side values from eq. 5.25
U 1-d array holding second derivatives from eq. 5.25
D2 1-d array of second derivatives including boundary conditions.

To reduce the length of the main program, two library subroutines are called as follows

NULL Puts zeros into a 2-d array
SOLVE Solves a set of N simultaneous equations using Gaussian elimination with partial pivoting (see Chapter 2)

All the subroutines listed are fully described in Appendices 1 and 2.

To illustrate use of the program, the same four-point interpolation problem solved previously by Program 5.1 is presented. The data and output from Program 5.2 are given in Figs 5.5(a) and (b) respectively. It may be noted that unlike the data for Lagrangian interpolation, the x-data

```
      PROGRAM P52
C
C      PROGRAM 5.2   INTERPOLATION USING CUBIC SPLINES
C
C      ALTER NEXT LINE TO CHANGE PROBLEM SIZE
C
      PARAMETER (MAX=10)
C
      REAL X(0:MAX),Y(0:MAX),U(MAX),F(MAX),D2(0:MAX)
      REAL DX(MAX),DY(MAX),K(MAX,MAX)
C
      WRITE (6,*) ('***** CUBIC SPLINE INTERPOLATION ******')
      WRITE (6,100)
      READ (5,*) NP
      N = NP - 1
      DO 1 I = 0,N
          READ (5,*) X(I),Y(I)
    1 WRITE (6,102) X(I),Y(I)
      READ (5,*) XI
      DO 2 I = 1,N
          DX(I) = X(I) - X(I-1)
    2 DY(I) = Y(I) - Y(I-1)
      CALL NULL(K,MAX,N-1,N-1)
      DO 3 I = 1,N - 1
          K(I,I) = 2.* (DX(I)+DX(I+1))
    3 F(I) = 6.* (DY(I)/DX(I)+DY(I+1)/DX(I+1))
      DO 4 I = 2,N - 1
          K(I-1,I) = DX(I)
    4 K(I,I-1) = DX(I-1)
      CALL SOLVE(K,MAX,U,F,1)
      DO 5 I = 1,N - 1
    5 D2(I) = U(I)
      D2(0) = 0.
      D2(N) = 0.
      I = 1
    6 IF (XI.LE.X(I)) GO TO 7
      I = I + 1
      GO TO 6
    7 YI = (D2(I-1)* (X(I)-XI)**3+D2(I)* (XI-X(I-1))**3)/ (6.*DX(I)) +
     +     (Y(I-1)/DX(I)-D2(I-1)*DX(I)/6.)* (X(I)-XI) +
     +     (Y(I)/DX(I)-D2(I)*DX(I)/6.)* (XI-X(I-1))
      WRITE (6,101)
      WRITE (6,102) XI,YI
  100 FORMAT (/,'DATA POINTS')
  101 FORMAT (/,'INTERPOLATED POINT USING CUBIC SPLINES')
  102 FORMAT (2E12.4)
      STOP
      END
```

Number of data points	NP	
	4	
Coordinates of data points	(X(I), Y(I), I=1, NP)	
	1.0	1.0
	3.0	5.0
	5.0	9.0
	6.0	10.0
Interpolation point	XI	
	4.5	

Fig. 5.5(a) Input data for Program 5.2

```
***** CUBIC SPLINE INTERPOLATION ******

DATA POINTS
   .1000E+01    .1000E+01
   .3000E+01    .5000E+01
   .5000E+01    .9000E+01
   .6000E+01    .1000E+02

INTERPOLATED POINT USING CUBIC SPLINES
   .4500E+01    .7531E+01
```

Fig. 5.5(b) Results from Program 5.2

must be given in ascending (or descending) order. As shown in Fig. 5.5(b) the cubic spline interpolated between $x=3$ and $x=5$ leads to the value $y(4.5)=7.531$. This may be compared with $y(4.5)=8.175$ obtained using Program 5.1 and Lagrangian interpolation.

5.3 Numerical differentiation

Numerical differentiation involves estimating derivatives from a set of data points. It should be stated straight away that numerical differentiation is often an unreliable process which can be highly sensitive to small fluctuation in data. This is in contrast to numerical integration covered in the next chapter, which is an inherently stable process.

As an example, consider the case of a particle moving with uniform acceleration. Under experimental conditions, the velocity is measured as a function of time at discrete intervals. The results of these readings are plotted in Fig. 5.6(a). The response is essentially a straight line, but due to inevitable small fluctuation in the readings the line is a little ragged. If these data are differentiated to give accelerations, it is clear that the small ripples have had a dramatic effect on the calculated derivative as shown in Fig. 5.6(b). The ripples have virtually no influence on the displacement plot however, which is obtained by integration of the velocity data as shown in Fig. 5.6(c).

Although there can be problems with numerical differentiation of measured data, quite often we may wish to obtain derivatives from discrete data which have been computed from some other numerical process. In cases where the discrete data are 'well behaved', numerical differentiation can lead to acceptable estimates of derivatives.

The most obvious way of obtaining derivatives from discrete data is to find the interpolating polynomial and differentiate it.

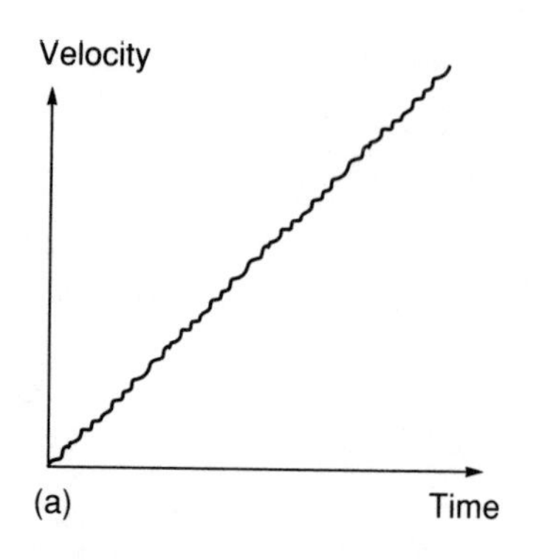

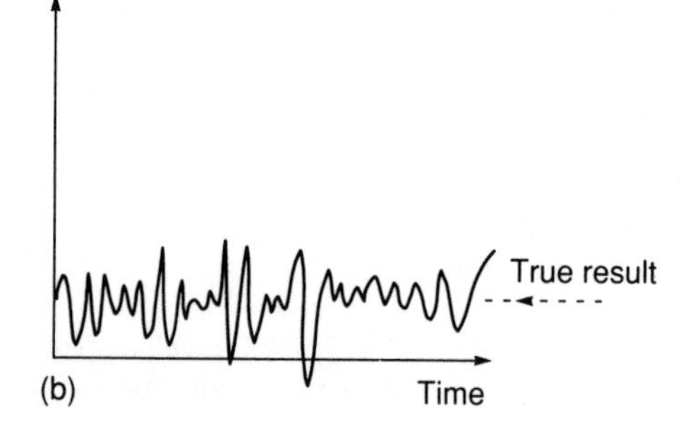

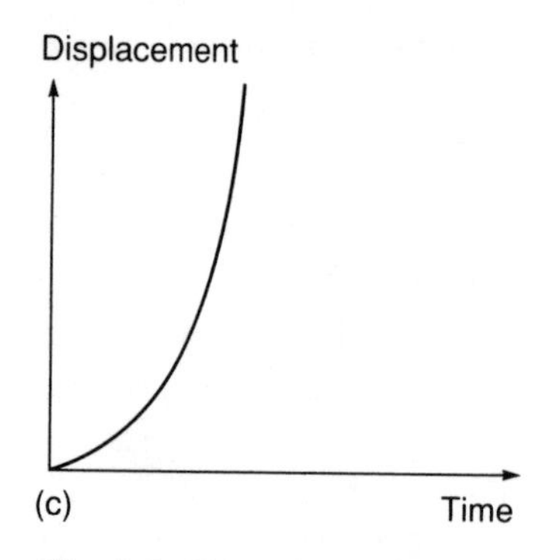

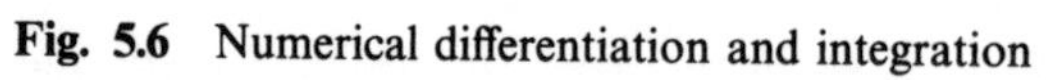
Fig. 5.6 Numerical differentiation and integration

Consider the case of three points, i.e.

$$(x_0, y_0), (x_1, y_1) \quad \text{and} \quad (x_2, y_2)$$

where the x's are equally spaced at h apart.

The interpolating polynomial is given by

$$Q_2(x) = C_0 + C_1(x - x_0) + C_2(x - x_0)(x - x_1) \tag{5.26}$$

when, from eqs 5.10, 5.11 and 5.15

$$C_0 = y_0$$

$$C_1 = \frac{y_1 - y_0}{h} \tag{5.27}$$

$$C_2 = \frac{y_2 - 2y_1 + y_0}{2h^2}$$

Differentiation of eq. 5.26 gives

$$Q_2'(x) = C_1 + C_2[2x - (x_0 + x_1)] \tag{5.28}$$

which when evaluated at x_0 leads to

$$Q_2'(x_0) = y_0' = \frac{1}{2h}(-3y_0 + 4y_1 - y_2) \tag{5.29}$$

Equation 5.29 is a 'second order' approximation to the first derivative at x_0 in terms of 'forward differences'. The term 'forward' in this context refers to the fact that the derivative is estimated from points that are greater than or equal to x_0.

'Backward difference' formulae estimate the required derivative using points that are less than or equal to x_0. A convenient notation in this context is to give points that are less than x_0 negative subscripts. Thus if we are presented with three points, i.e.

(x_{-2}, y_{-2}), (x_{-1}, y_{-1}) and (x_0, y_0)

where the x's are equally spaced at h apart, the backward difference formula for the first derivative at x_0 is given by

$$y_0' = \frac{1}{2h}(3y_0 - 4y_{-1} + y_{-2}) \tag{5.30}$$

It may be noted from eqs 5.29 and 5.30 that forward and backward difference formulae involving odd numbered derivatives, have coefficients of the same magnitude, but with the signs reversed.

To complete the picture we consider 'central difference' formulae where the required derivative is estimated from points symmetrically placed to the left and right of x_0. Thus, if we are given the three points

(x_{-1}, y_{-1}), (x_0, y_0) and (x_1, y_1)

formation of the second order interpolating polynomial and differentiation at x_0 leads to

$$y_0' = \frac{1}{2h}(y_1 - y_{-1}) \tag{5.31}$$

The nature of the three difference formulae is summarised in Fig. 5.7.

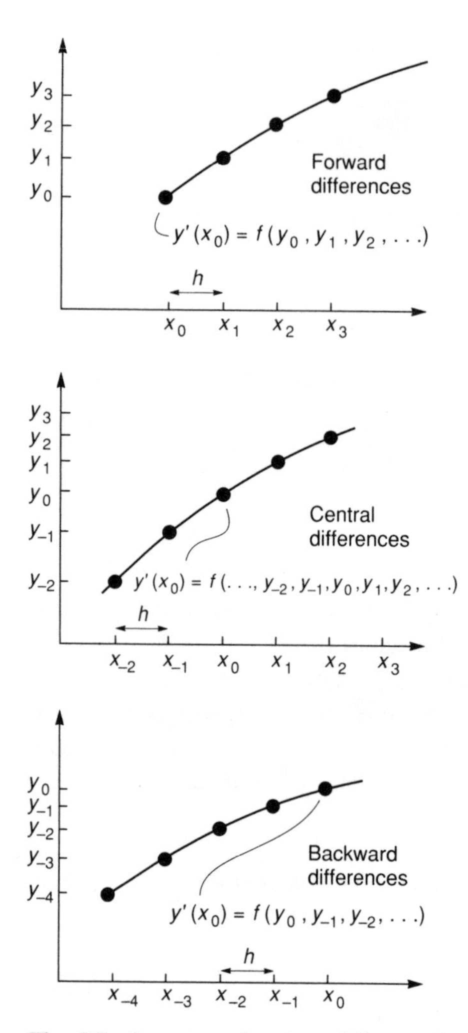

Fig. 5.7 Structure of various difference formulae

So far, only the calculation of first derivatives has been mentioned, but higher derivatives can also be generated from discrete data. For example, a forward difference version of the second derivative at x_0 can be obtained by further differentiation of eq. 5.26, i.e.

$$Q_2''(x) = 2C_2 = \text{constant}. \tag{5.32}$$

therefore

$$y_0'' = \frac{1}{h^2}(y_0 - 2y_1 + y_2) \tag{5.33}$$

Further differentiation of $Q_2(x)$ would be useless because the third and higher derivatives equal zero.

In general, if we are given $n+1$ data points equally spaced in x, the interpolating polynomial will be of order n, and the nth derivative will be the highest that can be deduced.

For example, if we wish to deduce a fourth derivative from discrete data, a minimum of five points will be required.

A summary of difference formulae is presented in Tables 5.2, 5.3 and 5.4.

Example 5.5

The following data rounded to four decimal places, is taken from the function $y=\sin x$ with x expressed in radians

x	y
0.4	0.3894
0.6	0.5646
0.8	0.7174
1.0	0.8415
1.2	0.9320

Table 5.2. Forward difference derivative formulae

Derivative	w_0	w_1	w_2	w_3	w_4	n	C	E
y'_0	−1	1				2	$\frac{1}{h}$	$-\frac{1}{2}hy''$
	−3	4	−1			3	$\frac{1}{2h}$	$\frac{1}{3}h^2y'''$
	−11	18	−9	2		4	$\frac{1}{6h}$	$-\frac{1}{4}h^3y^{iv}$
	−25	48	−36	16	−3	5	$\frac{1}{12h}$	$\frac{1}{5}h^4y^{v}$
y''_0	1	−2	1			3	$\frac{1}{h^2}$	$-hy'''$
	2	−5	4	−1		4	$\frac{1}{h^2}$	$\frac{11}{12}h^2y^{iv}$
	35	−104	114	−56	11	5	$\frac{1}{12h^2}$	$-\frac{5}{6}h^3y^{v}$
y'''_0	−1	3	−3	1		4	$\frac{1}{h^3}$	$-\frac{3}{2}hy^{iv}$
	−5	18	−24	14	−3	5	$\frac{1}{2h^3}$	$\frac{7}{4}h^2y^{v}$
y^{iv}_0	1	−4	6	−4	1	5	$\frac{1}{h^4}$	$-2hy^{v}$

General form to obtain a kth derivative

$$y_0^{(k)}=C(w_0y_0+w_1y_1+\cdots+w_{n-1}y_{n-1})+E$$

NB: The derivative occurring in the error term is taken at an unspecified value of $x=\xi$ in the range $x_0<\xi<x_{n-1}$.

Table 5.3. Central difference derivative formulae

Derivative	w_{-3}	w_{-2}	w_{-1}	w_0	w_1	w_2	w_3	n	C	E
			-1	0	1			3	$\frac{1}{2h}$	$-\frac{1}{6}h^2y'''$
y_0'		1	-8	0	8	-1		5	$\frac{1}{12h}$	$\frac{1}{30}h^4y^{\text{v}}$
	-1	9	-45	0	45	-9	1	7	$\frac{1}{60h}$	$-\frac{1}{140}h^6y^{\text{vii}}$
			1	-2	1			3	$\frac{1}{h^2}$	$-\frac{1}{12}h^2y^{\text{iv}}$
y_0''		-1	16	-30	16	-1		5	$\frac{1}{12h^2}$	$\frac{1}{90}h^4y^{\text{vi}}$
	2	-27	270	-490	270	-27	2	7	$\frac{1}{180h^2}$	$-\frac{1}{560}h^6y^{\text{viii}}$
y_0'''		-1	2	0	-2	1		5	$\frac{1}{2h^3}$	$-\frac{1}{4}h^2y^{\text{v}}$
	1	-8	13	0	-13	8	-1	7	$\frac{1}{8h^3}$	$\frac{7}{120}h^4y^{\text{vii}}$
y_0^{iv}		1	-4	6	-4	1		5	$\frac{1}{h^4}$	$-\frac{1}{6}h^2y^{\text{vi}}$
	-1	12	-39	56	-39	12	-1	7	$\frac{1}{6h^4}$	$\frac{7}{240}h^4y^{\text{viii}}$

General form to obtain a kth derivative

$$y_0^{(k)} = C(w_{-m}y_{-m} + w_{-m+1}y_{-m+1} + \cdots + w_0y_0 + \cdots + w_{m-1}y_{m-1} + w_my_m) + E$$

where $m = \dfrac{n-1}{2}$

NB: The derivative occurring in the error term is taken at an unspecified value of $x=\xi$ in the range $x_{-m} < \xi < x_m$.

Use forward differences to estimate $y'(0.4)$, backward differences to estimate $y'(1.2)$ and central differences to estimate $y'(0.8)$.

Solution 5.5

The data are equally spaced in x with the constant interval given by $h=0.2$. To illustrate the effect of taking more points on the accuracy of the solution, results are presented in tabular form. The error is also given as a percentage.

As an example, from Table 5.2, the four-point forward difference formula for $y'(0.4)$ gives the following expression

$$y'(0.4) \simeq \frac{1}{6(0.2)}[-11(0.3894) + 18(0.5646) - 9(0.7174) + 2(0.8415)] = 0.9215$$

Table 5.4. Backward difference derivative formulae

Derivative	w_{-4}	w_{-3}	w_{-2}	w_{-1}	w_0	n	C	E
				-1	1	2	$\frac{1}{h}$	$\frac{1}{2}hy''$
			1	-4	3	3	$\frac{1}{2h}$	$\frac{1}{3}h^2y'''$
y_0'		-2	9	-18	11	4	$\frac{1}{6h}$	$\frac{1}{4}h^3y^{iv}$
	3	-16	36	-48	25	5	$\frac{1}{12h}$	$\frac{1}{5}h^4y^{v}$
			1	-2	1	3	$\frac{1}{h^2}$	hy'''
y_0''		-1	4	-5	2	4	$\frac{1}{h^2}$	$\frac{11}{12}h^2y^{iv}$
	11	-56	114	-104	35	5	$\frac{1}{12h^2}$	$\frac{5}{6}h^3y^{v}$
y_0'''		-1	3	-3	1	4	$\frac{1}{h^3}$	$\frac{3}{2}hy^{iv}$
	3	-14	24	-18	5	5	$\frac{1}{2h^3}$	$\frac{7}{4}h^2y^{v}$
y_0^{iv}	1	-4	6	-4	1	5	$\frac{1}{h^4}$	$2hy^{v}$

General form to obtain a kth derivative

$y_0^{(k)} = C(w_0y_0 + w_{-1}y_{-1} + \cdots + w_{-n+1}y_{-n+1}) + E$

NB: The derivative occurring in the error term is taken at any unspecified value of $x=\xi$ in the range $x_{-n+1} < \xi < x_0$.

The exact solution is given by $\cos(0.4) = 0.9211$, hence the percentage error is given by

$$E = \frac{(0.9215 - 0.9211)}{0.9211} \times 100 = 0.04\%$$

N	$y'(0.4)$	Error %	Exact	
2	0.8760	-4.90	0.9211	
3	0.9320	1.18		FORWARD
4	0.9215	0.04		
5	0.9198	-0.15		

N	$y'(0.8)$	Error %	Exact	
3	0.6923	-0.64	0.6967	CENTRAL
5	0.6969	0.03		

N	$y'(1.2)$	Error %	Exact	
2	0.4524	24.88	0.3624	
3	0.3685	1.70		BACKWARD
4	0.3603	−0.56		
5	0.3621	−0.08		

Observation of the errors generated by the different formulae are of interest on several counts. In the forward difference calculation of $y'(0.4)$ the greatest accuracy was apparently obtained with the four-point formula. The larger error recorded in the 'more accurate' five-point formula occurred as a result of the tabulated data having been truncated after four decimal places.

For a given number of data points, the central difference formulae are more accurate than their forward or backward difference counterparts. This is borne out by the error terms in Tables 5.2, 5.3 and 5.4, and also by the percentage errors computed in this example.

The percentage error recorded in the two-point backward difference calculation of $y'(1.2)$ was very significant at nearly 25 per cent. This was due to the large value of the second derivative in the vicinity of the required solution. As shown in Table 5.4, the error term for this formula is given by

$$\tfrac{1}{2}hy''=\tfrac{1}{2}\times 0.2\times \sin(1.2)=0.093$$

which approximately equals the difference between the exact (0.3624) and computed (0.4524) values.

5.3.1 Errors in difference formulae

The error terms in Tables 5.2, 5.3 and 5.4 can be derived from the Taylor series. Consider a set of function values y_0, y_1, y_2, etc., which correspond to values of x which are equally spaced distance h apart. We can use the Taylor series to estimate y_1 and y_2 from y_0 and its derivatives at x_0, thus

$$y_1=y_0+hy'_0+\frac{h^2}{2}y''_0+\frac{h^3}{6}y'''_0+\cdots \tag{5.34}$$

and
$$y_2=y_0+2hy'_0+2h^2y''_0+\frac{4h^3}{3}y'''_0+\cdots \tag{5.35}$$

We can eliminate y''_0 from eqs 5.34 and 5.35, and rearrange to give

$$y'_0=\frac{1}{2h}(-3y_0+4y_1-y_2)+\frac{1}{3}h^2y'''_0 \tag{5.36}$$

which is the three-point forward difference formula for a first derivative from Table 5.2, together with the principal error term. Using this method, the error term for all the formulae quoted can be obtained.

5.4 Curve fitting

If we are seeking a function to follow closely a large number of data points measured in an experiment, it is often more practical to seek a function which represents a 'best fit' to the data rather than one which passes through all points exactly. Various strategies are possible for minimising the error between the individual data points and the approximating function. The method of least squares is the best known as this leads to 'regression' formulae used in statistical computation.

5.4.1 Least squares

Given n data points (x_1, y_1), (x_2, y_2) ... (x_n, y_n), the required 'best fit' function can be written in the form

$$f(x)=C_1 f_1(x)+C_2 f_2(x)+\cdots+C_k f_k(x) \tag{5.37}$$

where $f_j(x)$, $j=1,2$ ... k are chosen functions of x and the C_j, $j=1,2$ ... k are constants which are initially unknown.

The sum of the squares of the difference between $f(x)$ and the actual values of y is given by

$$E=\sum_{i=1}^{n}[f(x_i)-y_i]^2 \tag{5.38}$$

$$=\sum_{i=1}^{n}[C_1 f_1(x_i)+C_2 f_2(x_i)+\cdots+C_k f_k(x_i)-y_i]^2 \tag{5.39}$$

This error term can be minimised by taking the partial first derivative of E with respect to each of the constants, C_j, $j=1,2,\ldots,k$ and putting the result to zero, thus

$$\begin{aligned}
\frac{\partial E}{\partial C_1}&=2\sum_{i=1}^{n}\{[C_1 f_1(x_i)+C_2 f_2(x_i)+\cdots+C_k f_k(x_i)-y_i]\, f_1(x_i)\}=0\\
\frac{\partial E}{\partial C_2}&=2\sum_{i=1}^{n}\{[C_1 f_1(x_i)+C_2 f_2(x_i)+\cdots+C_k f_k(x_i)-y_i]\, f_2(x_i)\}=0\\
&\;\;\vdots\\
\frac{\partial E}{\partial C_k}&=2\sum_{i=1}^{n}\{[C_1 f_1(x_i)+C_2 f_2(x_i)+\cdots+C_k f_k(x_i)-y_i]\, f_k(x_i)\}=0
\end{aligned} \tag{5.40}$$

This symmetric system of k equation is linear, and can be written in matrix form as

$$\begin{bmatrix} \sum_{i=1}^{n} f_1(x_i)f_1(x_i) & \sum_{i=1}^{n} f_1(x_i)f_2(x_i) & \dots & \sum_{i=1}^{n} f_1(x_i)f_k(x_i) \\ \sum_{i=1}^{n} f_2(x_i)f_1(x_i) & \sum_{i=1}^{n} f_2(x_i)f_2(x_i) & \dots & \sum_{i=1}^{n} f_2(x_i)f_k(x_i) \\ \vdots & \vdots & & \vdots \\ \sum_{i=1}^{n} f_k(x_i)f_1(x_i) & \sum_{i=1}^{n} f_k(x_i)f_2(x_i) & \dots & \sum_{i=1}^{n} f_k(x_i)f_k(x_i) \end{bmatrix} \begin{Bmatrix} C_1 \\ C_2 \\ \vdots \\ C_k \end{Bmatrix} = \begin{Bmatrix} \sum_{i=1}^{n} f_1(x_i)y_i \\ \sum_{i=1}^{n} f_2(x_i)y_i \\ \vdots \\ \sum_{i=1}^{n} f_k(x_i)y_i \end{Bmatrix} \quad (5.41)$$

which can be solved for $C_1, C_2, \dots, C_k$ using any suitable algorithm (see Chapter 2).

Example 5.6

Use the method of least squares to derive the linear regression formula for the n data points given by (x_1, y_1), (x_2, y_2), $(x_3, y_3), \dots (x_n, y_n)$.

Solution 5.6

If a linear equation is to be fitted to the data, the following function involving two unknown constants could be used:

$$f(x) = C_1 + C_2 x$$

Following the general form of eq. 5.37, $f_1(x) = 1$ and $f_2(x) = x$.

From eq. 5.41, the following system of two equations is to be solved:

$$\begin{bmatrix} n & \sum_{i=1}^{n} x_i \\ \sum_{i=1}^{n} x_i & \sum_{i=1}^{n} x_i^2 \end{bmatrix} \begin{Bmatrix} C_1 \\ C_2 \end{Bmatrix} = \begin{Bmatrix} \sum_{i=1}^{n} y_i \\ \sum_{i=1}^{n} x_i y_i \end{Bmatrix}$$

hence
$$C_1 = \frac{\sum_{i=1}^{n} y_i - C_2 \sum_{i=1}^{n} x_i}{n}$$

and
$$C_2 = \frac{n \sum_{i=1}^{n} x_i y_i - \sum_{i=1}^{n} x_i \sum_{i=1}^{n} y_i}{n \sum_{i=1}^{n} x_i^2 - \left(\sum_{i=1}^{n} x_i \right)^2}$$

After substitution of C_1 and C_2 into the function $f(x)$, the classical linear regression formula is obtained.

It may be noted that the correlation coefficient for the data is given by

$$r=\frac{n\sum_{i=1}^{n} x_i y_i-\sum_{i=1}^{n} x_i \sum_{i=1}^{n} y_i}{\left\{\left[n\sum_{i=1}^{n} x_i^2-\left(\sum_{i=1}^{n} x_i\right)^2\right]\left[n\sum_{i=1}^{n} y_i^2-\left(\sum_{i=1}^{n} y_i\right)^2\right]\right\}^{1/2}}$$

Program 5.3. Curve fitting using 'least squares'

This program fits a function to a set of data points using least squares optimisation. The function must be of the form given by eq. 5.37. Input data consist of the number of data points (NP) and the number of coefficients to be optimised (NC). This is followed by the coordinates of the data points (x, y) and the value of x at which an estimate of y is required.

A PARAMETER statement defines the constants MAX and IC, which must be greater than or equal to the number of data points (NP) and coefficients (NC) respectively.

```
      PROGRAM P53
C
C      PROGRAM 5.3  CURVE FITTING USING 'LEAST SQUARES'
C
C      ALTER NEXT LINE TO CHANGE PROBLEM SIZE
C
      PARAMETER (MAX=10,IC=5)
C
      REAL X(MAX),Y(MAX),F(IC),C(IC),LS(IC,IC),R(IC)
C
      WRITE (6,*) ('******* LEAST SQUARES CURVE FITTING *******')
      WRITE (6,100)
      READ (5,*) NP,NC
      DO 1 I = 1,NP
          READ (5,*) X(I),Y(I)
    1 WRITE (6,101) X(I),Y(I)
      READ (5,*) XI
      CALL NULL(LS,IC,NC,NC)
      CALL NULVEC(R,NC)
      DO 2 I = 1,NP
          CALL FUNC(F,X(I))
          DO 3 J = 1,NC
              R(J) = R(J) + F(J)*Y(I)
              DO 3 K = 1,NC
    3     LS(J,K) = LS(J,K) + F(J)*F(K)
    2 CONTINUE
      CALL SOLVE(LS,IC,C,R,NC)
      CALL FUNC(F,XI)
      YI = 0.
      DO 4 I = 1,NC
    4 YI = YI + C(I)*F(I)
      WRITE (6,*)
      WRITE (6,*) ('OPTIMISED FUNCTION COEFFICIENTS')
      WRITE (6,101) (C(I),I=1,NC)
      WRITE (6,*)
      WRITE (6,*) ('INTERPOLATED POINT')
      WRITE (6,101) XI,YI
  100 FORMAT (/,'DATA POINTS')
  101 FORMAT (5E12.4)
      STOP
      END
C
      SUBROUTINE FUNC(F,X)
C
```

```
C       THIS SUBROUTINE PROVIDES THE VALUES OF THE REQUIRED
C       FUNCTIONS AT A PARTICULAR VALUE OF X
C
      REAL F(*)
C
      F(1) = 1.
      F(2) = ALOG(X)
      RETURN
      END
```

With the exception of simple integer counters, the variables can be summarised as follows:

NP	Number of data points
NC	Number of curve fitting coefficients
X	1-d arrays holding the coordinates
Y	of the data points
F	1-d array holding the function values
C	1-d array holding optimised coefficients
LS	2-d array holding function product summations
R	1-d array holding right hand sides
XI	Value of x at which estimate of y is required
YI	Estimated value of y

Library subroutines have been called as follows:

NULL	puts zeros into a 2-d array
NULVEC	puts zeros into a 1-d array
SOLVE	solves a system of simultaneous equation using Gaussian elimination with partial pivoting.

All these subroutines are described fully in Appendices 1 and 2.

To illustrate use of the program, an equation of the form

$$y = C_1 + C_2 \log_e x$$

is to be fitted to the following five data points:

x	y
29	1.6
50	23.5
74	38.0
103	46.4
118	48.9

The function $f_1(x) = 1$ and $f_2(x) = \log_e x$ have been included in a user-supplied subroutine FUNC, the contents of which will vary depending on the type of curve fitting to be performed.

The data and output from Program 5.3 are given in Figs 5.8(a) and (b) respectively. As shown in Fig. 5.8(b), the 'least squares' optimisation leads to the constants

$$C_1 = -111.1 \text{ and } C_2 = 34.02$$

Using these constants, the value of y when $x = 80$ is given to be 37.95.

The relation between the original points and the optimised curve in this example of 'logarithmic regression' is shown in Fig. 5.9.

Number of data points	NP 5
Number of curve fitting functions	NC 2
Coordinates of data points	(X(I), Y(I)), I = 1, NP) 29.0 1.6 50.0 23.5 74.0 38.0 103.0 46.4 118.0 48.9
Interpolation point	XI 80.0

Fig. 5.8(a) Input data for Program 5.3

```
 ******* LEAST SQUARES CURVE FITTING *******

DATA POINTS
   .2900E+02    .1600E+01
   .5000E+02    .2350E+02
   .7400E+02    .3800E+02
   .1030E+03    .4640E+02
   .1180E+03    .4890E+02

 OPTIMISED FUNCTION COEFFICIENTS
  -.1111E+03    .3402E+02

 INTERPOLATED POINT
   .8000E+02    .3795E+02
```

Fig. 5.8(b) Results from Program 5.3

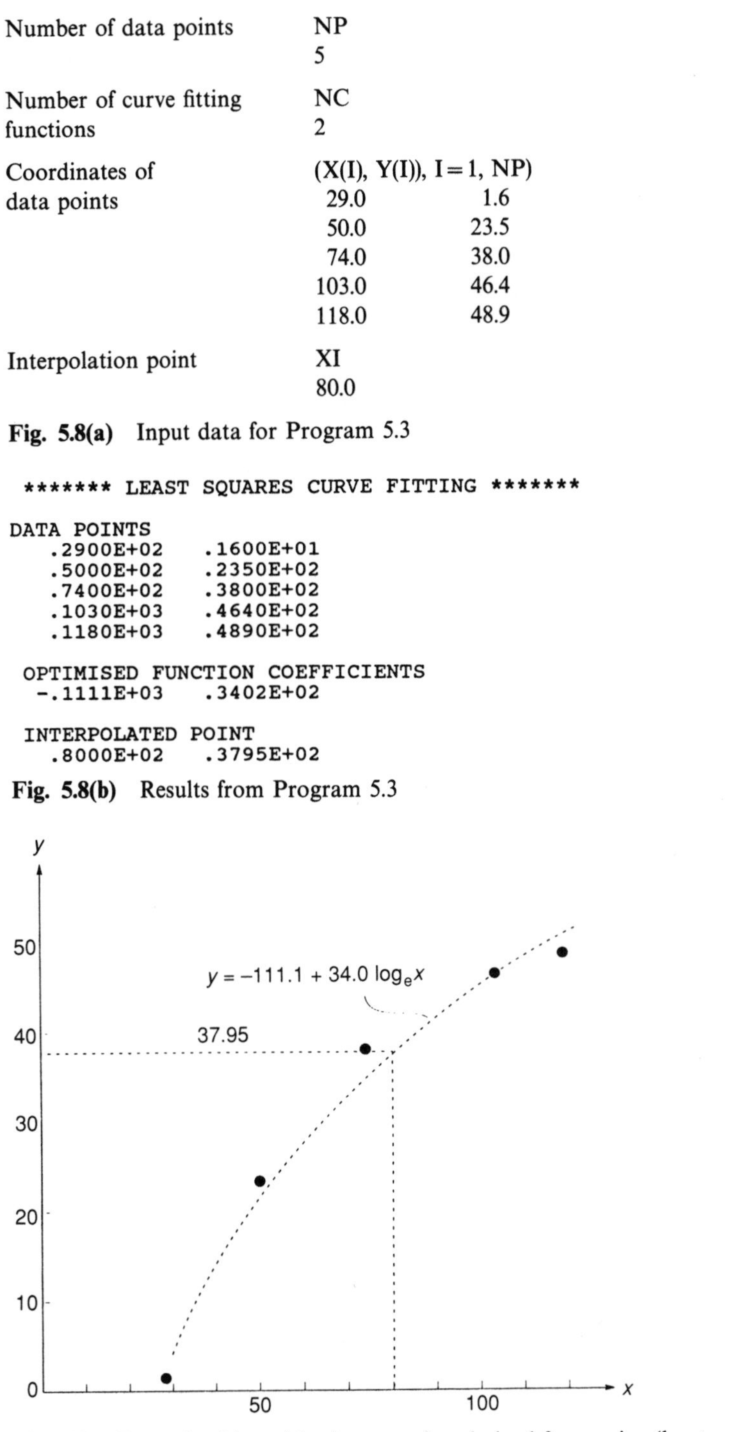

Fig. 5.9 Example of logarithmic regression derived from using 'least squares'

5.5 Exercises

1 Given the data

θ	$F(\theta)$
0	1.00
$\pi/4$	2.12
$\pi/2$	0.00

express the Lagrangian polynomials as function of θ, and hence estimate the value of $F(\pi/3)$.
Answer: $F(\pi/3)=1.773$

2 Use Lagrangian polynomials to obtain an interpolating polynomial for the data

x	y
0.0	0.1
0.1	0.1005
0.2	0.1020
0.3	0.1046

and use it to estimate the value of y when $x=0.4$. *Note:* This question involves extrapolation which is less reliable then interpolation.
Answer: $y(0.4)=0.1084$

3 Given the cubic

$$f(x)=x^3-2x^2+3x+1$$

derive an interpolating polynomial $g(x)$ that coincides with $f(x)$ at $x=1, 2$ and 3.
Show that $\int_1^3 f(x)\,dx=\int_1^3 g(x)\,dx$.

4 Derive a polynomial which passes through the points

x	y
0	1
1	1
2	15
3	61

using (a) Lagrangian polynomials (b) divided differences.

5 Given the sampling points:

$$x_1=-\sqrt{(\tfrac{3}{5})},\ x_2=0,\ x_3=\sqrt{(\tfrac{3}{5})}$$

form the Lagrangian polynomials $L_1(x)$, $L_2(x)$ and $L_3(x)$ and show that

$$\int_{-1}^{1} L_1(x)\,dx=\int_{-1}^{1} L_3(x)\,dx=\tfrac{5}{9},\quad \int_{-1}^{1} L_2(x)\,dx=\tfrac{8}{9}.$$

6 The following data are based on the function $y=\sin x$ where x is in degrees:

x	y
10	0.17365
15	0.25882
20	0.34202
25	0.42262
30	0.50000

Set up a table of forward differences, and by including more terms each time, estimate $\sin(28°)$ using first, second, third and fourth order interpolating polynomials.
Answer: First order 0.4690, second order 0.4694, third order 0.4695, fourth order 0.4695.

7 Rework Exercise 4 to obtain the fourth order interpolating polynomial, if the additional point (4,100) is included.

8 Rework Exercise 1 using a cubic spline function and hence estimate $F(\pi/3)$.
Answer: $F(\pi/3)=1.506$

9 Rework Exercise 2 using a cubic spline function and hence estimate the value of y when $x=0.15$.
Answer: $y(0.15)=0.1011$

10 Rework Exercise 3 to derive a cubic spline function which coincides with $f(x)$ at $x=1, 2$ and 3. Compute the value of the cubic spline and $f(x)$ at $x=1.5$ and $x=2.5$.
Answer: spline $(1.5)=3.5$, $f(1.5)=4.375$; spline $(2.5)=11.5$, $f(2.5)=11.625$

11 Rework Exercise 4 using cubic spline functions to estimate the value of y when $x=2.5$.
Answer: 38.00

12 Rework Exercise 6 using a cubic spline function to estimate the value of y when $x=17$.
Answer: 0.2759

13 Given the data of Exercise 1 estimate $F'(0)$, $F'(\pi/4)$ and $F'(\pi/2)$ making use of as much of the data as possible.
Answer: $F'(0)\simeq 3.49$, $F'(\pi/4)\simeq -0.64$, $F'(\pi/2)=-4.76$

14 Using the data of Exercise 6 estimate: $y'(30)$ using

(a) 2-point backward differences
(b) 3-point backward differences
(c) 4-point backward differences
(d) 5-point backward differences

Compare your results with the exact solution, 0.86603
Answer: (a) 0.887, (b) 0.868, (c) 0.866, (d) 0.866

15 Use the data of Exercise 6 to estimate $y'(20)$ using

(a) 3-point central differences
(b) 5-point central differences

Compare your results with the exact solution, 0.93969
Answer: (a) 0.939, (b) 0.940

16 Use the data of Exercise 6 to estimate $y''(10)$ using

(a) 3-point forward differences
(b) 4-point forward differences
(c) 5-point forward differences

Compare your results with the exact solution, -0.17365
Answer: (a) -0.259, (b) -0.176, (c) -0.175

17 Use the data of Exercise 6 to estimate $y'''(10)$ and $y^{iv}(10)$, using 5 points in each case. Compare your results with the exact solution of -0.98481 and 0.17365 respectively.
Answer: -0.971, 0.172

18 Prandtl's bearing capacity factor N_c is related to the friction angle ϕ as summarised in the table below

ϕ	N_c
0	5.14
5	6.49
10	8.35
15	10.98
20	14.83
25	20.72
30	30.14

Use least squares to fit first, second and third order polynomials to this data, and use these to estimate N_c when $\phi = 28°$.
Answer: First order 24.02, second order 26.08, third order 26.01

19 Use least squares to fit a straight line to the data

x	1.3	2.1	3.4	4.3	4.7	6.0	7.0
y	2.6	3.5	3.2	3.0	4.1	4.6	5.2

and hence estimate the value of y when $x = 2.5$ and $x = 5.5$.
Answer: $y(2.5) \simeq 3.092$, $y(5.5) \simeq 4.302$

20 Use least squares to fit a power equation of the form

$y = ax^b$

to the data

x	2.5	3.5	5.0	6.0	7.5	10.0	12.5
y	4.8	3.5	2.1	1.7	1.2	0.9	0.7

and hence estimate y when $x = 9.0$.
Note: You will need to arrange the above equation into a linear form before solving.
Answer: $a = 15.35$, $b = -1.233$, $y(9.0) \simeq 1.02$

21 Use last squares to fit an exponential equation of the form

$y = ae^{bx}$

to the data

x	0.6	1.0	2.3	3.1	4.4	5.8	7.2
y	3.6	2.5	2.2	1.5	1.1	1.3	0.9

and hence estimate the value of y when $x = 2.0$.
Note: You will need to arrange the above equation into a linear form before solving.
Answer: $a = 3.23$, $b = -0.188$, $y(2.0) \simeq 2.22$

5.6 Further reading

Abramowitz, M. and Stegun, I.A. (1964). *Handbook of Mathematical Functions*, U.S. Government Printing Office, Washington D.C.
Ahlberg, J.H., Nilson, E.N. and Walsh, J.L. (1967). *The Theory of Splines and their Application*, Academic Press, New York.
Bickley, W.G. (1941). Formulae for numerical differentiation, *Mathematical Gazette*, **25**, 19–27.
Chapra, S.C. and Canale, R.P. (1988). *Numerical Methods for Engineers*, 2nd edn, McGraw-Hill, New York.
Cheney, E.W. and Kincaid, D. (1980). *Numerical Mathematics and Computing*, Brooks/Cole, Monterey, California.
Conte, S.D. and de Boor, C. (1980). *Elementary Numerical Analysis*, 3rd edn, McGraw-Hill, New York.
Hildebrand, F.D. (1956). *Introduction to Numerical Analysis*, McGraw-Hill, New York.
Schumaker, L.L. (1981). *Spline Functions: Basic Theory*, Wiley, New York.
Shoup, T.E. (1979). *A Practical Guide to Computer Methods for Engineers*, Prentice-Hall, Englewood Cliffs, New Jersey.

6

Numerical Integration

6.1 Introduction

Numerical integration or 'quadrature' is used whenever analytical approaches are either inconvenient or impossible to perform. Numerical integration is also suitable for integrating discrete data, such as those measured in an experiment. Initially in this chapter, we will consider methods for numerical integration of a function of one variable, i.e.

$$\int_a^b f(x)\,dx \tag{6.1}$$

although area integrals of functions of more than one variable will also be considered subsequently.

We seek general algorithms or 'rules' for numerical integration which are suitable for programming on a computer. Our methods will usually not depend on the type of function being integrated, although some 'customised' approaches will be discussed.

Although numerical integration can often yield exact solutions, especially if the function under consideration is a simple polynomial, more often, our 'solution' will only be approximate. This is especially true when integrating combinations of transcendental functions (e.g. sine, cosine, logarithm, etc.) for which no analytical solution can be found. In addition, once our approximate 'solution' has been found, it will be important to have an idea of the magnitude of the errors.

The first step in numerical integration methods is to replace the function to be integrated, $f(x)$, by a simple polynomial $Q_{n-1}(x)$ of degree $n-1$ which coincides with $f(x)$ at n points x_i where $i=1,2,\ldots,n-1,n$.

Thus

$$\int_a^b f(x)\,dx \simeq \int_a^b Q_{n-1}(x)\,dx \tag{6.2}$$

where

$$Q_{n-1}(x)=a_0+a_1x+a_2x^2+\cdots+a_{n-1}x^{n-1} \tag{6.3}$$

and the a_i are constant coefficients.

Polynomials such as $Q_{n-1}(x)$ can be easily integrated analytically, and also exactly integrated numerically. Hence, provided approximation 6.2 is acceptable, we should be able to obtain reasonable estimates of the required integral.

Computers are usually efficient at addition and multiplication, so numerical integration 'rules' can be expressed as a summation of the form

$$\int_a^b Q_{n-1}(x)\,\mathrm{d}x = \sum_{i=1}^{n} w_i Q_{n-1}(x_i) \tag{6.4}$$

hence
$$\int_a^b f(x)\,\mathrm{d}x \simeq \sum_{i=1}^{n} w_i f(x_i) \tag{6.5}$$

The x_i's are termed 'sampling points', being those places where the function $f(x)$ must be evaluated, and the w_i's are constant 'weighting coefficients'. Equation 6.5 forms the basis for all our numerical integration methods in a single variable.

Usually, we do not need to know much about $Q_{n-1}(x)$ except to bear in mind that it is the function that is really being integrated from eq. 6.4. Of course if $f(x)$ also happens to be a polynomial, it too can be integrated exactly if n is sufficiently large.

All that distinguishes one of our numerical integration methods from another is the location and number of sampling points within, or even outside the range of integration. Methods in which the sampling points are equally spaced are called 'Newton–Cotes' rules, and these are considered first. It is later shown that 'Gaussian' rules, in which the sampling points are optimally spaced, lead to considerable improvements in accuracy and efficiency.

Other aspects of numerical integration dealt with in this chapter include special integrals, where the integrand contains an exponential function or a certain type of singularity. Also covered are some integration rules in which sampling points stray outside the range of integration. Finally, a section is included on area integrals in which functions of two variables are integrated numerically over quadrilateral regions.

6.2 Newton–Cotes rules

6.2.1 Introduction

These methods are probably the best known, and most popular for 'hand' calculation. Newton–Cotes rules are distinguished by the property that the sampling points are equally spaced within the range of integration and coincide with the limits of integration. It is usual to define the constant distance between adjacent sampling points as h.

In the methods that follow, the approximating polynomial $Q_{n-1}(x)$ is always integrated exactly, so the accuracies of the different methods depend on how closely $Q_{n-1}(x)$ fits the actual function $f(x)$ under consideration.

6.2.2 One-point rule (rectangle rule, $n=1$)

This trivial method is rarely used in practice but is included here for completeness. The function $f(x)$ is approximated by $Q_0(x)$ which coincides at one point only, namely the lower limit of integration as shown in Fig. 6.1. The method is not usually going to produce very accurate solutions, and will only be exact if $f(x)$ is itself a zeroth order polynomial i.e. a line parallel to the x-axis.

$$\text{Rectangle rule:} \quad \int_{x_1}^{x_2} f(x)\,dx \simeq hf(x_1) \tag{6.6}$$

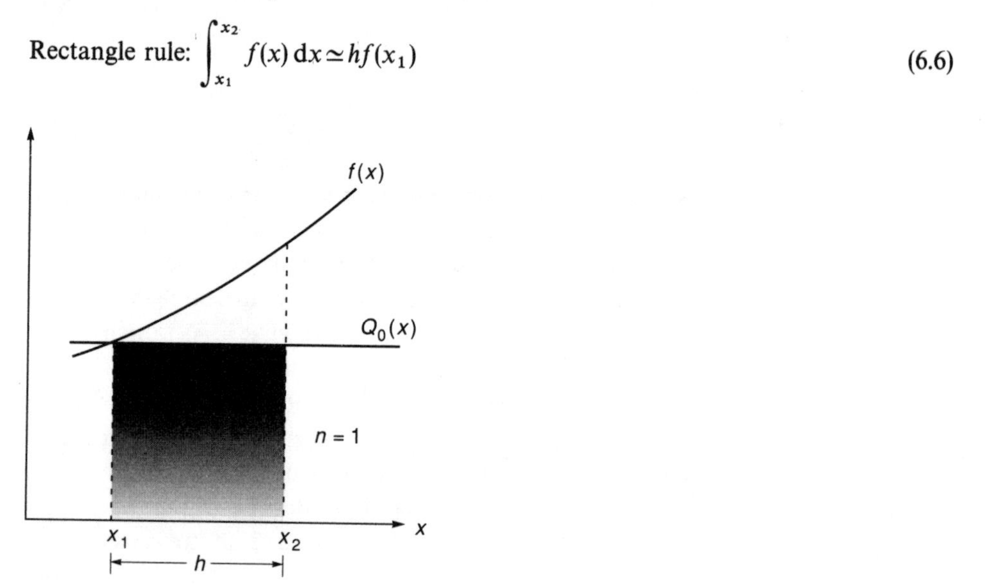

Fig. 6.1 Rectangle rule

6.2.3 Two-point rule (trapezium rule, $n=2$)

This is a popular approach in which $f(x)$ is approximated by a first order polynomial $Q_1(x)$ which coincides at the limits of integration. As shown in Fig. 6.2, the integral is approximated by the area of a trapezium.

$$\text{Trapezium rule:} \quad \int_{x_1}^{x_2} f(x)\,dx \simeq \tfrac{1}{2}hf(x_1)+\tfrac{1}{2}hf(x_2) \tag{6.7}$$

The formula will be exact for $f(x)$ of degree 1 or less.

Example 6.1

Estimate the value of

$$I=\int_{\pi/4}^{\pi/2} \sin x\,dx$$

using the trapezium rule.

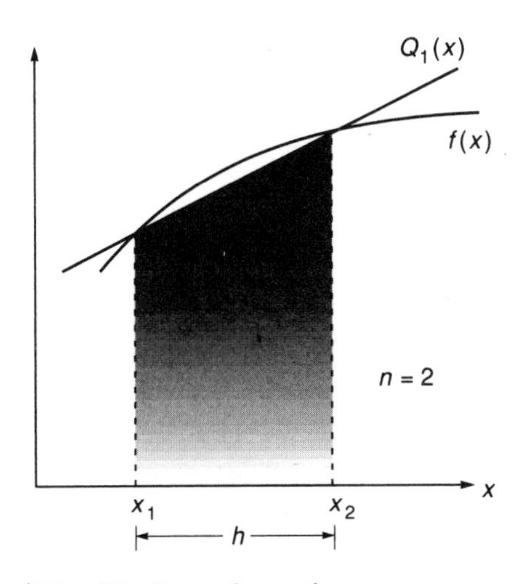

Fig. 6.2 Trapezium rule

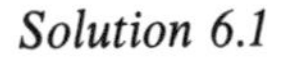

Solution 6.1

$$h=\frac{\pi}{4}$$

$$\therefore \quad I \simeq \frac{1}{2}\frac{\pi}{4}\left[\sin\left(\frac{\pi}{4}\right)+\sin\left(\frac{\pi}{2}\right)\right]$$

$$=0.6704 \text{ (cf. exact solution 0.7071)}$$

More accurate estimates can be found by using 'repeated' rules as will be described in a later section.

6.2.4 Three-point rule ('Simpson's' rule, $n=3$)

Another popular method is where $f(x)$ is approximated by a second order polynomial $Q_2(x)$ which coincides at three points, namely the ends and middle of the range of integration as shown in Fig. 6.3. It is apparent that Simpson's rule should exactly reproduce the hatched area under $Q_2(x)$. In order to derive the formula from first principles, we use a procedure called polynomial substitution where we require the general summation expression, namely

$$\int_{x_1}^{x_3} f(x)\,\mathrm{d}x = w_1 f(x_1) + w_2 f(x_2) + w_3 f(x_3) \tag{6.8}$$

to be exact for $f(x)$ of degree 0, 1, and 2. This gives three equations in the unknown weighting coefficients w_1, w_2 and w_3.

The algebra is simplified quite considerably at this stage if we transform the axes of the problem so that the new y-axis coincides with the midpoint of the range.

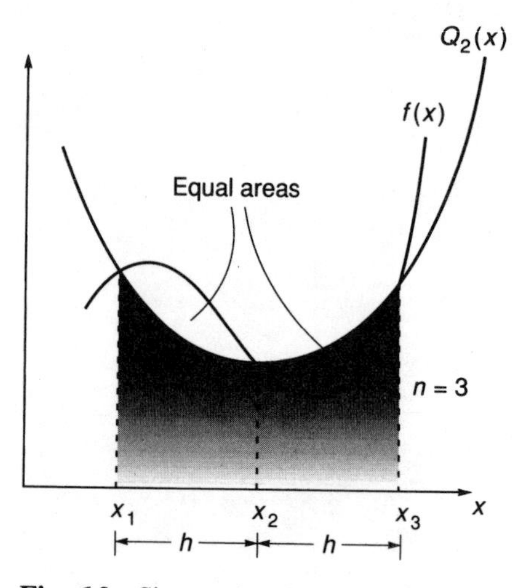

Fig. 6.3 Simpson's rule

Thus

$$\int_{x_1}^{x_3} f(x)\,\mathrm{d}x \equiv \int_{-h}^{h} F(x)\,\mathrm{d}x \tag{6.9}$$

where $\quad F(x) \equiv f(x + x_1 + h)$ (6.10)

and $\quad 2h = x_3 - x_1$ (6.11)

Hence, let $F(x) = 1$

then $$\int_{-h}^{h} \mathrm{d}x = 2h = w_1 + w_2 + w_3 \tag{6.12}$$

let $\quad F(x) = x$

$$\int_{-h}^{h} x\,\mathrm{d}x = 0 = w_1(-h) + w_2(0) + w_3(h) \tag{6.13}$$

let $\quad F(x) = x^2$

then $$\int_{-h}^{h} x^2\,\mathrm{d}x = \tfrac{2}{3}h^3 = w_1(-h)^2 + w_2(0)^2 + w_3(h)^2 \tag{6.14}$$

Solution of eqs 6.12, 6.13 and 6.14 leads to

$$w_1 = w_3 = \tfrac{1}{3}h$$
$$w_2 = \tfrac{4}{3}h \tag{6.15}$$

hence

$$\int_{-h}^{h} F(x)\,\mathrm{d}x \simeq \tfrac{1}{3}hF(-h) + \tfrac{4}{3}hF(0) + \tfrac{1}{3}hF(h) \tag{6.16}$$

or, after returning to the original function and limits we get

$$\int_{x_1}^{x_3} f(x)\,dx \simeq \tfrac{1}{3}hf(x_1)+\tfrac{4}{3}hf(x_2)+\tfrac{1}{3}hf(x_3) \tag{6.17}$$

where x_2 is midway between x_1 and x_3.

Example 6.2

Estimate the value of

$$I=\int_{\pi/4}^{\pi/2} \sin x\,dx$$

using Simpson's rule.

Solution 6.2

$$h=\frac{\pi}{8}$$

$$\therefore \quad I\simeq\frac{1}{3}\frac{\pi}{8}\left[\sin\left(\frac{\pi}{4}\right)+4\sin\left(\frac{3\pi}{8}\right)+\sin\left(\frac{\pi}{2}\right)\right]$$

$$=0.7072 \text{ (cf. exact solution 0.7071)}$$

It is instructive to re-derive the one and two-point rules given by eqs 6.6 and 6.7 to gain confidence in the generality of polynomial substitution.

It turns out that Simpson's rule given by eq. 6.17 is exact for $f(x)$ up to degree 3, although we did not need to assume this in the derivation. This anomaly, and other issues of accuracy will be discussed in a later section.

6.2.5 Higher order Newton–Cotes rules ($n>3$)

There is no limit to the number of sampling points that could be incorporated in a Newton–Cotes rule. For example, following the pattern already established in the previous section, a four-point rule fits a cubic to $f(x)$ and is exact for $f(x)$ of degree 3 or less. A five-point rule fits a quartic to $f(x)$ and is exact for $f(x)$ of degree 5 or less.

These high order methods are rarely used in practice as repeated lower order methods (see Section 6.2.8) are often preferred.

6.2.6 Accuracy of Newton–Cotes rules

Rules with an even number of sampling points (i.e. $n=2,4,\ldots$) will exactly integrate polynomials of degree up to one less than the number of sampling points (i.e. $n-1$). Rules with an odd number of sampling points (i.e. $n=3$, 5, etc.) are more efficient, in that they integrate exactly polynomials of degree up to the number of sampling points (i.e. n). The reason for this difference is explained by Fig. 6.3 for Simpson's rule, where a cubic function $f(x)$ will be integrated exactly, because the errors introduced above and below the approximating polynomial $Q_2(x)$ cancel out.

In order to assess the level of approximation in any Newton–Cotes formula, we need to consider the equivalent Taylor series expansion of the function to be integrated. The largest term ignored in the truncated series is known as the 'dominant error term'.

6.2.6.1 *Error in rectangle rule*

Consider the Taylor series expansion of $f(x)$ about the lower limit of integration x_1, i.e.

$$f(x)=f(x_1)+(x-x_1)f'(x_1)+\frac{(x-x_1)^2}{2}f''(x_1)+\cdots \tag{6.18}$$

Integration of this equation gives

$$\int_{x_1}^{x_2} f(x)\,dx=\left[xf(x_1)+\tfrac{1}{2}(x-x_1)^2 f'(x_1)+\tfrac{1}{6}(x-x_1)^3 f''(x_1)+\cdots\right]_{x_1}^{x_2} \tag{6.19}$$

leading to

$$\int_{x_1}^{x_2} f(x)\,dx=hf(x_1)\;\Big|\;\overbrace{+\tfrac{1}{2}h^2 f'(x_1)+\tfrac{1}{6}h^3 f''(x_1)+\cdots}^{\text{Truncated by rectangle rule}} \tag{6.20}$$

Hence the rectangle rule has a dominant error term of the form $\tfrac{1}{2}h^2 f'(x_1)$.

6.2.6.2 *Error in trapezium rule*

Consider the Taylor series expansion about x_1 to obtain $f(x_2)$, i.e.

$$f(x_2)=f(x_1)+hf'(x_1)+\tfrac{1}{2}h^2 f''(x_1)+\cdots \tag{6.21}$$

Multiplication by $\tfrac{1}{2}h$ and rearrangement gives

$$\tfrac{1}{2}h^2 f'(x_1)=\tfrac{1}{2}hf(x_2)-\tfrac{1}{2}hf(x_1)-\tfrac{1}{4}h^3 f''(x_1)-\cdots \tag{6.22}$$

and after substitution into eq. 6.20 we get

$$\int_{x_1}^{x_2} f(x)\,dx=\tfrac{1}{2}hf(x_1)+\tfrac{1}{2}hf(x_2)\;\Big|\;\overbrace{-\tfrac{1}{12}h^3 f''(x_1)-\cdots}^{\text{Truncated by trapezium rule}} \tag{6.23}$$

hence the trapezium rule has a dominant error term of the form

$-\tfrac{1}{12}h^3 f''(x_1)$.

6.2.6.3 *Error in Simpson's rule*

An alternative approach to finding the dominant error term can be used, if the highest order of polynomial for which a certain rule is exact is known in advance.

For example, Simpson's rule exactly integrates cubics, but will only approximately integrate quartics, hence we can deduce that the dominant error term will contain a fourth derivative as all lower derivatives must disappear, thus

$$\int_{-h}^{h} f(x)\,dx = \tfrac{1}{3}hf(-h) + \tfrac{4}{3}hf(0) + \tfrac{1}{3}hf(h) + \alpha f^{(iv)}(x) \tag{6.24}$$

where α is a constant coefficient.

Letting $f(x) = x^4$, we get

$$\int_{-h}^{h} x^4\,dx = \tfrac{2}{5}h^5 = \tfrac{1}{3}h^5 + 0 + \tfrac{1}{3}h^5 + 24\alpha \tag{6.25}$$

$$\text{therefore } \alpha = \frac{h^5}{24}\left(\frac{2}{5} - \frac{2}{3}\right) = -\frac{1}{90}h^5 \tag{6.26}$$

Hence Simpson's rule has a dominant error term of the form $-\frac{1}{90}h^5 f^{(iv)}(x)$.

6.2.7 Summary of Newton–Cotes rules

The weighting coefficients often have common factors, so the general form of Newton–Cotes rules, together with the dominant error term, can be written as follows:

$$\int_{a}^{b} f(x)\,dx \simeq C_0 h \sum_{i=1}^{n} W_i f(x_i) + C_1 h^{k+1} f^{(k)}(\xi) \tag{6.27}$$

where ξ represents a value of x in the range of integration, and h represents the distance between adjacent sampling points. A summary of the values of C_0, C_1, the W_i's and k for the first five formulae in the Newton–Cotes series is given in Table 6.1.

Table 6.1. Summary of Newton–Cotes rules

n	C_0	W_1	W_2	W_3	W_4	W_5	C_1	k	Name
1	1	1					$\frac{1}{2}$	1	Rectangle
2	$\frac{1}{2}$	1	1				$-\frac{1}{12}$	2	Trapezium
3	$\frac{1}{3}$	1	4	1			$-\frac{1}{90}$	4	Simpson
4	$\frac{3}{8}$	1	3	3	1		$-\frac{3}{80}$	4	4-point
5	$\frac{2}{45}$	7	32	12	32	7	$-\frac{8}{945}$	6	5-point

It should be noted that the weighting coefficients are always symmetrical about the midpoint of the range of integration and their sum must always equal the range of integration.

Although the dominant error term is helpful in assessing the accuracy of these methods, the reality is often more complicated. Clearly as h gets smaller, h^{k+1} gets even smaller, and errors should diminish. However, the error term also depends on $f^{(k)}(\xi)$ and care must be taken with certain functions to ensure that the higher derivatives are not becoming excessively large.

6.2.8 Repeated rules

If the range to be integrated is large, rather than trying to fit a high order polynomial over the full range, it is often better to use a lower order polynomial repeated several times. This strategy is widely used in more advanced engineering numerical methods such as the 'finite element' method.

6.2.8.1 Repeated trapezium rule

As shown in Fig. 6.4, the range of integration $[A, B]$ is split into k strips, each of width h. It is not essential to have equally wide strips, but it is algebraically simpler to do so. The area of each small trapezium is found, and the total sum calculated. It can be seen from Fig. 6.4 that the method amounts to replacing a smooth continuous function $f(x)$ by a series of short straight lines. The more strips taken, the more closely the actual shape of the function is reproduced.

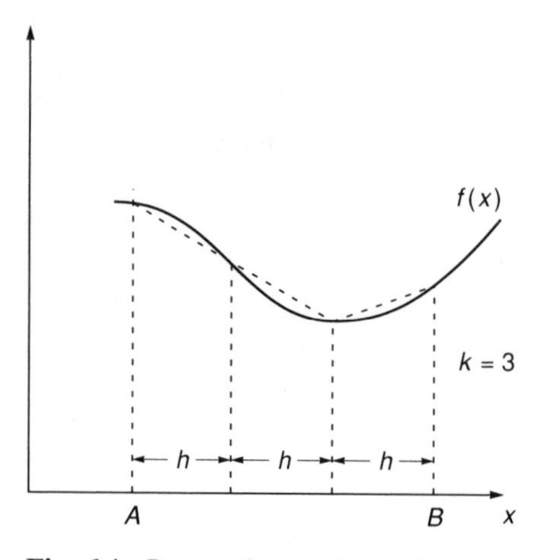

Fig. 6.4 Repeated trapezium rule

If there are k strips, then

$$h=\frac{B-A}{k} \tag{6.28}$$

and

$$\int_A^B f(x)\,dx \simeq \tfrac{1}{2}h[f(A)+2f(A+h)+2f(A+2h)+\cdots+2f(B-h)+f(B)] \tag{6.29}$$

or alternatively

$$\int_A^B f(x)\,dx \simeq \tfrac{1}{2}h[f(A)+f(B)]+h\sum_{i=1}^{k-1} f(A+ih) \tag{6.30}$$

Example 6.3

Estimate the value of

$$I=\int_{\pi/4}^{\pi/2} \sin x \, dx$$

using the repeated trapezium rule with $k=3$.

Solution 6.3

$$h=\frac{\pi}{12}$$

$$I \simeq \frac{1}{2}\frac{\pi}{12}\left[\sin\left(\frac{\pi}{4}\right)+\sin\left(\frac{\pi}{2}\right)\right]+\frac{\pi}{12}\left[\sin\left(\frac{\pi}{3}\right)+\sin\left(\frac{5\pi}{12}\right)\right]$$

$$=0.7031 \text{ (cf. exact solution 0.7071)}$$

It may be remembered that a single application of the trapezium rule gave 0.6704.

6.2.8.2 Repeated Simpson's rule

A single application of Simpson's rule requires three sampling points, so the repeated rule must have an even number of strips as shown in Fig. 6.5. Each pair of strips must have the same width, but widths may differ from pair to pair. The repeated Simpson's rule fits a parabola over each set of three sampling points, leading to the following expression assuming k (even) strips of equal width h:

$$\int_A^B f(x)\,dx \simeq \tfrac{1}{3}h[f(A)+4f(A+h)+2f(A+2h)+4f(A+3h)+\cdots+2f(B-2h)+ \\ +4f(B-h)+f(B)] \tag{6.31}$$

Fig. 6.5 Repeated Simpson's rule

or alternatively

$$\int_A^B f(x)\,dx \simeq \tfrac{1}{3}h[f(A)+f(B)] + \tfrac{4}{3}h \sum_{i=1,3,5\ldots}^{k-1} f(A+ih) + \tfrac{2}{3}h \sum_{i=2,4,6\ldots}^{k-2} f(A+ih) \tag{6.32}$$

Example 6.4

Estimate the value of

$$I = \int_{\pi/4}^{\pi/2} \sin x \, dx$$

using the repeated Simpson's rule with $k=4$.

Solution 6.4

$$h = \frac{\pi}{16}$$

$$I \simeq \frac{1}{3}\frac{\pi}{16}\left[\sin\left(\frac{\pi}{4}\right)+\sin\left(\frac{\pi}{2}\right)\right] + \frac{4}{3}\frac{\pi}{16}\left[\sin\left(\frac{5\pi}{16}\right)+\sin\left(\frac{7\pi}{16}\right)\right]$$

$$+\frac{2}{3}\frac{\pi}{16}\left[\sin\left(\frac{3\pi}{8}\right)\right] = 0.7071 \text{ (cf. exact solution 0.7071)}$$

Program 6.1. Repeated Newton–Cotes rules
The first program in this chapter estimates the value of the integral

$$I = \int_A^B f(x)\,dx$$

using a repeated Newton–Cotes rule of the user's choice. Input data consist of the limits of integration A and B, the number of repetitions (NR) of the rule across the range, and the number of sampling points (NNCP) in each application of the rule. In the method presented here, the distance between adjacent sampling points (H) is assumed constant.

```
      PROGRAM P61
C
C      PROGRAM 6.1   REPEATED NEWTON-COTES RULE
C
      REAL WEIGHT(5)
C
      READ (5,*) A,B,NNCP,NR
      CALL NEWCOT(WEIGHT,NNCP)
      WR = (B-A)/NR
      H = WR
      IF (NNCP.GE.2) H = WR/ (NNCP-1)
      AREA = 0.
      DO 10 I = 1,NR
          BR = A + (I-1)*WR
```

```
          DO 10 J = 1,NNCP
               XS = BR + (J-1)*H
               W = WEIGHT(J)
   10 AREA = AREA + W*H*F(XS)
      WRITE (6,*) ('****** REPEATED NEWTON-COTES RULE *****')
      WRITE (6,100) A,B
      WRITE (6,101) NNCP
      WRITE (6,102) NR
      WRITE (6,103) AREA
  100 FORMAT (/,'  LIMITS OF INTEGRATION                        ',2F12.4)
  101 FORMAT ('  NUMBER OF SAMPLING POINTS PER STRIP        ',I7)
  102 FORMAT ('  NUMBER OF REPETITIONS                      ',I7)
  103 FORMAT ('  COMPUTED RESULT                            ',1F12.4)
      STOP
      END
C
      FUNCTION F(X)
C
C      THIS FUNCTION PROVIDES THE VALUE OF F(X)
C      AND WILL VARY FROM ONE PROBLEM TO THE NEXT
C
      F = SIN(X)*SIN(X) + X
      RETURN
      END
```

The function $f(x)$ to be integrated is evaluated by function F which will be changed by the user from one problem to the next.

The weighting coefficients for the chosen rule are provided by library subroutine NEWCOT in the form of column array WEIGHT of length NNCP (maximum 5). All subroutines used in this book are described and listed in Appendices 1 and 2.

With the exception of simple integer counters, the variables can be summarised as follows:

A, B	Limits of integration
NNCP	Number of sampling points in rule
NR	Number of rule repetitions
WR	Range of each rule application (assumed constant)
H	Distance between sampling points
AREA	Running total of integral
BR	Beginning of each range
XS	Sampling point
W	Weighting coefficient
WEIGHT	Column array holding weighting coefficients

To illustrate use of the program, the following problem is to be solved using the trapezium rule, repeated five times.

$$I = \int_{0.25}^{0.75} (\sin^2 x + x)\, dx$$

The input data and output from Program 6.1 are given in Figs 6.6(a) and (b) respectively. The function $\sin^2 x + x$ has been written into the user-supplied function F.

As shown in Fig. 6.6(b) with NR = 5, the method gives a value for the above integral of 0.3709 which compares quite well with the analytical solution of 0.3705. In problems of this type, especially when no analytical solution is available, it is good practice to repeat the calculation with more strips to ensure that the solution is converging with sufficient accuracy. For example, if the above problem is recalculated with NR = 10, the computed result improves to 0.3706.

	A	B
Limits of integration	0.25	0.75
	NNCP	
Number of sampling points	2	
	NR	
Number of repetitions	5	

Fig. 6.6(a) Data for Program 6.1

```
****** REPEATED NEWTON-COTES RULE *****

 LIMITS OF INTEGRATION                       .2500        .7500
 NUMBER OF SAMPLING POINTS PER STRIP        2
 NUMBER OF REPETITIONS                      5
 COMPUTED RESULT                             .3709
```

Fig. 6.6(b) Results from Program 6.1

6.2.9 Remarks on Newton–Cotes rules

The choice of a suitable method for numerical integration is never completely straightforward. If the analyst is free to choose the position of the sampling points, then Newton–Cotes methods should probably not be used at all, as the Gaussian rules covered in the next section are more efficient, especially for computer applications. The Newton–Cotes rules do have the great advantage of simplicity however, especially the lower order members of the family such as the trapezium and Simpson's rule.

Generally speaking, frequently repeated low order methods are preferred to high order methods, and the repeated trapezium rule presented in Program 6.1 will give acceptable solutions to many problems.

The choice of the interval h in a repeated rule presents a further difficulty. Ideally, h should be 'small' but not so small that excessive computer time is required, or that the accuracy is affected by computer word-length limitations. When using Program 6.1 to integrate a function, it is recommended that two or three solutions are obtained using an increasing number of strips. This will indicate whether the result is converging. Lack of convergence suggests a 'badly behaved' function which could contain a singularity requiring special treatment.

The repeated trapezium rule is also suitable for integrating tabulated data from an experiment where, for example, the interval at which measurements are made is not constant. In such cases, a formula similar to eq. 6.29 would be used, but with h allowed to vary from strip to strip.

6.3 Gauss–Legendre rules

6.3.1 Introduction

Newton–Cotes formulae were easy to use, because their sampling points were evenly spaced within the range of integration, and the weighting coefficients easy to remember (at least up to Simpson's rule!).

Gaussian rules recognise that the sampling points should be optimally spaced within the range of integration to achieve maximum accuracy. These locations are still symmetrical about the middle of the range of integration, but the positions and weights are not so easy to remember. This latter point may be a slight hindrance to 'hand' calculation using Gaussian methods, but is no problem at all for computer programs where information can be stored in a subroutine library.

In Gaussian rules the summation notation adopted previously is still applicable, i.e.

$$\int_a^b f(x)\,\mathrm{d}x \simeq \sum_{i=1}^{n} w_i f(x_i) \tag{6.33}$$

When using polynomial substitution however, both the w_i's and the x_i's are initially treated as unknowns. The resulting $2n$ equations will result in rules that are exact for $f(x)$ up to degree $2n-1$. This represents a considerable improvement over the equivalent Newton–Cotes rule, which, for the same number of sampling points, will only be exact for $f(x)$ up to degree $n-1$ or n (if n is odd).

There are various types of Gaussian quadrature, but the most important are called Gauss–Legendre rules, and it is these which will be considered first.

Rather than limits of a and b as indicated in eq. 6.33 the development of these rules is greatly simplified if limits of ± 1 are used. This in no way limits the generality of the method because the actual limits can be easily retrieved by a simple transformation.

6.3.2 One-point rule (midpoint rule, $n=1$)

This familiar method samples the function at the midpoint of the range as shown in Fig. 6.7. The function $f(x)$ is approximated by $Q_0(x)$ which runs parallel to the x-axis. The formula given by eq. 6.34 will be exact for $f(x)$ up to degree 1.

$$\int_{-1}^{1} f(x)\,\mathrm{d}x \simeq 2f(0) \tag{6.34}$$

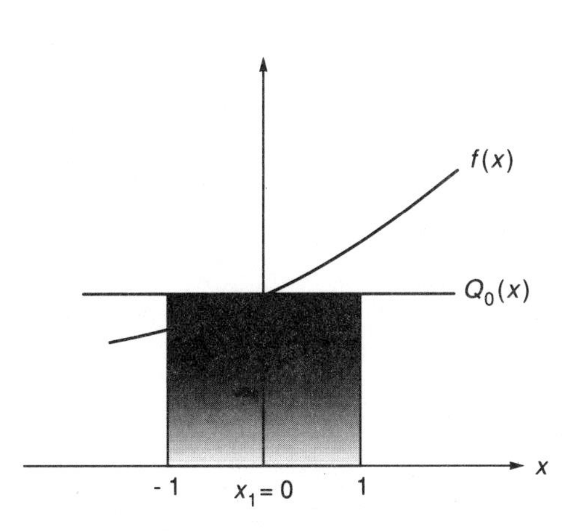

6.3.3 Two-point rule ($n=2$)

The formula will be of the form

$$\int_{-1}^{1} f(x)\,dx \simeq w_1 f(x_1) + w_2 f(x_2) \tag{6.35}$$

and the four unknowns w_1, w_2, x_1 and x_2 can be found by polynomial substitution, thus let $f(x)=1$

$$\int_{-1}^{1} dx = 2 = w_1 + w_2 \tag{6.36}$$

let $f(x)=x$

$$\int_{-1}^{1} x\,dx = 0 = w_1 x_1 + w_2 x_2 \tag{6.37}$$

let $f(x)=x^2$

$$\int_{-1}^{1} x^2\,dx = \tfrac{2}{3} = w_1 x_1^2 + w_2 x_2^2 \tag{6.38}$$

let $f(x)=x^3$

$$\int_{-1}^{1} x^3\,dx = 0 = w_1 x_1^3 + w_2 x_2^3 \tag{6.39}$$

From eqs 6.37 and 6.39 we can write

$$\frac{w_1}{w_2} = \frac{-x_2}{x_1} = \frac{-x_2^3}{x_1^3} \tag{6.40}$$

hence $\quad x_2 = \pm x_1 \qquad (6.41)$

Assuming the sampling points do not coincide, we must have $x_2 = -x_1$, and from eq. 6.37

$$w_1 = w_2 = 1 \tag{6.42}$$

hence from eq. 6.38

$$x_1^2 = x_2^2 = \tfrac{1}{3} \tag{6.43}$$

and $\quad x_1 = -\dfrac{1}{\sqrt{3}}, \quad x_2 = \dfrac{1}{\sqrt{3}} \qquad (6.44)$

Substitution of these values into eq. 6.35 gives the final form of the two-point rule, thus

$$\int_{-1}^{1} f(x)\,dx \simeq f\left(-\frac{1}{\sqrt{3}}\right) + f\left(\frac{1}{\sqrt{3}}\right) \tag{6.45}$$

which will be exact for $f(x)$ up to degree 3. A typical function and the location of

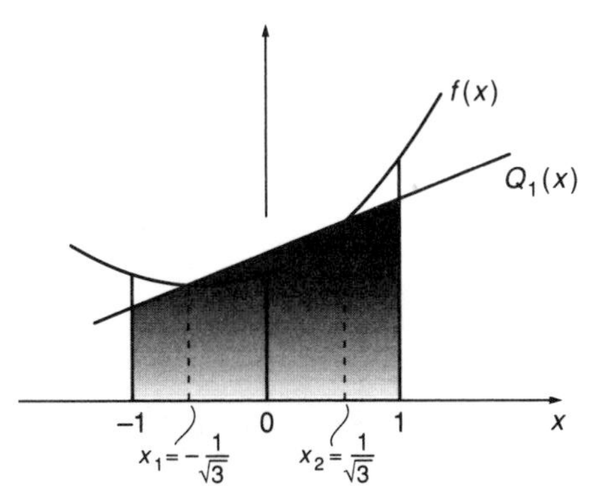

Fig. 6.8 Two-point Gauss–Legendre rule

sampling points is shown in Fig. 6.8. Although the approximating polynomial $Q_1(x)$ is never computed explicitly, it is interesting to note that the area beneath a cubic function is exactly the same as the area beneath a straight line coinciding at the two sampling points derived above.

Example 6.5

Estimate the value of

$$I=\int_{-1}^{1}(x^3+2x^2+1)\,\mathrm{d}x$$

using the two-point Gauss–Legendre rule.

Solution 6.5

$$I=\left(-\frac{1}{3\sqrt{3}}+\frac{2}{3}+1\right)+\left(\frac{1}{3\sqrt{3}}+\frac{2}{3}+1\right)$$

$$=\frac{10}{3}\left(\text{cf. exact solution }\frac{10}{3}\right)$$

The two-point Gauss–Legendre rule integrates cubics exactly.

6.3.4 Three-point rule ($n=3$)

This formula will be of the form

$$\int_{-1}^{1}f(x)\,\mathrm{d}x\simeq w_1f(x_1)+w_2f(x_2)+w_3f(x_3) \tag{6.46}$$

and will be exact for $f(x)$ up to degree 5 $(2n-1)$. Although there are six 'unknowns' in eq. 6.46 which would require polynomial substitution of $f(x)$ up to degree 5, advantage can be taken of symmetry to reduce the number of equations. Symmetries are always apparent in Gauss–Legendre rules about the midpoint of the range of integration, hence it can be stated from inspection that

$$\begin{aligned} w_1 &= w_3 \\ x_1 &= -x_3 \\ x_2 &= 0 \end{aligned} \tag{6.47}$$

Thus we are faced with only three unknowns, namely w_1, w_2 and x_1. It is left as an exercise for the reader to show that the final form of the three-point rule is as follows

$$\int_{-1}^{1} f(x)\,dx \simeq \tfrac{5}{9} f(-\sqrt{\tfrac{3}{5}}) + \tfrac{8}{9} f(0) + \tfrac{5}{9} f(\sqrt{\tfrac{3}{5}}) \tag{6.48}$$

The locations of the sampling points for this integration rule are indicated in Fig. 6.9.

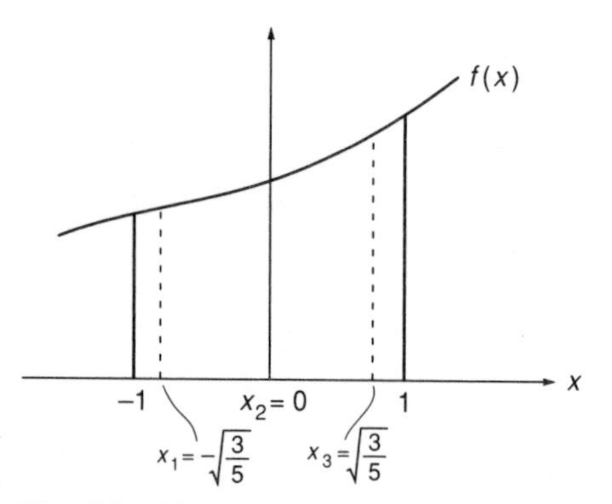

Fig. 6.9 Three-point Gauss–Legendre rule

6.3.5 Changing the limits of integration

All the Gauss–Legendre rules described in this section used limits of integration ± 1. Naturally, we will frequently wish to integrate functions between other limits. One method of dealing with this is to transform the actual problem to one which does have limits of ± 1.

For example

$$\int_{a}^{b} f(x)\,dx \equiv \int_{-1}^{1} F(t)\,dt \tag{6.49}$$

where

$$x = \frac{(b-a)t + (b+a)}{2} \tag{6.50}$$

$$dx = \frac{(b-a)}{2} dt \tag{6.51}$$

Once the problem has been transformed in this way, the familiar Gauss–Legendre weights and sampling points can be applied.

Example 6.6

Estimate the value of

$$I = \int_1^4 x \cos x \, dx$$

using the Gauss–Legendre three-point rule.

Solution 6.6

From eqs 6.50 and 6.51

$$x = \frac{3t+5}{2}$$

$$dx = \tfrac{3}{2} dt$$

hence the transformed problem becomes

$$I = \frac{3}{4} \int_{-1}^{1} (3t+5) \cos[\tfrac{1}{2}(3t+5)] \, dt$$

and substitution of weights and sampling points gives

$$I \simeq \tfrac{3}{4} [\![\tfrac{5}{9}\{(-3\sqrt{\tfrac{3}{5}}+5) \cos[\tfrac{1}{2}(-3\sqrt{\tfrac{3}{5}}+5)] + (3\sqrt{\tfrac{3}{5}}+5) \cos[\tfrac{1}{2}(3\sqrt{\tfrac{3}{5}}+5)]\} + \tfrac{8}{9}(5 \cos\tfrac{5}{2})]\!]$$

$$= -5.0611 \text{ (cf. exact solution } -5.0626)$$

An alternative approach and the one used by Program 6.2, for dealing with limits of integration different from ± 1, is to proportion the sampling points symmetrically about the midpoint of the range. This method involves no change of variable and the original limits of integration are maintained. Care must be taken over the positioning of the sampling points however, and the weights from the standard formulae must all be multiplied by half the range of integration.

Example 6.7

Estimate the value of

$$I = \int_1^4 x \cos x \, dx$$

using the Gauss–Legendre three-point rule. Proportion the sampling points without changing the variable.

Solution 6.7

As shown in Fig. 6.10, the standard sampling points $-\sqrt{\tfrac{3}{5}}, 0, \sqrt{\tfrac{3}{5}}$ are multiplied by half the range of integration, i.e. 1.5, and positioned relative to the midpoint of the range i.e. 2.5. The standard weights are also augmented by a factor of 1.5, thus

$$I \simeq \tfrac{3}{2}\{\tfrac{5}{9}[(\tfrac{5}{2}-\tfrac{3}{2}\sqrt{\tfrac{3}{5}})\cos(\tfrac{5}{2}-\tfrac{3}{2}\sqrt{\tfrac{3}{5}})+(\tfrac{5}{2}+\tfrac{3}{2}\sqrt{\tfrac{3}{5}})\cos(\tfrac{5}{2}+\tfrac{3}{2}\sqrt{\tfrac{3}{5}})]+\tfrac{8}{9}(\tfrac{5}{2}\cos\tfrac{5}{2})\}$$
$$= -5.0611$$

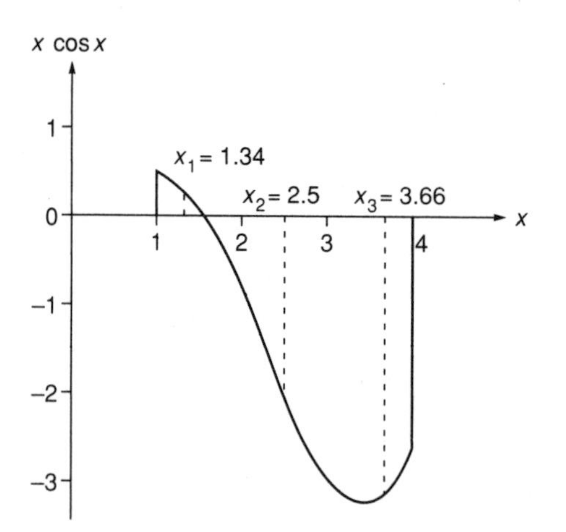

Fig. 6.10 Sampling points from Example 6.7

Program 6.2. Repeated Gauss–Legendre rules
The second program in this chapter estimates the value of the integral

$$I = \int_A^B f(x)\,dx$$

using a repeated Gauss–Legendre rule of the user's choice. Input data consist of the limits of integration A and B, the number of repetitions (NR) of the rule across the range, and the number of sampling points in each application of the rule (NGP).

The function $f(x)$ to be integrated, is evaluated by user-supplied function F which will be changed from one problem to the next. The weighting coefficients and sampling points are provided by library subroutine GAULEG in the form of array SAMP. The first column of SAMP holds sampling points, and the second column holds weighting coefficients, in each case of length NGP (maximum 7). Many of the variables have the same meaning as in Program 6.1, but new ones are defined below:

NGP	Number of sampling points in rule
HR	Half the range of each rule application
CR	Centre of each range
X	Sampling point for limits of ± 1
SAMP	2-d array holding sampling points and weighting coefficients.

The PARAMETER statement defines ISAMP $\geq$ NGP.

```
      PROGRAM P62
C
C      PROGRAM 6.2  REPEATED GAUSS-LEGENDRE RULE
C
C      ALTER NEXT LINE TO CHANGE PROBLEM SIZE
C
      PARAMETER (ISAMP=7)
C
      REAL SAMP(ISAMP,2)
C
      READ (5,*) A,B,NGP,NR
      CALL GAULEG(SAMP,ISAMP,NGP)
      WR = (B-A)/NR
      HR = 0.5*WR
      AREA = 0.
      DO 10 I = 1,NR
          CR = A + (I-1)*WR + HR
          DO 10 J = 1,NGP
              X = SAMP(J,1)
              W = SAMP(J,2)
              XS = CR + X*HR
   10 AREA = AREA + W*HR*F(XS)
      WRITE (6,*) ('****** REPEATED GAUSS-LEGENDRE RULE *****')
      WRITE (6,100) A,B
      WRITE (6,101) NGP
      WRITE (6,102) NR
      WRITE (6,103) AREA
  100 FORMAT (/,'  LIMITS OF INTEGRATION                   ',2F12.4)
  101 FORMAT ('  NUMBER OF GAUSS POINTS PER STRIP       ',I7)
  102 FORMAT ('  NUMBER OF REPETITIONS                  ',I7)
  103 FORMAT ('  COMPUTED RESULT                        ',F12.4)
      STOP
      END
C
      FUNCTION F(X)
C
C      THIS FUNCTION PROVIDES THE VALUE OF F(X)
C      AND WILL VARY FROM ONE PROBLEM TO THE NEXT
C
      F = COS(X)*X
      RETURN
      END
```

The problem of Example 6.7 is considered once more, to demonstrate use of the program. This time however, the three-point Gauss–Legendre rule is repeated twice over the range 1 to 4.

The input data and output from Program 6.2 are given in Figs 6.11(a) and (b). It is clear that application of the three-point rule twice over the range has improved the result from -5.0611 for a single application, to -5.0626 which agrees with the exact solution to 4DP.

	A	B
Limits of integration	1.0	4.0
	NGP	
Number of sampling points	3	
	NR	
Number of repetitions	2	

Fig. 6.11(a) Data for Program 6.2

```
****** REPEATED GAUSS-LEGENDRE RULE *****

 LIMITS OF INTEGRATION                        1.0000        4.0000
 NUMBER OF GAUSS POINTS PER STRIP             3
 NUMBER OF REPETITIONS                        2
 COMPUTED RESULT                             -5.0626
```

Fig. 6.11(b) Results from Program 6.2

6.3.6 Accuracy of Gauss–Legendre rules

Gauss–Legendre rules with n sampling points will exactly integrate polynomials of degree up to $2n-1$. In order to find the dominant error term in any of the methods, the Taylor series expansion must be considered in relation to the formula.

We know for example that the two-point Gauss-Legendre rule will integrate cubics exactly, but not quartics, hence the dominant error term must include a fourth derivative of the function.

Thus we can write

$$\int_{-1}^{1} f(x)\,\mathrm{d}x = f\left(-\frac{1}{\sqrt{3}}\right) + f\left(\frac{1}{\sqrt{3}}\right) + \alpha f^{(\mathrm{iv})}(x) \tag{6.52}$$

Substitution of $f(x)=x^4$ gives

$$\int_{-1}^{1} x^4\,\mathrm{d}x = \tfrac{2}{5} = \tfrac{1}{9} + \tfrac{1}{9} + 24\alpha \tag{6.53}$$

therefore $\alpha = \frac{1}{135}$. (6.54)

In the case of Gauss–Legendre rules, if the general expression including the dominant error term is written as follows:

$$\int_{-1}^{1} f(x)\,\mathrm{d}x = \sum_{i=1}^{n} w_i f(x_i) + \alpha f^{(2n)}(\xi) \tag{6.55}$$

where ξ is a value of x within the range of integration, it can be shown that

$$\alpha = \frac{2^{2n+1}(n!)^4}{(2n+1)[(2n)!]^3} \tag{6.56}$$

6.3.7 Summary of Gauss–Legendre rules

With reference to eq. 6.55, Table 6.2 presents a summary of the weights and sampling points for the first four formulae in the Gauss–Legendre series. All the coefficients up to $n=7$ are stored by library subroutine GAULEG as described in Program 6.2.

It should be remembered that these weights and sampling points are based on limits of integration of ± 1.

The Gauss–Legendre rules are so called, because the sampling points in the range -1 to $+1$ are the roots of the Legendre family of polynomials given by

$$P_n(x) = \frac{1}{2^n n!}\frac{\mathrm{d}^n}{\mathrm{d}x^n}(x^2-1)^n = 0 \tag{6.57}$$

Table 6.2. Sampling points and weights for Gauss–Legendre integration

n	x_i	w_i
1	0.0	2.0
2	−0.57735027	1.0
	+0.57735027	1.0
3	−0.77459667	0.55555556
	0.0	0.88888889
	+0.77459667	0.55555556
4	−0.86113631	0.34785485
	−0.33998104	0.65214515
	0.33998104	0.65214515
	0.86113631	0.34785485

6.4 Special integration rules

6.4.1 Introduction

The methods of numerical integration discussed so far have usually involved rather straightforward functions with finite limits of integration. In this section, some more specialised integration rules are described.

6.4.2 Gauss–Laguerre rules

This method is specially designed for numerical integration of functions of the form

$$\int_0^\infty e^{-x} f(x)\,dx \simeq \sum_{i=1}^{n} w_i f(x_i) \tag{6.58}$$

The term e^{-x} is sometimes referred to as the 'weighting function' on $f(x)$. It should be noted that the sampling points and weights for this method take account of the exponential term and the infinite range of integration, hence only the function $f(x)$ needs to be evaluated at each sampling point in the summation terms on the right-hand side of eq. 6.58.

It can be shown that sampling points for this method are the roots of the Laguerre family of polynomials given by

$$L_n(x) = e^x \frac{d^n}{dx^n}(e^{-x}x^n) = 0 \tag{6.59}$$

A summary of some weights and sampling points for this method is given in Table 6.3.

It may be noted that polynomial substitution of $f(x) = 1$ into eq. 6.58 yields the

Table 6.3. Sampling points and weights for Gauss–Laguerre integration

n	x_i	w_i
1	1.0	1.0
2	0.58578644	0.85355339
	3.41421356	0.14644661
3	0.41577456	0.71109301
	2.29428036	0.27851773
	6.28994508	0.01038926
4	0.32254769	0.60315410
	1.74576110	0.35741869
	4.53662030	0.03888791
	9.39507091	0.00053929

properties that

$$\sum_{i=1}^{n} w_i = 1$$

$$\sum_{i=1}^{n} w_i x_i = 1 \qquad (6.60)$$

It is interesting to observe how rapidly the weighting coefficients diminish as the order of the method is increased.

Example 6.8

Estimate

$I = \int_0^\infty e^{-x} \sin x \, dx$ using three-point Gauss–Laguerre integration

Solution 6.8

$$I \simeq 0.71109 \sin(0.41577) + 0.27852 \sin(2.29428) + 0.01038 \sin(6.28995)$$
$$= 0.496 \text{ (cf. exact solution 0.5).}$$

Program 6.3. Gauss–Laguerre integration
This program evaluates integrals of the form

$$I = \int_0^\infty e^{-x} f(x) \, dx$$

using Gauss–Laguerre integration. Input data require only the number of integration points

```
      PROGRAM P63
C
C      PROGRAM 6.3  GAUSS-LAGUERRE RULE
C
C      ALTER NEXT LINE TO CHANGE PROBLEM SIZE
C
      PARAMETER (ISAMP=5)
C
      REAL SAMP(ISAMP,2)
C
      READ (5,*) NGL
      CALL GAULAG(SAMP,ISAMP,NGL)
      AREA = 0.
      DO 10 I = 1,NGL
          X = SAMP(I,1)
          W = SAMP(I,2)
   10 AREA = AREA + W*F(X)
      WRITE (6,*) ('****** GAUSS-LAGUERRE RULE *****')
      WRITE (6,101) NGL
      WRITE (6,103) AREA
  101 FORMAT (/,'  NUMBER OF GAUSS POINTS                   ',I7)
  103 FORMAT ('  COMPUTED RESULT                          ',F12.4)
      STOP
      END
C
      FUNCTION F(X)
C
C      THIS FUNCTION PROVIDES THE VALUE OF F(X)
C      AND WILL VARY FROM ONE PROBLEM TO THE NEXT
C
      F = SIN(X)
      RETURN
      END
```

NGL. The weights and sampling points are provided by library subroutine GAULAG in the form of array SAMP. The function $f(x)$ is evaluated by user-supplied function F in the usual way.

The program is quite trivial, and involves a single loop in which the summation of eq. 6.58 is performed. The PARAMETER restriction is that ISAMP $\geq$ NGL.

The problem from Example 6.8, i.e.

$$I=\int_0^\infty e^{-x}\sin x\,dx$$

has been recalculated using the program with NGP = 5. The data and output are given in Figs 6.12(a) and (b) respectively, and an improved solution of 0.4989 is indicated.

```
Number of sampling points  NGL
                            5
```

Fig. 6.12(a) Data for Program 6.3

```
****** GAUSS-LAGUERRE RULE *****

 NUMBER OF GAUSS POINTS                 5
 COMPUTED RESULT                          .4989
```

Fig. 6.12(b) Results from Program 6.3

6.4.3 Gauss–Chebyshev rules

This method is specially designed for integration of functions of the form,

$$\int_{-1}^{1} \frac{f(x)}{(1-x)^{1/2}}\,dx \simeq \sum_{i=1}^{n} w_i f(x_i) \tag{6.61}$$

In this case, the 'weighting function' is $(1-x)^{-1/2}$, and contains singularities at $x=\pm 1$. As might be expected from the symmetry of the limits of integration in eq. 6.61, the weights and sampling points will also be symmetrical about the middle of the range. It can be shown that the general equation is of the form

$$\int_{-1}^{1} \frac{f(x)}{(1-x)^{1/2}}\,dx \simeq \frac{\pi}{n}\sum_{i=1}^{n} f\left(\cos\frac{2i-1}{2n}\pi\right) + \frac{2\pi}{2^{2n}(2n)!} f^{(2n)}(\xi) \tag{6.62}$$

and it is left to the reader to compute the required sampling points. The weighting coefficients are seen to involve simple multiples of π.

6.4.4 Fixed weighting coefficients

Using the polynomial substitution technique, many different variations on the methods already described are possible. In the Newton–Cotes approaches, the sampling points were prescribed 'a priori' and the weighting coefficient treated as unknowns. In the Gauss–Legendre approaches, both sampling points and weighting coefficients were initially treated as unknowns.

A further variation occurs when the weighting coefficients are all equal, leading to a summation formula of the form

$$\int_{-1}^{1} f(x)\,dx \simeq \frac{2}{n}\sum_{i=1}^{n} f(x_i) \tag{6.63}$$

Clearly when $n=2$, the formula is identical to the corresponding Gauss–Legendre rule with $n=2$. Consider the case when $n=3$, whence

$$\int_{-1}^{1} f(x)\,dx \simeq \tfrac{2}{3}[f(x_1)+f(x_2)+f(x_3)] \tag{6.64}$$

By inspection, $x_1=-x_3$ and $x_2=0$, so only one unknown remains to be found.

Let $f(x)=x^2$, then

$$\int_{-1}^{1} x^2\,dx = \tfrac{2}{3} = \tfrac{2}{3}({x^2}_1+0+{x^2}_1) \tag{6.65}$$

$$\therefore \quad {x^2}_1 = \tfrac{1}{2} \text{ and } x_1 = \pm 0.7071$$

hence (6.64) becomes

$$\int_{-1}^{1} f(x)\,dx \simeq \tfrac{2}{3}[f(-0.7071)+f(0)+f(0.7071)] \tag{6.66}$$

Higher order methods of this type can be found in a similar way.

6.4.5 Hybrid methods

An alternative, hybrid approach is to prescribe some of the sampling points, but leave the rest as unknowns. For example, there may be a good reason to sample at the limits of integration, leading to formulae of the following type, called Lobatto quadrature

$$\int_{-1}^{1} f(x)\,dx \simeq w_1 f(-1) + \sum_{i=2}^{n-1} w_i f(x_i) + w_n f(1) \tag{6.67}$$

From inspection, $w_1 = w_n$ and the remaining weights and sampling points will also be symmetrical about the middle of the range.

It is left to the reader to show that the four-point rule of this type will be of the following form:

$$\int_{-1}^{1} f(x)\,dx \simeq \tfrac{1}{6}f(-1) + \tfrac{5}{6}f(-\sqrt{0.2}) + \tfrac{5}{6}f(\sqrt{0.2}) + \tfrac{1}{6}f(1) \tag{6.68}$$

6.4.6 Sampling points outside the range of integration

All the methods of numerical integration described so far in this chapter involved sampling points within the range of integration. Although the majority of problems will be of this type, there exists an important class of formula in the solution of differential equations called 'predictors' which require sampling points outside the range of integration.

For example, consider an integration formula of the type

$$\int_{x_3}^{x_4} f(x)\,dx \simeq w_1 f(x_1) + w_2 f(x_2) + w_3 f(x_3) \tag{6.69}$$

where the points x_1 to x_4 are equally spaced at h apart as shown in Fig. 6.13. Only the hatched area is required, but the fact that x_1 and x_2 lie outside the range of integration makes no difference to the method of finding the weighting coefficients.

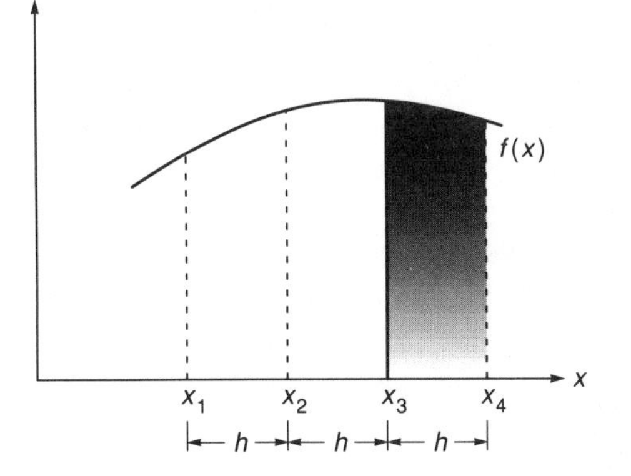

Fig. 6.13 Four-point Adams–Bashforth method

Using polynomial substitution

let $f(x)=1$

$$\int_{x_3}^{x_4} dx = h = w_1 + w_2 + w_3 \tag{6.70}$$

let $f(x)=x$

$$\int_{x_3}^{x_4} x\, dx = \tfrac{1}{2}(x_4^2 - x_3^2) = w_1 x_1 + w_2 x_2 + w_3 x_3 \tag{6.71}$$

let $f(x)=x^2$

$$\int_{x_3}^{x_4} x^2\, dx = \tfrac{1}{3}(x_4^3 - x_3^3) = w_1 x_1^2 + w_2 x_2^2 + w_3 x_3^2 \tag{6.72}$$

Solution of these three equations is greatly simplified if the substitutions $x_1=0$ and $h=1$ are made, leading to the final equation

$$\int_{x_3}^{x_4} f(x)\, dx \simeq \frac{h}{12}[5f(x_1) - 16f(x_2) + 23f(x_3)] \tag{6.73}$$

which will be exact for $f(x)$ up to degree 2.

Higher order 'predictors' of this type can be derived by similar means, for example a four-point formula is given by

$$\int_{x_4}^{x_5} f(x)\, dx \simeq \frac{h}{24}[-9f(x_1) + 37f(x_2) - 59f(x_3) + 55f(x_4)] \tag{6.74}$$

and is exact for $f(x)$ up to degree 3. Formulae such as those given by eqs 6.73 and 6.74 are called Adams–Bashforth predictors, and will be encountered again in Chapter 7 on solution of ordinary differential equations.

6.5 Multiple integrals

6.5.1 Introduction

In engineering analysis we are frequently required to integrate functions of more than one variable over an area or volume. Analytical methods for performing multiple integrals will be possible in a limited number of cases, but in this section we will consider numerical integration techniques.

Consider integration of a function of two variables over the two-dimensional region R as shown in Fig. 6.14. The function $f(x, y)$ could be considered to represent a third dimension coming out of the page at right angles over the region R.

The required integral is

$$I = \int_R \int f(x, y)\, dx\, dy \tag{6.75}$$

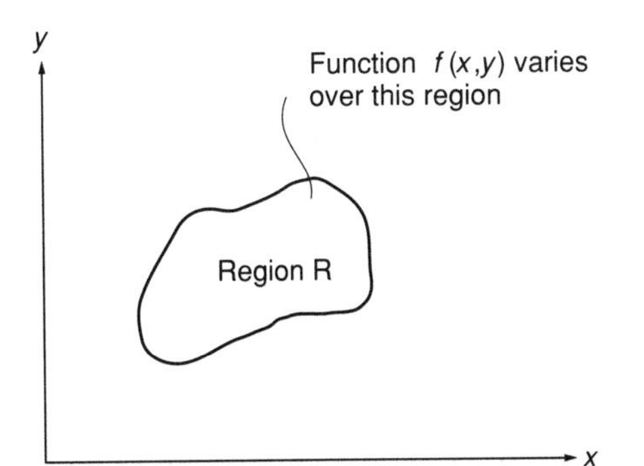

Fig. 6.14 Two-dimensional range of integration

which might be visualised as representing the volume of material bounded by the function $f(x, y)$ and the region R in the x–y plane.

Following our previous techniques, we seek numerical integration formulae involving a summation of products of the function evaluated at certain sampling points, and weighting coefficients.

Hence, the expression given by eq. 6.76

$$\int_R \int f(x, y)\,\mathrm{d}x\,\mathrm{d}y \simeq \sum_{i=1}^{n} \sum_{j=1}^{n} w_i w_j f(x_i y_j) \tag{6.76}$$

forms the basis of our methods for 'double' integrals.

Clearly, a problem arises in defining explicitly the limits of integration for an irregular region such as that shown in Fig. 6.17. In practice, it may be sufficient to subdivide the irregularly shaped region into a number of simpler shapes, over which numerical integration can be easily performed. The final result over the full region R, would then be obtained by adding together the solutions obtained over each subregion. This approach is analogous to the 'repeated' rules covered earlier in the chapter.

Initially, we consider integration over rectangular regions which lie parallel to the Cartesian coordinate directions, as this greatly simplifies the definition of the limits. Later on, the concepts are extended to integration over general quadrilateral regions in the x–y plane. Although the examples given in the remainder of this section are always double integrals, the methods described are readily extrapolated to higher order multiple integrals.

6.5.2 Integration over a rectangular area

Consider, in Fig. 6.15, integration of a function $f(x, y)$ over the rectangular region shown. As the boundaries of the rectangle lie parallel to the Cartesian coordinate directions, the variables can be uncoupled, and any of the methods described previously can be applied directly.

For example, the trapezium rule applied in each direction would lead to four

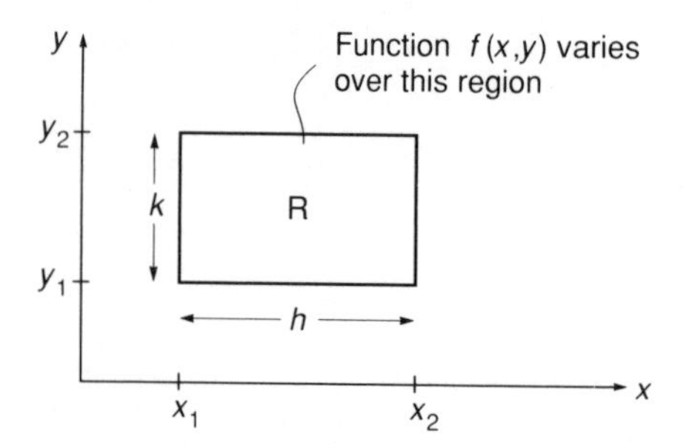

Fig. 6.15 Rectangular region of integration

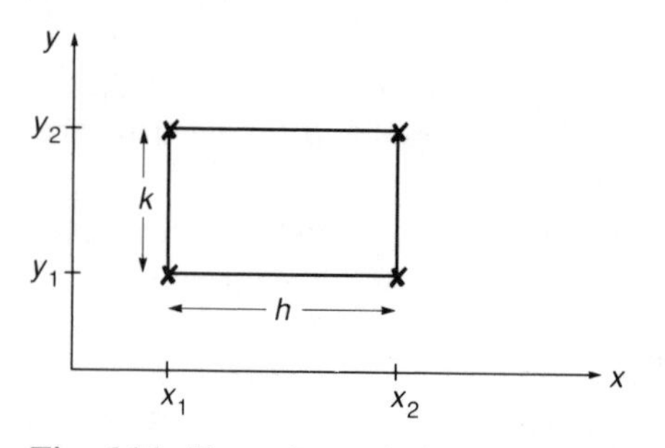

Fig. 6.16 Trapezium rule in two dimensions (Example 6.9)

sampling points at each corner of the rectangle as shown in Fig. 6.16, i.e.

$$\int_{y_1}^{y_2}\int_{x_1}^{x_2} f(x, y)\,\mathrm{d}x\,\mathrm{d}y \simeq \tfrac{1}{4}hk[f(x_1 y_1)+f(x_2 y_1)+f(x_1 y_2)+f(x_2 y_2)] \tag{6.77}$$

It should be noted that the weights in each direction, i.e. $\frac{1}{2}h$ and $\frac{1}{2}k$, have been multiplied together at each sampling point as indicated in the general expression (6.76).

Example 6.9

Estimate

$$I = \int_1^3 \int_1^2 f(x, y)\,\mathrm{d}x\,\mathrm{d}y$$

where $f(x, y) = xy(1+x)$ using the trapezium rule.

Solution 6.9

$h=1$ and $k=2$, hence from eq. 6.77

$I \simeq \frac{1}{2}[f(1, 1)+f(2, 1)+f(1, 3)+f(2, 3)]$

$= \frac{1}{2}(2+6+6+18) = 16$ (cf. exact solution 15.333)

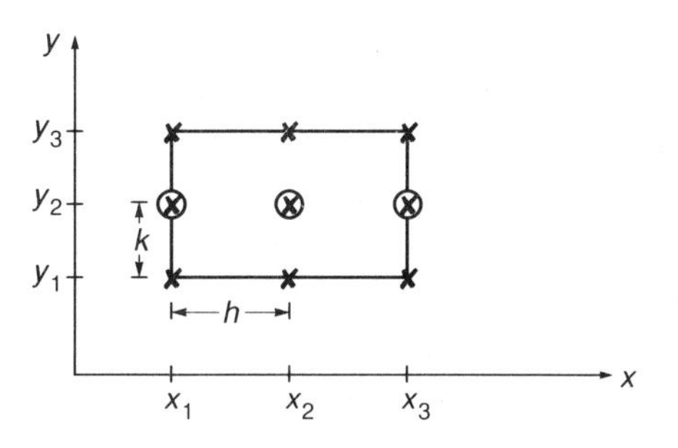

Fig. 6.17 Simpson's/midpoint rule in two dimensions (Example 6.10)

Simpson's rule applied in each direction, would lead to the nine sampling points indicated by crosses in Fig. 6.17, hence the formula

$$\int_{y_1}^{y_3}\int_{x_1}^{x_3} f(x, y)\,\mathrm{d}x\,\mathrm{d}y \simeq \tfrac{1}{9}hk[f(x_1, y_3)+4f(x_2, y_3)+f(x_3, y_3)+4f(x_1, y_2)+16f(x_2, y_2)$$
$$+4f(x_3, y_2)+f(x_1, y_1)+4f(x_2, y_1)+f(x_3, y_1)] \qquad (6.78)$$

Example 6.10

Estimate

$$I=\int_1^3\int_1^2 f(x, y)\,\mathrm{d}x\,\mathrm{d}y$$

where $f(x, y)=xy(1+x)$ using Simpson's rule.

Solution 6.10

$h=\frac{1}{2}$ and $k=1$, hence from eq. 6.78

$$I=\tfrac{1}{18}[f(1, 3)+4f(1.5, 3)+f(2, 3)+4f(1, 2)+16f(1.5, 2)$$
$$+4f(2, 2)+f(1, 1)+4f(1.5, 1)+f(2, 1)]$$
$$=\tfrac{1}{18}[6+45+18+16+120+48+2+15+6]$$
$$=15.3333$$

As expected, Simpson's rule gives the exact solution in this case because the function $f(x, y)$ is second order in x and linear in y. In this instance, it was rather inefficient to use Simpson's rule in both directions as the exact solution could also have been achieved with Simpson's rule in the x-direction and the midpoint rule (say) in the y-direction. Using just these three sampling points, circled in Fig. 6.17, the following expression is obtained

$$I=\tfrac{1}{3}[f(1, 2)+4f(1.5, 2)+f(2, 2)]$$
$$=\tfrac{1}{3}[4+30+12]=15.3333$$

Gauss–Legendre rules can also be applied to multiple integrals of this type, but care must be taken to find the correct locations of the sampling points. One approach would be to perform a coordination transformation so that the limits of integration in each direction become ± 1. This would enable the weights and sampling points from Table 6.2 to be used directly. The general topic of transformation is covered in the next section on integration over general quadrilateral areas.

Consider two-point Gauss–Legendre integration over the rectangular region shown in Fig. 6.18. In the x-direction the sampling points will be located at

$$\xi_1 = \left(x_1 + h - \frac{1}{\sqrt{3}}h\right) \quad \text{and} \quad \xi_2 = \left(x_1 + h + \frac{1}{\sqrt{3}}h\right) \tag{6.79}$$

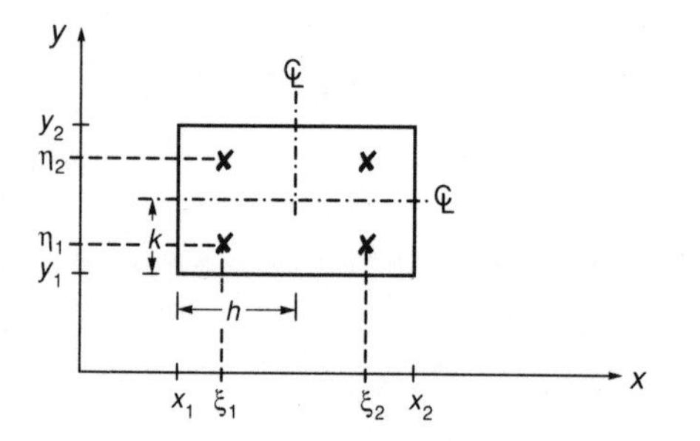

Fig. 6.18 Two-point Gauss–Legendre rule in two dimensions

with corresponding weighting coefficients $w_1 = w_2 = h$. Similarly, in the y-direction the sampling points will be located at

$$\eta_1 = \left(y_1 + k - \frac{1}{\sqrt{3}}k\right) \quad \text{and} \quad \eta_2 = \left(y_1 + k + \frac{1}{\sqrt{3}}k\right) \tag{6.80}$$

with corresponding weighting $w_1 = w_2 = k$. Hence the formula becomes

$$\int_{y_1}^{y_2}\int_{x_1}^{x_2} f(x, y)\,\mathrm{d}x\,\mathrm{d}y \simeq hk[f(\xi_1, \eta_1) + f(\xi_2, \eta_1) + f(\xi_1, \eta_2) + f(\xi_2, \eta_2)] \tag{6.81}$$

Example 6.11

Estimate

$$\int_{-1}^{0}\int_{0}^{2} x^3 y^4\,\mathrm{d}x\,\mathrm{d}y$$

using the two-point Gauss–Legendre rule in both directions.

Solution 6.11

$h=1$, $k=0.5$ and referring to eq. 6.81

(a) $\xi_1=0.4226$ (c) $\eta_1=-0.7887$

(b) $\xi_2=1.5774$ (d) $\eta_2=-0.2113$

$$\therefore \quad I \simeq 0.5\,[f(0.4226,\ -0.7887)+f(1.5774,\ -0.7887)+ f(0.4226,\ -0.2113)+f(1.5774,\ -0.2113)]$$
$$=0.7778 \text{ (cf. exact solution 0.8)}$$

In this instance, the two-point rule is capable of exact integration in the x-direction but not in the y-direction.

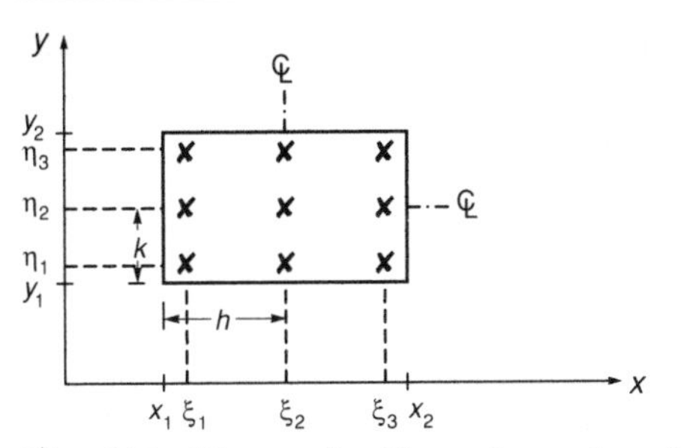

Fig. 6.19 Three-point Gauss–Legendre rule in two dimensions

By similar reasoning, the three-point Gauss–Legendre rule shown in Fig. 6.19 leads to the following terms:

$$\xi_1=x_1+h-\sqrt{(0.6h)},\ \xi_2=x_1+h,\ \xi_3=x_1+h+\sqrt{(0.6h)} \tag{6.82}$$
$$w_1=\tfrac{5}{9}h,\ w_2=\tfrac{8}{9}h,\qquad w_3=\tfrac{5}{9}h$$

$$\eta_1=y_1+k-\sqrt{(0.6k)},\ \eta_2=y_1+k,\ \eta_3=y_1+k+\sqrt{(0.6k)} \tag{6.83}$$
$$w_1=\tfrac{5}{9}k,\ w_2=\tfrac{8}{9}k,\qquad w_3=\tfrac{5}{9}k$$

Hence the formula becomes

$$\int_{y_1}^{y_2}\int_{x_1}^{x_2} f(x,y)\,\mathrm{d}x,\mathrm{d}y=\frac{hk}{81}[25f(\xi_1,\eta_1)+40f(\xi_2,\eta_1)+25f(\xi_3,\eta_1)+ +40f(\xi_1,\eta_2)+64f(\xi_2,\eta_2)+40f(\xi_3,\eta_2)+ +25f(\xi_1,\eta_3)+40f(\xi_2,\eta_3)+25f(\xi_3,\eta_3)] \tag{6.84}$$

Example 6.12

Estimate

$$I=\int_{-1}^{0}\int_{0}^{2} x^3y^4\,\mathrm{d}x\,\mathrm{d}y$$

using the three-point Gauss–Legendre rule in both directions.

Solution 6.12

$h=1$, $k=0.5$ and referring to eqs 6.82–6.84

$\xi_1 = 0.2254 \quad \xi_2 = 1 \quad \xi_3 = 1.7746$

$\eta_1 = -0.8873 \quad \eta_2 = -0.5 \quad \eta_3 = -0.1127$

hence $I=0.8$ (cf. exact solution 0.8).

The exact solution to the integration of the quartic terms in y is obtained by the three-point rule, which leads to nine sampling points in the region of integration. In fact it was unnecessary to use three-point integration in the x-direction in Example 6.12, as two-point integration would have been sufficient. However, in most computer applications it is usual to employ the same order of integration in all directions.

6.5.3. Integration over a general quadrilateral area

We now turn our attention to numerical integration of a function $f(x, y)$ over a general quadrilateral area such as that shown in Fig. 6.20. Even analytical integration of simple functions over this region is quite tedious due to the variable limits. It should also be noted that the required order of integration to obtain exact solutions over such regions is higher than for regions whose boundaries are parallel to the Cartesian coordinate directions.

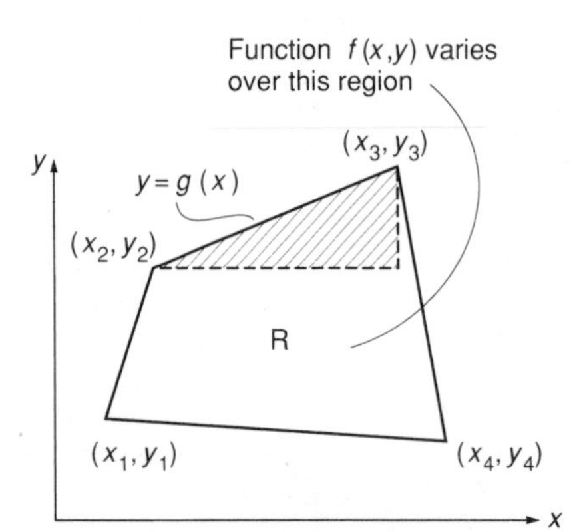

Fig. 6.20 General quadrilateral region of integration

Consider integration of $f(x, y)=x^n y^m$ over the hatched triangular region in Fig. 6.20, i.e.

$$I=\int_{x_2}^{x_3}\int_{y_2}^{g(x)} x^n y^m \,\mathrm{d}y\,\mathrm{d}x \tag{6.85}$$

where $g(x)$ is a linear function given by

$$g(x)=ax+b \tag{6.86}$$

Performing the inner integral first with respect to y, and substituting the limits leads to

$$I=\frac{1}{m+1}\int_{x_2}^{x_3} x^n[(ax+b)^{m+1}-y_2^{m+1}]\,dx \tag{6.87}$$

The remaining outer integral involves integration of a polynomial in x of the $(m+n+1)$th order. If we are to obtain an exact solution, our numerical integration formula must be capable of exactly integrating an $(m+n+1)$th order polynomial. If on the other hand, the region of integration was a rectangle of the kind considered in the previous section, the order of integration required to give an exact solution would relate only to a polynomial of order m or n (whichever was the greater).

We will concentrate on Gauss–Legendre methods for numerical integration of problems such as that described in Fig. 6.20. Before we proceed, it is convenient to perform a coordinate transformation which results in every point in the region R being mapped onto a square region of side length 2 as shown in Fig. 6.21. The one-to-one

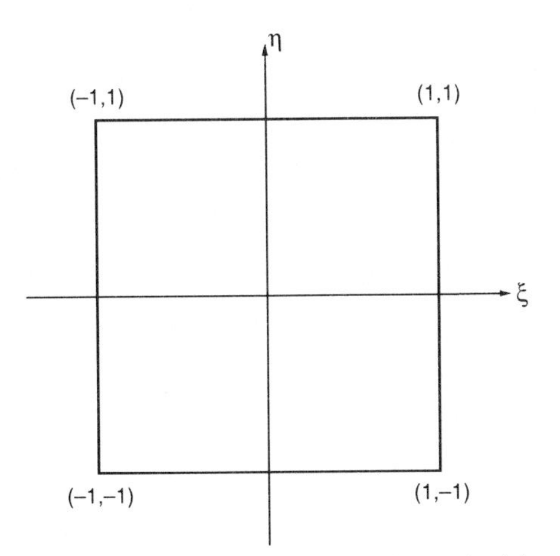

Fig. 6.21 Transformed quadrilateral with coordinates of ± 1

correspondence between points in the two regions is achieved by the transformation relationships

$$\begin{aligned} x(\xi,\eta)&=N_1x_1+N_2x_2+N_3x_3+N_4x_4\\ y(\xi,\eta)&=N_1y_1+N_2y_2+N_3y_3+N_4y_4 \end{aligned} \tag{6.88}$$

where

$$\begin{aligned} N_1&=\tfrac{1}{4}(1-\xi)(1-\eta)\\ N_2&=\tfrac{1}{4}(1-\xi)(1+\eta)\\ N_3&=\tfrac{1}{4}(1+\xi)(1+\eta)\\ N_4&=\tfrac{1}{4}(1+\xi)(1-\eta) \end{aligned} \tag{6.89}$$

The N-functions are often called 'shape functions', and one of their properties is that they equal either 1 or 0 at each corner. Thus (x_1, y_1) is mapped to $(-1, -1)$, (x_2, y_2) to $(-1, 1)$ and so on. In addition, the line joining (x_2, y_2) to (x_3, y_3) is mapped onto line $\eta = 1$ in the transformed space, etc.

It can be shown from coordinate transformation theory, that the integrals in the original and transformed spaces are related as follows

$$\int_R\int f(x, y)\,\mathrm{d}x\,\mathrm{d}y \equiv \int_{-1}^{1}\int_{-1}^{1} Jf[x(\xi, \eta), y(\xi, \eta)]\,\mathrm{d}\xi\,\mathrm{d}\eta \tag{6.90}$$

where J is called the 'Jacobian', and is a scaling factor relating $\mathrm{d}x\,\mathrm{d}y$ to $d\xi\,\mathrm{d}\eta$.

The Jacobian is a function of position within the region of integration and is given as the determinant of the 'Jacobian matrix', thus

$$J = \det \begin{vmatrix} \dfrac{\partial x}{\partial \xi} & \dfrac{\partial y}{\partial \xi} \\ \dfrac{\mathrm{d}x}{\mathrm{d}\eta} & \dfrac{\mathrm{d}y}{\mathrm{d}\eta} \end{vmatrix} \tag{6.91}$$

The Jacobian matrix is readily computed from eqs 6.88, thus

$$\begin{aligned} \frac{\partial x}{\partial \xi} &= \frac{\partial N_1}{\partial \xi}x_1 + \frac{\partial N_2}{\partial \xi}x_2 + \frac{\partial N_3}{\partial \xi}x_3 + \frac{\partial N_4}{\partial \xi}x_4 \\ \frac{\partial x}{\partial \eta} &= \frac{\partial N_1}{\partial \eta}x_1 + \frac{\partial N_2}{\partial \eta}x_2 + \frac{\partial N_3}{\partial \eta}x_3 + \frac{\partial N_4}{\partial \eta}x_4 \quad \text{etc.} \end{aligned} \tag{6.92}$$

where from eqs 6.89

$$\begin{aligned} \frac{\partial N_1}{\partial \xi} &= -\tfrac{1}{4}(1-\eta) \\ \frac{\partial N_2}{\partial \eta} &= \tfrac{1}{4}(1-\xi) \quad \text{and so on.} \end{aligned} \tag{6.93}$$

It should be noted that the shape functions from eq. 6.89 are only smooth functions of ξ and η provided all interior angles of the untransformed quadrilateral are less than 180°.

Program 6.4. Integration over a general quadrilateral area
This program evaluates integrals of the form

$$I = \int_R\int f(x, y)\,\mathrm{d}x\,\mathrm{d}y$$

where R is a general quadrilateral area in x–y space with all internal angles less than 180°. The

```
      PROGRAM P64
C
C      PROGRAM 6.4 AREA INTEGRAL OVER A QUADRILATERAL USING
C      GAUSS-LEGENDRE QUADRATURE
C
C      ALTER NEXT LINE TO CHANGE PROBLEM SIZE
C
      PARAMETER (ISAMP=7)
C
      REAL SAMP(ISAMP,2),COORD(4,2),FUN(4),JAC(2,2),DER(2,4)
C
      DO 10 I = 1,4
   10 READ (5,*) COORD(I,1),COORD(I,2)
      READ (5,*) NGP
      CALL GAULEG(SAMP,ISAMP,NGP)
      VOL = 0.
      DO 20 I = 1,NGP
          DO 20 J = 1,NGP
              CALL BILIN(DER,FUN,SAMP,ISAMP,I,J)
              CALL MATMUL(DER,2,COORD,4,JAC,2,2,4,2)
              DET = JAC(1,1)*JAC(2,2) - JAC(1,2)*JAC(2,1)
              X = 0.
              Y = 0.
              DO 30 K = 1,4
                  X = X + FUN(K)*COORD(K,1)
   30             Y = Y + FUN(K)*COORD(K,2)
   20 VOL = VOL + DET*SAMP(I,2)*SAMP(J,2)*F(X,Y)
      WRITE (6,*) ('****** AREA INTEGRAL OVER A QUADRILATERAL ******')
      WRITE (6,*) ('************ GAUSS-LEGENDRE IN 2-D *************')
      WRITE (6,*)
      DO 40 I = 1,4
   40 WRITE (6,100) COORD(I,1),COORD(I,2)
      WRITE (6,101) NGP
      WRITE (6,103) VOL
  100 FORMAT ('  X,Y COORDINATES                          ',2F12.4)
  101 FORMAT (/,'  NO. OF GAUSS POINTS IN EACH DIRECTION ',I7)
  103 FORMAT (/,'  COMPUTED RESULT                         ',F12.4)
      STOP
      END
C
      FUNCTION F(X,Y)
C
C      THIS FUNCTION PROVIDES THE VALUE OF F(X,Y)
C      AND WILL VARY FROM ONE PROBLEM TO THE NEXT
C
      F = X**2*Y**2
      RETURN
      END
```

method used is Gauss–Legendre integration in two dimensions, with the same number of sampling points in each direction.

Input data consist of the x- and y-coordinates of the four corners in a *clockwise* sense read into array COORD, and the number of sampling points in each direction read into NGP. The function $f(x, y)$ is evaluated by function F, which is problem dependent. The Gaussian weights and sampling points are provided by library subroutine GAULEG. Library subroutine BILIN computes at each Gauss point, the shape functions held in FUN, and their derivatives with respect to ξ and η held in the first and second rows respectively of array DER. Library subroutine MATMUL multiplies matrix DER by matrix COORD to generate the Jacobian matrix JAC following eqs 6.91 and 6.92. The determinant of the Jacobian matrix DET is computed, and the x, y-coordinates of each Gauss point evaluated from eqs 6.88. A running total of the contribution to the integral from each Gauss point is kept by VOL.

A summary of variable names and their meanings is given below.

NGP	Number of Gauss points in each direction
VOL	Running total of integral
DET	Determinant of Jacobian matrix
X, Y	x- and y-coordinates of Gauss point
SAMP	2-d array holding sampling points and weighting coefficients
COORD	2-d array holding corner coordinates
FUN	1-d array holding shape functions
JAC	2-d array holding Jacobian matrix
DER	2-d array holding derivatives of shape functions

The PARAMETER restriction is that ISAMP$\geq$NGP.

The example solved here involves integration of the function $f(x, y)=x^2y^2$, given by user-supplied function F, over the quadrilateral region given in Fig. 6.22. Following eqs 6.85, 6.86 and

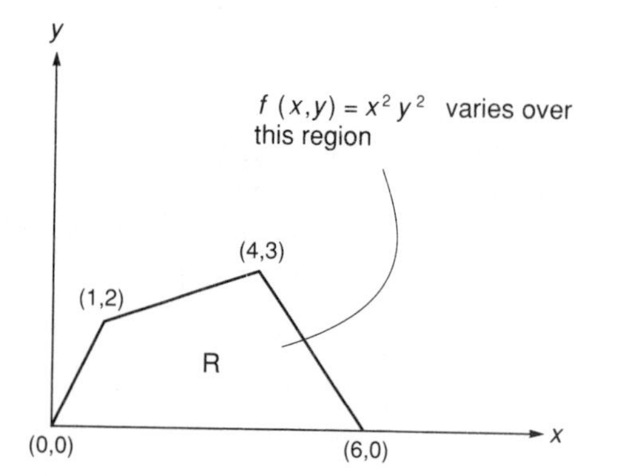

Fig. 6.22 Example problem for Program 6.4

	x	y
Corner coordinates of	0.0	0.0
quadrilateral	1.0	2.0
(clockwise sense)	4.0	3.0
	6.0	0.0
Number of Gauss points	NGP	
in each dimension	3	

Fig. 6.23(a) Input Data for Program 6.4

```
****** AREA INTEGRAL OVER A QUADRILATERAL ******
************ GAUSS-LEGENDRE IN 2-D *************

X,Y COORDINATES                              .0000       .0000
X,Y COORDINATES                             1.0000      2.0000
X,Y COORDINATES                             4.0000      3.0000
X,Y COORDINATES                             6.0000       .0000

NO. OF GAUSS POINTS IN EACH DIRECTION          3

COMPUTED RESULT                           225.7944
```

Fig. 6.23(b) Results from Program 6.4

6.87, we need the capability to integrate a fifth order polynomial, hence NGP is set to 3. The input data and output from Program 6.4 are given in Figs 6.23(a) and (b) respectively. It is left for the reader to check by analytical means that the numerically obtained solution of 225.7944 is exact.

6.6 Exercises

1 Calculate the area of a quarter of a circle of radius a by the following methods, and compare with the exact result of $0.7854a^2$.

(a) Rectangle rule
(b) Trapezium rule
(c) Simpson's rule
(d) 4-point Newton–Cotes
(e) 5-point Newton–Cotes

Answer: (a) a^2, (b) $0.5a^2$, (c) $0.7440a^2$, (d) $0.7581a^2$, (e) $0.7727a^2$.

2 Attempt Exercise 1 using the following methods repeated twice.

(a) Rectangle rule
(b) Trapezium rule
(c) Simpson's rule

Answer: (a) $0.9330a^2$, (b) $0.6830a^2$, (c) $0.7709a^2$.

3 Calculate the area of a quarter of an ellipse whose semi-axes are a and b $(a=2b)$ by the following methods, and compare with the exact result of $\frac{1}{2}\pi b^2$.

$$\left(\text{Equation of ellipse } \frac{x^2}{a^2}+\frac{y^2}{b^2}=1.\right)$$

(a) Rectangle rule
(b) Trapezium rule
(c) Simpson's rule
(d) 4-point Newton–Cotes
(e) 5-point Newton–Cotes

Answer: (a) $2b^2$, (b) b^2, (c) $1.4880b^2$, (d) $1.5161b^2$, (e) $1.5454b^2$.

4 Attempt Exercise 3 using the following methods repeated twice:

(a) Rectangle rule
(b) Trapezium rule
(c) Simpson's rule

Answer: (a) $1.8660b^2$, (b) $1.3660b^2$, (c) $1.5418b^2$.

5 Determine the weights w_1, w_2 and w_3 in the integration formula

$$\int_0^{2h} f(x)\,dx \simeq w_1 f(0) + w_2 f(h) + w_3 f(2h)$$

which ensure that it is exact for all polynomials $f(x)$ of degree 2 or less. Show that the formula is in fact also exact for $f(x)$ of degree 3.

6 How many repetitions of the trapezium rule are necessary in order to compute

$$\int_0^{\pi/3} \sin x \, dx$$

accurate to three decimal places?
Answer: 11 repetitions.

7 Compute the volume of a hemisphere by numerical integration using the lowest order Newton–Cotes method that would give an exact solution.
Answer: Simpson's rule gives $\frac{2}{3}\pi r^3$.

8 Estimate

$$\int_0^3 (x^3 - 3x^2 + 2) \, dx \text{ using}$$

(a) Simpson's rule and (b) the trapezium rule repeated three times. Which method is the most accurate in this case and why?
Answer: (a) -0.75 (exact), (b) 0.0!

9 From the tabulated data given, estimate the area between the function $y(x)$, and the lines $x = 0.2$ and $x = 0.6$

x	y
0.2	1.221403
0.3	1.349859
0.4	1.491825
0.5	1.648721
0.6	1.822119

using the following methods repeated twice:

(a) Simpson's rule
(b) Trapezium rule
(c) Midpoint rule

If the exact solution is given by $y(0.6) - y(0.2)$ comment on the numerical solution obtained.
Answer: (a) 0.6007, (b) 0.6027, (c) 0.5997.

10 Calculate the area of a quarter of a circle of radius a by the following methods and compare with the exact result of $0.7854a^2$.

(a) Midpoint rule
(b) 2-point Gauss–Legendre
(c) 3-point Gauss–Legendre

Answer: (a) $0.8660a^2$, (b) $0.7961a^2$, (c) $0.7890a^2$.

11 Attempt Exercise 10 using the same methods repeated twice.
Answer: (a) $0.8148a^2$, (b) $0.7891a^2$, (c) $0.7867a^2$.

12 Use polynomial substitution to find the coefficients w_0, w_1 and w_2, and the sampling points x_0, x_1, and x_2 in the Gauss–Legendre formula

$$I \simeq w_0 f(x_0) + w_1 f(x_1) + w_2 f(x_2)$$

You may assume symmetry of weights and sampling points.

13 Derive the two-point Gauss–Legendre integration rule, and use it to estimate the area enclosed by the ellipse

$$\frac{x^2}{4} + \frac{y^2}{9} = 1$$

Compare your solution with that obtained using one application of the midpoint rule.
Answer: 19.1067, 20.7846, 18.8496 (exact).

14 Estimate

$$\int_{0.3}^{0.8} e^{-2x} \tan x \, dx$$

using the one-point Gauss–Legendre formula repeated once, twice and three times.
Answer: 0.1020, 0.1002, 0.0999.

15 Use Gauss–Legendre integration to find the exact value of

$$\int_0^1 (x^7 + 2x^2 - 1) \, dx$$

Answer: 4-point gives −0.2083.

16 Estimate the value of

$$\int_1^3 \frac{dx}{(x^4+1)^{1/2}}$$

using (a) Midpoint rule
(b) 2-point Gauss–Legendre
(c) 3-point Gauss–Legendre

Answer: (a) 0.4851, (b) 0.5918, (c) 0.5951.

17 Attempt Exercise 16 using the same methods repeated twice.
Answer: (a) 0.5641, (b) 0.5947, (c) 0.5942.

18 Use Gauss–Legendre integration with four sampling points to estimate the value of

(a) $\displaystyle\int_{-2}^{2} \frac{dx}{1+x^2}$ (b) $\displaystyle\int_0^1 x \exp(-3x^2) \, dx$

Answer: (a) 2.1346, (b) 0.1584.

19 Estimate the value of

$$\int_0^\infty e^{-x} \cos x \, dx$$

using (a) 1-point Gauss–Laguerre
(b) 3-point Gauss–Laguerre
(c) 5-point Gauss–Laguerre

Answer: (a) 0.5403, (b) 0.4765, (c) 0.5005.

20 Determine approximate values of the integral

$$\int_0^\infty \frac{e^{-x}}{x+4}\,dx \simeq 0.206346$$

by using Gauss–Laguerre quadrature employing, two, three, and four sampling points.
Answer: 0.2059, 0.2063, 0.2063.

21 Use two-point Gauss–Legendre integration to estimate

$$\int_1^2 \int_3^4 f(x, y)\,dy\,dx$$

where $f(x) =$ (a) xy, (b) x^2y, (c) x^3y, (d) x^4y

Check your estimates against analytical solutions.
Answer: (a) 5.25, (b) 8.1667, (c) 13.125, (d) 21.6806 (approx).

22 Estimate the value of

$$\int_0^1 \int_0^1 \exp(-x^2)y^2\,dy\,dx$$

using (a) Gauss–Legendre 1-point method
(b) Gauss–Legendre 2-point method

Answer: (a) 0.1947, (b) 0.2489.

23 Use the minimum number of sampling points to find the exact value of

$$\int_1^2 \int_0^3 xy^3\,dy\,dx$$

(*Note:* Program 6.4 always uses the same number of sampling points in each direction, which is inefficient in this case).
Answer: 1-point in x, 2-points in y, 30.375.

24 Estimate the value of

$$\int_{-2}^0 \int_0^1 e^x \sin y\,dx\,dy$$

using 3-point Gauss–Legendre quadrature.
Answer: 0.3975.

25 Find the exact solution to

$$\int_R x^2y\,dR$$

where R is a region bounded by the (x, y) coordinates (0,0), (0.5, 1) (1.5, 1.5), (2, 0.5).
Answer: 1.5375 using 3 Gauss points in each direction.

26 Estimate the triple integral

$$\int_0^1 \int_1^2 \int_0^{0.5} e^{xyz}\, dx\, dy\, dz$$

using both 1- and 2-point Gauss–Legendre integration in each direction.
(*Note:* Program 6.4 is only available for double integrals.)
Answer: 0.6031, 0.6127.

6.7 Further reading

Abramowitz, M. and Stegun, I.A. (1964). *Handbook of Mathematical Functions*, U.S. Government Printing Office, Washington D.C.

Burden, R.L. and Faires, J.D. (1985). *Numerical Analysis*, 3rd edn, Prindle, Weber and Schmidt, Boston, Mass.

Davis, P.J. and Rabinowitz, P. (1984). *Methods of Numerical Integration*, 2nd edn, Academic Press, New York.

Froberg, C.E. (1985). *Numerical Mathematics*, Benjamin/Cummings, Menlo Park, California.

Goldstine, H.H. (1977). *A History of Numerical Analysis*, Springer–Verlag, New York.

Johnson L.W. and Riess, R.D. (1982). *Numerical Analysis*, 3rd edn, Addison-Wesley, Reading Mass.

Ralston, A. (1965). *A First Course in Numerical Analysis*, McGraw-Hill, New York.

Smith, I.M. and Griffiths, D.V. (1988). *Programming the Finite Element Method*, 2nd edn, Wiley, New York.

Stroud, A.H. and Secrest, D. (1966). *Gaussian Quadrature Formulas*, Prentice-Hall, Englewood Cliffs, New Jersey.

7

Numerical Solutions of Ordinary Differential Equations

7.1 Introduction

Differential equations express relationships between variables in terms of their derivatives, in contrast to the algebraic equations covered in Chapters 2 and 3 where no derivatives occurred. The need to solve differential equations arises in a great many problems of mathematical modelling, because physical laws in engineering and science are often expressed in terms of the derivatives of variables rather that just the variables themselves.

The solution to a differential equation can sometimes be arrived at by analytical integration. For example, the simplest type of differential equation is of the form

$$\frac{dy}{dx} = f(x) \tag{7.1}$$

where $f(x)$ is a given function of x, and $y(x)$ is the required solution. Provided $f(x)$ can be integrated, the solution to eq. 7.1 is of the form

$$y = \int f(x)\,dx + C \tag{7.2}$$

where C is an arbitrary constant. In order to find the value of C, some additional piece of information is required, such as an initial value of y corresponding to a particular value of x.

Differential equations can be presented in many different forms however, often involving functions of x *and* y. Before describing different solution techniques therefore, we need to define some important classes of differential equations, as this may influence our method of tackling a particular problem.

7.2 Definitions and types of differential equations

Differential equations fall into two distinct categories depending on the number of independent variables they contain. If there is only one independent variable, the derivatives will be 'ordinary', and the equation will be called an 'ordinary differential equation'. If more than one independent variable exists, the derivatives will be 'partial', and the equation will be called a 'partial differential equation'.

Although ordinary differential equations with only one independent variable may be considered to be a special case of partial differential equations, it is best to consider the solution techniques for the two classes quite separately. The remainder of this chapter is devoted to the solution of ordinary differential equations.

The 'order' of an ordinary differential equation corresponds to the highest derivative that appears in the equation, thus

$$y''-3y'+4=y \qquad \text{is second order} \tag{7.3}$$

where we use the notation $y'=\dfrac{\mathrm{d}y}{\mathrm{d}x}$, $y''=\dfrac{\mathrm{d}^2y}{\mathrm{d}x^2}$ etc.

In the same way

$$\frac{\mathrm{d}^4y}{\mathrm{d}x^4}+\left(\frac{\mathrm{d}y}{\mathrm{d}x}\right)^2=1 \qquad \text{is fourth order} \tag{7.4}$$

A 'linear' equation is one which contains no products of the dependent variable or its derivatives, thus eq. 7.3 is linear whereas 7.4 is nonlinear, due to the squared term. A general nth order linear equation is given as

$$A_n(x)\frac{\mathrm{d}^ny}{\mathrm{d}x^n}+A_{n-1}(x)\frac{\mathrm{d}^{n-1}y}{\mathrm{d}x^{n-1}}+\cdots+A_1(x)\frac{\mathrm{d}y}{\mathrm{d}x}+A_0(x)y=R(x) \tag{7.5}$$

where the A_i's and R are functions of x.

Sometimes the degree of an equation is also referred to, and this represents the power to which the highest derivative is raised, thus eqs 7.3 and 7.4 are both first degree. A consequence of this is that all linear equations are first degree, but not all first degree equations are linear, for example

$$y''+2y'+y=0 \qquad \text{is second order, first degree, linear} \tag{7.6}$$

but

$$y''+2y'+y^2=0 \qquad \text{is second order, first degree, nonlinear} \tag{7.7}$$

Nonlinear equations are harder to solve analytically and may have multiple solutions. Numerical solution of nonlinear boundary value problems will require iterative approaches as will be shown in a later section.

The higher the order of a differential equation, the more additional information must be supplied in order to obtain a solution. For example, eqs 7.8–7.10 are all equivalent statements mathematically, but the second order equation requires two additional pieces of information to be equivalent to the first order equation which only requires one additional piece of information.

$$y''=y+x+1, \quad y(0)=0, \quad y'(0)=0 \tag{7.8}$$

$$y'=y+x, \quad y(0)=0 \tag{7.9}$$

$$y=\mathrm{e}^x-x-1 \tag{7.10}$$

The third equation, 7.10, is purely algebraic and represents the solution to the two differential equations.

In general, to obtain a solution to an nth order ordinary differential equation such as that given in eq. 7.5, n additional pieces of information will be required.

The way in which this additional information is supplied greatly influences the method of numerical solution. If all the information is given at the same value of the independent variable, such as in eqs 7.8 and 7.9, the problem is termed an 'initial value problem'. If the information is provided at different values of the independent variable, such as in the second order system given by eq. 7.11, the problem is termed a 'boundary value problem':

$$y''+\frac{1}{x}y'-\frac{1}{x^2}y=\frac{6}{x^2}, \quad y(1)=1, \quad y'(1.5)=-1 \tag{7.11}$$

All boundary value problems will be at least of second order. It should also be noted that as first order equations only require one piece of additional information, all first order equations may be treated as initial value problems.

The solution techniques for initial and boundary value problems usually differ substantially, so they will be considered separately.

7.3 Initial value problems

We will limit our discussion for now to the numerical solution of first order equations subject to an initial condition, thus

$$\frac{dy}{dx}=f(x, y), \quad \text{with } y(x_0)=y_0 \tag{7.12}$$

where $f(x, y)$ is any function of x and y. The equation in this general form could be linear or nonlinear depending on the nature of the function $f(x, y)$. It may be noted that if the equation is linear it can often be solved analytically by means of an integrating factor.

If the function $f(x, y)$ is nonlinear, the analytical approach is greatly limited, and may even prove impossible. Only a limited number of nonlinear differential equations can be solved analytically.

All the numerical techniques for solving equations such as eq. 7.12 involve starting at the initial condition (x_0, y_0) and stepping along the x-axis. At each step, a new value of y is estimated. As more steps are taken the form of the required solution $y(x)$ is obtained.

The effect on y of a change in x is given by eq. 7.13

$$y_{i+1}-y_i=\int_{x_i}^{x_{i+1}}\frac{dy}{dx}\,dx \tag{7.13}$$

where (x_i, y_i) is the initial condition, and y_{i+1} is the new estimate of y corresponding to

$x = x_{i+1}$. The step length in x is under the user's control and is usually defined as h, thus

$$x_{i+1} - x_i = h \tag{7.14}$$

Rearrangement of 7.13 leads to

$$y_{i+1} = y_i + \int_{x_i}^{x_{i+1}} \frac{dy}{dx}\, dx \tag{7.15}$$

or alternatively

$$\begin{bmatrix}\text{new value}\\ \text{of } y\end{bmatrix} = \begin{bmatrix}\text{old value}\\ \text{of } y\end{bmatrix} + \begin{bmatrix}\text{numerical integration}\\ \text{process}\end{bmatrix} \tag{7.16}$$

Equation 7.16 gives the general form for all numerical solution techniques for initial value problems. In the previous chapter we integrated under a curve of y vs. x to compute an area, but in this chapter we will be integrating under a curve of (dy/dx) vs. x to compute the change in y. Many of the methods of numerical integration described in the previous chapter still apply. The main difference here is that the function to be integrated may depend on both x and y; thus we will need to modify our methods slightly to account for this.

There are two main approaches for performing the integration required by eq. 7.15:

(a) One-step methods, which use information from only one preceding point, (x_i, y_i) to estimate the next point (x_{i+1}, y_{i+1})

(b) Multi-step methods or predictor–corrector methods, which use information about several previous points, (x_i, y_i), (x_{i-1}, y_{i-1}) ... etc. to estimate the next point (x_{i+1}, y_{i+1}). These methods have the ability to refine the initial prediction iteratively by repeated application of a 'corrector' formula.

One-step methods are self-starting with only the initial condition, whereas multi-step methods require several consecutive values of x and y to get started. These initial values can be provided by a one-step method if not provided in the initial data.

We will concentrate initially on the numerical solution of a single first order equation. It will be shown subsequently in Section 7.3.1.6 that this is not a limitation, because a higher order equation can be broken down into systems of first order equations which can be solved by the same methods.

7.3.1 One-step methods

One-step methods are so called, because information about one previous step only is used to generate the solution at the next step. This makes the one-step methods relatively simple to implement in a computer program. There are several one-step methods of increasing complexity, and as is often the case in numerical methods, the more work that has to be done at each step, the greater the accuracy. The trade-off to be sought is between increasing the work per step and decreasing the number of steps to span a given range.

7.3.1.1 Euler's method

The simplest one-step method has limited accuracy and is certainly not recommended. However, it forms the basis for all subsequent methods. The numerical integration of eq. 7.15 is performed using the rectangle rule (see Section 6.2.2) where the derivative is 'sampled' at the initial condition. Given the differential equation

$$y' = f(x, y) \quad \text{with } y(x_0) = y_0 \tag{7.17}$$

Euler's method estimates the new value of y using the expression

$$y_1 = y_0 + hf(x_0, y_0) \tag{7.18}$$

where $\quad h = x_1 - x_0 \qquad (7.19)$

This is the simplest statement of eq. 7.16. The method is then repeated using the 'new' initial conditions (x_1, y_1) to estimate y_2 and so on. For a general step i, we get

$$y_{i+1} = y_i + hf(x_i, y_i) \tag{7.20}$$

where $\quad h = x_{i+1} - x_i \qquad (7.21)$

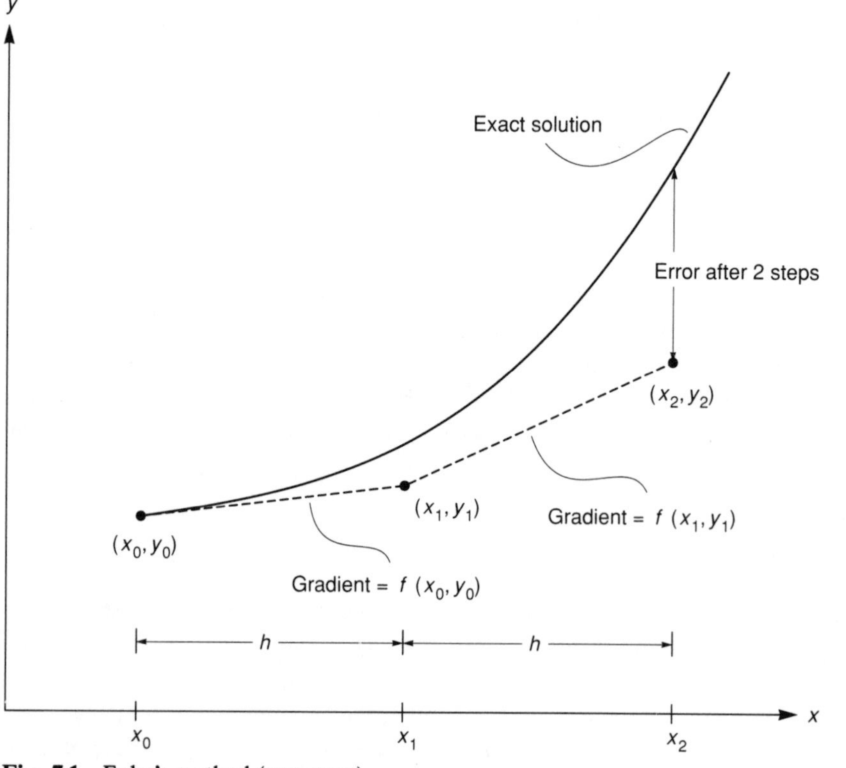

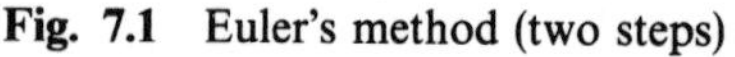
Fig. 7.1 Euler's method (two steps)

Figure 7.1 shows how Euler's method operates. The approximate solution is assumed to follow a straight line corresponding to the tangent at the initial point. Clearly an error is introduced by this assumption unless the actual solution happens to be linear. Errors tend to accumulate at each step by this method as shown in Fig. 7.1. The local error at each step can be reduced by making h smaller, but there is a limit to how small h can be made from efficiency and machine accuracy considerations.

It should be noted that Euler's method from eq. 7.18 is equivalent to the Taylor series truncated after two terms, i.e.

$$y(x_0+h)=y(x_0)+hy'(x_0) \;\vdots\; \underbrace{+\frac{h^2}{2!}y''(x_0)+\frac{h^3y'''}{3!}(x_0)+\cdots}_{\text{Truncated by Euler's method}} \tag{7.22}$$

Example 7.1

Given

$$y'=\frac{x+y}{x} \quad \text{with } y(2)=2,$$

estimate $y(2.5)$ using Euler's method. Let (a) $h=0.25$ and (b) $h=0.1$.

Solution 7.1

(a) Two steps will be required if $h=0.25$

$$y(2.25)\simeq 2+0.25\left(\frac{2+2}{2}\right)=2.5$$

$$y(2.5)\simeq 2.5+0.25\left(\frac{2.25+2.5}{2.25}\right)=3.028$$

(b) Five steps will be required if $h=0.1$

$$y(2.1)\simeq 2+0.1\left(\frac{2+2}{2}\right)=2.2$$

$$y(2.2)\simeq 2.2+0.1\left(\frac{2.1+2.2}{2.1}\right)=2.405$$

$$y(2.3)\simeq 2.405+0.1\left(\frac{2.2+2.405}{2.2}\right)=2.614$$

$$y(2.4)\simeq 2.614+0.1\left(\frac{2.3+2.614}{2.3}\right)=2.828$$

$$y(2.5) \simeq 2.828 + 0.1\left(\frac{2.4 + 2.828}{2.4}\right) = 3.046$$

The exact solution is given by $y = x[1 + \ln(x/2)]$ hence $y(2.5) = 3.058$.

7.3.1.2 Modified Euler's method

A logical refinement to Euler's method is to include more terms in the Taylor series of eq. 7.22. The modified Euler method is analogous to the trapezium rule described in Section 6.2.3 and is equivalent to the Taylor series truncated after three terms.

Given the differential equation and initial condition

$$y' = f(x, y) \quad \text{with } y(x_0) = y_0 \tag{7.23}$$

the modified Euler method estimates the new value of y using the sequence

$$K_0 = hf(x_0, y_0) \tag{7.24}$$

$$K_1 = hf(x_0 + h, y_0 + K_0) \tag{7.25}$$

$$y_1 = y_0 + \tfrac{1}{2}(K_0 + K_1) \tag{7.26}$$

where K_0 is the change in y based on the slope at the beginning of the step, and K_1 is the change in y based on the slope at the end of the step.

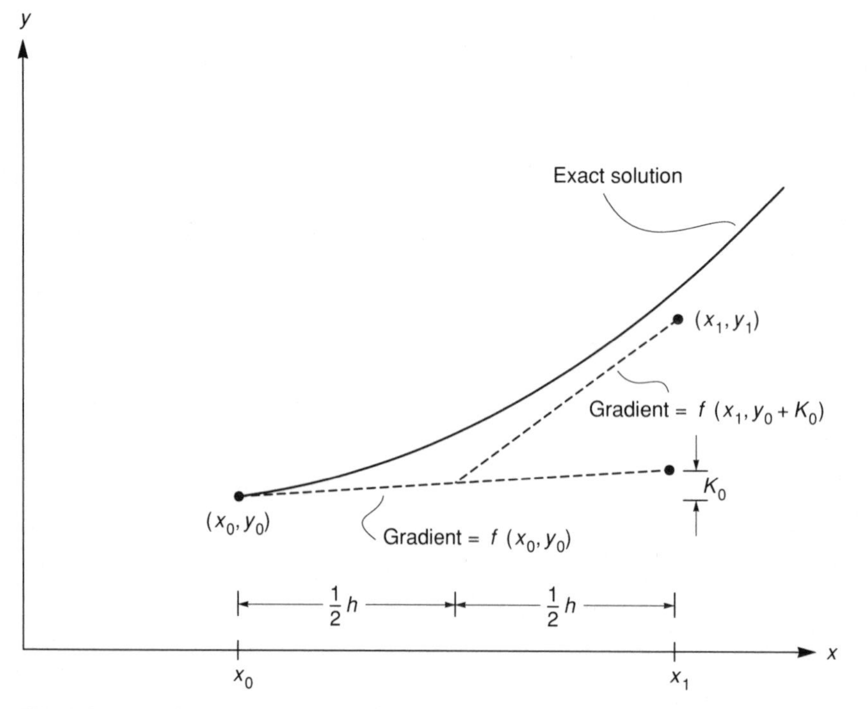

Fig. 7.2 Modified Euler method

As shown in Fig. 7.2, the approximate solution follows the exact solution more closely than the simple Euler method because the integration of eq. 7.15 is performed by sampling the derivative $f(x, y)$ at the beginning *and* end of the step. The calculation of K_0 in eq. 7.24 is equivalent to a simple Euler step, and this is used to compute K_1 leading to a 'modified' Euler approach.

Example 7.2

Given

$$y' = \frac{x+y}{x} \quad \text{with } y(2)=2,$$

estimate $y(2.5)$ using the modified Euler method. Let $h=0.25$.

Solution 7.2

Two steps will be required if $h=0.25$

$$\left.\begin{aligned} K_0 &= 0.25\left(\frac{2+2}{2}\right)=0.500 \\ K_1 &= 0.25\left(\frac{2.25+2.5}{2.25}\right)=0.528 \\ y(2.25) &= 2+\tfrac{1}{2}(0.500+0.528)=2.514 \end{aligned}\right\} \text{Step 1}$$

$$\left.\begin{aligned} K_0 &= 0.25\left(\frac{2.25+2.514}{2.25}\right)=0.529 \\ K_1 &= 0.25\left(\frac{2.5+3.043}{2.5}\right)=0.554 \\ y(2.5) &= 2.514+\tfrac{1}{2}(0.529+0.554)=3.056 \end{aligned}\right\} \text{Step 2}$$

(cf. exact solution 3.058)

7.3.1.3 Midpoint method

Another simple one-step method is based on the midpoint rule of integration. It may be recalled from Section 6.3.2 that this is also the one-point Gauss–Legendre rule.

Given the differential equation and initial condition

$$y' = f(x, y) \quad \text{with } y(x_0)=y_0 \tag{7.27}$$

the midpoint method estimates the new value of y using the sequence

$$K_0 = hf(x_0, y_0) \tag{7.28}$$

$$K_1 = hf(x_0 + \tfrac{1}{2}h, y_0 + \tfrac{1}{2}K_0) \tag{7.29}$$

$$y_1 = y_0 + K_1 \tag{7.30}$$

As shown in Fig. 7.3, the gradient of the solution curve is estimated at the midpoint of the range and this 'average' gradient is assumed to act across the full range.

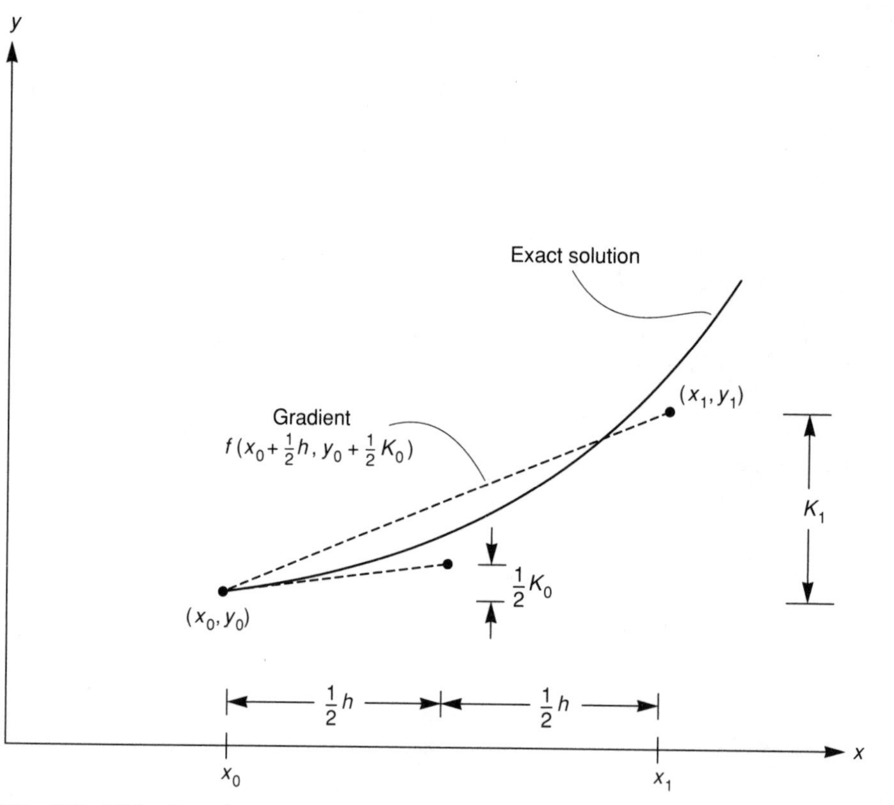

Fig. 7.3 Midpoint rule

Example 7.3

Given the differential equation

$$y' = \frac{x+y}{x} \quad \text{with } y(2) = 2$$

estimate $y(2.5)$ using the midpoint rule. Let $h = 0.25$.

Solution 7.3

Step 1:

$$K_0 = 0.25\left(\frac{2+2}{2}\right) = 0.50$$

$$K_1 = 0.25\left(\frac{2.125+2.25}{2.125}\right) = 0.5147$$

$$y(2.25) = 2 + 0.5147 = 2.5147$$

Step 2:

$$K_0 = 0.25\left(\frac{2.25+2.5147}{2.25}\right) = 0.5294$$

$$K_1 = 0.25\left(\frac{2.375+2.7794}{2.375}\right) = 0.5426$$

$$y(2.5) = 2.5147 + 0.5426 = 3.0573$$

(cf. exact solution 3.058)

7.3.1.4 Runge–Kutta methods

'Runge–Kutta' methods refer to a large family of one-step methods for numerical solution of initial value problems, and include the Euler and modified Euler methods, which are first and second order Runge–Kutta methods. The 'order' of a method indicates the highest power of h included in the equivalent truncated Taylor series expansion.

The general form of all Runge–Kutta methods for advancing from step 'i' to '$i+1$' is as follows:

$$y_{i+1} = y_i + \sum_{j=0}^{r-1} W_j K_j \Big/ \sum_{j=0}^{r-1} W_j \tag{7.31}$$

where the W_j are constant weighting coefficients and r is the order of the method. The K_j's are estimates of the change in y evaluated at r locations within the range h.

A third order Runge–Kutta method would be analogous to Simpson's rule but this will not be covered here. A surprisingly simple fourth order Runge–Kutta method has received widespread use, and is sometimes referred to as *the* Runge–Kutta method.

The Runge–Kutta fourth order method takes the following form. Given the differential equation and initial condition

$$y' = f(x, y) \quad \text{with } y(x_0) = y_0 \tag{7.32}$$

then

$$\begin{aligned} K_0 &= hf(x_0, y_0) \\ K_1 &= hf(x_0 + \tfrac{1}{2}h, y_0 + \tfrac{1}{2}K_0) \\ K_2 &= hf(x_0 + \tfrac{1}{2}h, y_0 + \tfrac{1}{2}K_1) \\ K_3 &= hf(x_0 + h, y_0 + K_2) \end{aligned} \tag{7.33}$$

and $y_1 = y_0 + \frac{1}{6}(K_0 + 2K_1 + 2K_2 + K_3)$ (7.34)

The simplicity and accuracy of this method make it the most popular of all one-step methods for numerical solution of linear or nonlinear first order differential equations. Derivation of eqs 7.33 and 7.34 is beyond the scope of the present work, however.

Example 7.4

Given the differential equation

$$y' = \frac{x+y}{x} \quad \text{with } y(2) = 2$$

estimate $y(2.5)$ using the fourth order Runge–Kutta method. Let $h = 0.5$.

Solution 7.4

$$K_0 = 0.5\left(\frac{2+2}{2}\right) = 1$$

$$K_1 = 0.5\left(\frac{2.25+2.5}{2.25}\right) = 1.056$$

$$K_2 = 0.5\left(\frac{2.25+2.528}{2.25}\right) = 1.062$$

$$K_3 = 0.5\left(\frac{2.5+3.062}{2.5}\right) = 1.112$$

$$\begin{aligned} y(2.5) &= 2 + \tfrac{1}{6}[1 + 2(1.056 + 1.062) + 1.112] \\ &= 3.058 \end{aligned}$$

(cf. exact solution 3.058)

Program 7.1. One-step methods for a single first order equation

The first program in this chapter obtains a numerical solution to the initial value problem

$$y' = f(x, y) \quad \text{with } y(x_0) = y_0$$

using a one-step method of the user's choice. Input data initially define the method to be used, where ITYPE = 1 gives Euler's method, ITYPE = 2 gives the modified Euler method, ITYPE = 3 gives the midpoint method, and ITYPE = 4 gives the fourth order Runge–Kutta method. Other data consist of the number of steps (NSTEPS), the step length (H) and the initial values of x and y (X and Y).

The function $f(x, y)$ is evaluated by the user-supplied function F which will be changed from one problem to the next. With the exception of simple integer counters, the variables can be summarised as follows:

ITYPE	Defines method to be used (see above)
NSTEPS	Number of steps

```
      PROGRAM P71
C
C      PROGRAM 7.1  ONE-STEP METHODS (SINGLE EQUATION)
C
C      ITYPE= 1 (EULER'S METHOD)    ITYPE= 2 (MODIFIED EULER'S METHOD)
C      ITYPE= 3 (MID-POINT RULE)    ITYPE= 4 (4TH ORDER RUNGE-KUTTA)
C
      READ (5,*) ITYPE,NSTEPS,H
      READ (5,*) X,Y
      WRITE (6,*) ('*********** ONE-STEP METHODS **********')
      WRITE (6,*)
      GO TO (1,2,3,4) ITYPE
    1 WRITE (6,*) ('************ EULERS METHODS ***********')
      WRITE (6,*)
      WRITE (6,*) ('     X             Y')
      DO 10 I = 0,NSTEPS
          WRITE (6,100) X,Y
          RK0 = H*F(X,Y)
          X = X + H
   10 Y = Y + RK0
      GO TO 50
    2 WRITE (6,*) ('******* MODIFIED EULERS METHOD ********')
      WRITE (6,*)
      WRITE (6,*) ('     X             Y')
      DO 20 I = 0,NSTEPS
          WRITE (6,100) X,Y
          RK0 = H*F(X,Y)
          RK1 = H*F(X+H,Y+RK0)
          X = X + H
   20 Y = Y + 0.5* (RK0+RK1)
      GO TO 50
    3 WRITE (6,*) ('*********** MID-POINT RULE ************')
      WRITE (6,*)
      WRITE (6,*) ('     X             Y')
      DO 30 I = 0,NSTEPS
          WRITE (6,100) X,Y
          RK0 = H*F(X,Y)
          RK1 = H*F(X+0.5*H,Y+0.5*RK0)
          X = X + H
   30 Y = Y + RK1
      GO TO 50
    4 WRITE (6,*) ('******** 4TH ORDER RUNGE-KUTTA ********')
      WRITE (6,*)
      WRITE (6,*) ('     X             Y')
      DO 40 I = 0,NSTEPS
          WRITE (6,100) X,Y
          RK0 = H*F(X,Y)
          RK1 = H*F(X+0.5*H,Y+0.5*RK0)
          RK2 = H*F(X+0.5*H,Y+0.5*RK1)
          RK3 = H*F(X+H,Y+RK2)
          X = X + H
   40 Y = Y + (RK0+2.*RK1+2.*RK2+RK3)/6.
   50 CONTINUE
  100 FORMAT (2E13.5)
      STOP
      END
C
      FUNCTION F(X,Y)
C
C      THIS FUNCTION PROVIDES THE VALUE OF F(X,Y)
C      AND WILL VARY FROM ONE PROBLEM TO THE NEXT
C
      F = (Y+X)**2
      RETURN
      END
```

H	Step length
X	Initial and updated value of x
Y	Initial and updated value of y
RK0 RK1 RK2 RK3	Intermediate values of $hf(x, y)$

To illustrate use of the program the following problem is to be solved using the fourth order Runge–Kutta method. Given the nonlinear equation

$$y' = (x+y)^2 \quad \text{with } y(0) = 1$$

estimate $y(0.5)$ using 5 steps of $h = 0.1$

The input data and output from Program 7.1 are given in Figs 7.4(a) and (b) respectively and the function $f(x, y) = (x+y)^2$ has been written into function F.

One-step method to be used	ITYPE 4	
Number and size of steps	NSTEPS 5	H 0.1
Initial values of x and y	X 0.0	Y 1.0

Fig. 7.4(a) Data for Program 7.1

```
*********** ONE-STEP METHODS **********

******** 4TH ORDER RUNGE-KUTTA ********

      X              Y
  .00000E+00    .10000E+01
  .10000E+00    .11230E+01
  .20000E+00    .13085E+01
  .30000E+00    .15958E+01
  .40000E+00    .20649E+01
  .50000E+00    .29078E+01
```

Fig. 7.4(b) Results from Program 7.1

As shown in Fig. 7.4(b), the method gives the approximate solution $y(0.5) \simeq 2.9078$ to four decimal places. For comparison, the analytical solution to this problem is given by

$$y = \tan\left(x + \frac{\pi}{4}\right) - x$$

leading to an exact solution of $y(0.5) = 2.9082$

7.3.1.5 Accuracy of one-step methods

In assessing the accuracy of one-step methods, the dominant error term gives useful information regarding the influence of the step size h. The dominant error term is the term with the lowest power of h not included in the truncated Taylor series. It can be assumed that this error term is the largest and therefore the most important, provided h is 'small'. The dominant error term results in a local error associated with a single step of the numerical process. The local errors accumulate step by step leading to a global error after several steps.

From eq. 7.22, Euler's method has a dominant error term of $\frac{1}{2}h^2f''(x_0)$. The h^2 term implies that if the step length is reduced by a factor of 2, the local error will be reduced by a factor of 4. The global error however, will only be reduced by a factor of 2 because twice as many of the smaller steps will be required to reach the same value of x, some distance from the initial condition at x_0.

The next method of increased sophistication is the modified Euler method, which is based on the trapezium rule of integration, i.e.

$$y(x_0+h) \simeq y(x_0) + \tfrac{1}{2}h[y'(x_0) + y'(x_0+h)] \tag{7.35}$$

Noting that the second derivative of y at x_0 in finite difference form can be written (see Section 7.4.1)

$$y''(x_0) \simeq \frac{y'(x_0+h) - y'(x_0)}{h} \tag{7.36}$$

we get after substitution into eq. 7.35, the following truncated Taylor series:

$$y(x_0+h) \simeq y(x_0) + hy'(x_0) + \tfrac{1}{2}h^2y''(x_0) \tag{7.37}$$

Hence, the dominant error term of the modified Euler method is of the form $\frac{1}{6}h^3y'''(x_0)$. Halving the step length in this instance will reduce the local error by a factor of 8, but due to the doubling of the number of steps required to reach the required solution, the global error is only reduced by a factor of 4.

In summary, if the dominant error term in a one-step method based on a truncated Taylor series involves h^{k+1}, then the global error at a particular value of x will be approximately proportional to h^k. Although one-step methods on their own do not usually give convenient error estimates, the property described above can be used to estimate the necessary step length to achieve a particular level of accuracy as will be shown in Example 7.5.

A more pragmatic approach to error analysis, and probably the one most often used in practice, is to repeat a particular calculation with a different step size. The sensitivity of the solution to the value of h will often give a good indication of the accuracy of the solution.

Example 7.5

Given that $dy/dx = (x+y)^2$ with $y(0) = 1$, estimate the value of $y(0.5)$ using the modified Euler method with

(a) 5 steps of 0.1

(b) 10 steps of 0.05

Use your solution to estimate the value of h required to obtain a solution accurate to five decimal places.

Solution 7.5

Using Program 7.1
5 steps of $h=0.1$ gives $y(0.5)=2.82541$
10 steps of $h=0.05$ gives $y(0.5)=2.88402$
The global error in this method is proportional to h^2, hence

$$y_{exact}-2.82541=C(0.1)^2$$
$$y_{exact}-2.88402=C(0.05)^2$$

where C is a constant of proportionality.

These equations can be solved for C to give

$$C=7.81477$$

In order for the solution to be accurate to five decimal places, the error must not be greater than 0.000005, hence

$$0.000005=7.81467h^2$$

and $\quad h=0.0008$

A final run of Program 7.1 with 625 steps of $h=0.0008$ gives $y(0.5)=2.90822$ which is accurate to five decimal places.

7.3.1.6 Reduction of high order equations

All the examples considered so far have involved the solution of first order equations. This does not prove to be a restriction to solving higher order equations, because it is easily shown that an nth order differential equation can be broken down into an equivalent system of n first order equations. In addition, if the n conditions required to obtain a particular solution to the nth order equation are all given at the same value of the independent variable, then the resulting set of initial value problems can be solved using the same methods as described previously.

Consider an nth order differential equation arranged so that all terms except the nth derivative term are placed on the right-hand side, thus

$$\frac{d^n y}{dx^n}=f\left(x, y, \frac{dy}{dx}, \frac{d^2y}{dx^2}, \ldots, \frac{d^{n-1}y}{dx^{n-1}}\right) \tag{7.38}$$

with initial conditions given as

$$y(x_0)=A_0, \frac{dy}{dx}(x_0)=A_1, \frac{d^2y}{dx^2}(x_0)=A_2, \ldots \frac{d^{n-1}y}{dx^{n-1}}(x_0)=A_{n-1} \tag{7.39}$$

We now replace all terms (except x) on the right-hand side of eq. 7.38 by simple variables, i.e.

$$\text{let} \quad y=y_0, \frac{dy}{dx}=y_1, \frac{d^2y}{dx^2}=y_2, \ldots \frac{d^{n-1}y}{dx^{n-1}}=y_{n-1} \tag{7.40}$$

and noting that $\frac{dy}{dx}=\frac{dy_0}{dx}, \frac{d^2y}{dx^2}=\frac{dy_1}{dx}, \dots \frac{d^{n-1}y}{dx^{n-1}}=\frac{dy_{n-2}}{dx}$ etc.,

the n first order equations together with their initial condition can be written thus:

$$
\begin{aligned}
&\frac{dy_0}{dx}=y_1 && y_0(x_0)=A_0 \\
&\frac{dy_1}{dx}=y_2 && y_1(x_0)=A_1 \\
&\vdots && \vdots \\
&\frac{dy_{n-2}}{dx}=y_{n-1} && y_{n-2}(x_0)=A_{n-2} \\
&\frac{dy_{n-1}}{dx}=f(x, y_0, y_1, \dots, y_{n-1}) && y_{n-1}(x_0)=A_{n-1}
\end{aligned}
\qquad (7.41)
$$

All the derivatives up to $n-1$ are simply treated as dependent variables. This process of reducing a higher order equation to several first order equations always results in $n-1$ equations with trivial right-hand sides, and one equation closely resembling the original differential equation.

Before attempting to solve any high order differential equations, they must first be arranged in the 'standard form' of eq. 7.41. This may not always be a straightforward process.

Example 7.6

Reduce the following third order equation to standard form:

$$\frac{d^3y}{dx^3}+2\frac{dy}{dx}=2e^x \quad \text{with } y(x_0)=A, \quad y'(x_0)=B, \quad y''(x_0)=C$$

Solution 7.6

Let $y=y_0, \frac{dy}{dx}=y_1, \frac{d^2y}{dx^2}=y_2$

then

$$
\left.
\begin{aligned}
&\frac{dy_0}{dx}=y_1 && y_0(x_0)=A \\
&\frac{dy_1}{dx}=y_2 && y_1(x_0)=B \\
&\frac{dy_2}{dx}=2e^x-2y_1 && y_2(x_0)=C
\end{aligned}
\right\} \text{Standard form}
$$

7.3.1.7 Solution of simultaneous first order equations

In general, a system of n first order equations will be of the form

$$\frac{dy_i}{dx}=f_i(x, y_0, y_1, \ldots, y_{n-1}) \qquad i=0, 1, 2, \ldots, n \tag{7.42}$$

with n initial conditions $y_i(x_0)=A_i$, $i=0, 1, 2, \ldots, n$

Consider the system of two equations given below

$$\begin{aligned} \frac{dy}{dx}&=f(x, y, z) \quad y(x_0)=y_0 \\ \frac{dz}{dx}&=g(x, y, z) \quad z(x_0)=z_0 \end{aligned} \tag{7.43}$$

We may advance the solution of y and z to new values at $x_1=x_0+h$ using any of the one-step or Runge–Kutta methods described previously.

In general our solutions will be advanced using expressions of the form

$$\begin{aligned} y(x_1)&=y(x_0)+K \\ z(x_1)&=z(x_0)+L \end{aligned} \tag{7.44}$$

where the nature of K or L depends on the method being applied. The modified Euler method leads to the expressions

$$K=\tfrac{1}{2}(K_0+K_1) \qquad \text{and} \qquad L=\tfrac{1}{2}(L_0+L_1) \tag{7.45}$$

where

$$\begin{aligned} K_0&=hf(x_0, y_0, z_0) \\ L_0&=hg(x_0, y_0, z_0) \\ K_1&=hf(x_0+h, y_0+K_0, z_0+L_0) \\ L_1&=hg(x_0+h, y_0+K_0, z_0+L_0), \end{aligned} \tag{7.46}$$

the midpoint method leads to the expressions

$$K=K_1 \text{ and } L=L_1 \tag{7.47}$$

where

$$\begin{aligned} K_0&=hf(x_0, y_0, z_0) \\ L_0&=hg(x_0, y_0, z_0) \\ K_1&=hf(x_0+\tfrac{1}{2}h, y_0+\tfrac{1}{2}K_0, z_0+\tfrac{1}{2}L_0) \\ L_1&=hg(x_0+\tfrac{1}{2}h, y_0+\tfrac{1}{2}K_0, z_0+\tfrac{1}{2}L_0), \end{aligned} \tag{7.48}$$

and the fourth order Runge–Kutta method leads to the expressions

$$K=\frac{K_0+2K_1+2K_2+K_3}{6} \qquad \text{and} \qquad L=\frac{L_0+2L_1+2L_2+L_3}{6} \tag{7.49}$$

where

$$
\begin{aligned}
K_0 &= hf(x_0, y_0, z_0)\\
L_0 &= hg(x_0, y_0, z_0)\\
K_1 &= hf(x_0+\tfrac{1}{2}h,\ y_0+\tfrac{1}{2}K_0,\ z_0+\tfrac{1}{2}L_0)\\
L_1 &= hg(x_0+\tfrac{1}{2}h,\ y_0+\tfrac{1}{2}K_0,\ z_0+\tfrac{1}{2}L_0)\\
K_2 &= hf(x_0+\tfrac{1}{2}h,\ y_0+\tfrac{1}{2}K_1,\ z_0+\tfrac{1}{2}L_1)\\
L_2 &= hg(x_0+\tfrac{1}{2}h,\ y_0+\tfrac{1}{2}K_1,\ z_0+\tfrac{1}{2}L_1)\\
K_3 &= hf(x_0+h,\ y_0+K_2,\ z_0+L_2)\\
L_3 &= hg(x_0+h,\ y_0+K_2,\ z_0+L_2)
\end{aligned}
\qquad (7.50)
$$

Example 7.7

Given the equation $d^2y/dx^2 = x - y$ with initial conditions $y(0)=1$ and $dy/dx(0)=0$, estimate $y(0.5)$ using
(a) midpoint rule ($h=0.25$)
(b) fourth order Runge–Kutta ($h=0.5$)

Solution 7.7

Firstly reduce the problem to two first order equations in standard form, i.e.

$$\frac{dy}{dx} = z \qquad y(0)=1$$

$$\frac{dz}{dx} = x - y \qquad z(0)=0$$

A matrix layout is useful for hand calculation.

(a) Midpoint rule ($h=0.25$)

$$\begin{Bmatrix} y \\ z \end{Bmatrix}_{i+\frac{1}{2}} = \begin{Bmatrix} y \\ z \end{Bmatrix}_i + \frac{1}{2}h\begin{Bmatrix} y' \\ z' \end{Bmatrix}_i$$

$$\begin{Bmatrix} y \\ z \end{Bmatrix}_{i+1} = \begin{Bmatrix} y \\ z \end{Bmatrix}_i + h\begin{Bmatrix} y' \\ z' \end{Bmatrix}_{i+\frac{1}{2}}$$

$$\begin{Bmatrix} y \\ z \end{Bmatrix}_{0.125} = \begin{Bmatrix} 1 \\ 0 \end{Bmatrix} + \frac{0.25}{2}\begin{Bmatrix} 0 \\ 0-1 \end{Bmatrix} = \begin{Bmatrix} 1 \\ -0.125 \end{Bmatrix}$$

$$\begin{Bmatrix} y \\ z \end{Bmatrix}_{0.25} = \begin{Bmatrix} 1 \\ 0 \end{Bmatrix} + 0.25\begin{Bmatrix} -0.125 \\ 0.125-1 \end{Bmatrix} = \begin{Bmatrix} 0.969 \\ -0.219 \end{Bmatrix}$$

$$\begin{Bmatrix} y \\ z \end{Bmatrix}_{0.375} = \begin{Bmatrix} 0.969 \\ -0.219 \end{Bmatrix} + \frac{0.25}{2}\begin{Bmatrix} -0.219 \\ 0.25-0.969 \end{Bmatrix} = \begin{Bmatrix} 0.941 \\ -0.309 \end{Bmatrix}$$

$$\begin{Bmatrix} y \\ z \end{Bmatrix}_{0.5} = \begin{Bmatrix} 0.969 \\ -0.219 \end{Bmatrix} + 0.25\begin{Bmatrix} -0.309 \\ 0.375-0.941 \end{Bmatrix} = \begin{Bmatrix} 0.892 \\ -0.360 \end{Bmatrix}$$

therefore $y(0.5)=0.892$

(b) Fourth order Runge–Kutta ($h=0.5$)

$$\frac{dy}{dx}=z=f(x, y, z), \qquad y(0)=1$$

$$\frac{dz}{dx}=x-y=g(x, y, z), \quad z(0)=0$$

$K_0=0.5(0) \quad = \quad 0$

$L_0=0.5(0-1)=-0.5$

$K_1=0.5(0-0.25)=-0.125$

$L_1=0.5[(0+0.25)-(1+0)]=-0.375$

$K_2=0.5(0-0.1875)=-0.094$

$L_2=0.5[(0+0.25)-(1-0.0725)]=-0.344$

$K_3=0.5(0-0.344)=-0.172$

$L_3=0.5[(0+0.5)-(1-0.095)]=-0.203$

therefore

$$\begin{Bmatrix} y \\ z \end{Bmatrix}_{0.5} = \begin{Bmatrix} y \\ z \end{Bmatrix}_{0} + \frac{1}{6}\left[\begin{Bmatrix} K_0 \\ L_0 \end{Bmatrix} + 2\begin{Bmatrix} K_1 \\ L_1 \end{Bmatrix} + 2\begin{Bmatrix} K_2 \\ L_2 \end{Bmatrix} + \begin{Bmatrix} K_3 \\ L_3 \end{Bmatrix}\right]$$

$$= \begin{Bmatrix} 1 \\ 0 \end{Bmatrix} + \frac{1}{6}\left[\begin{Bmatrix} 0 \\ -0.5 \end{Bmatrix} + 2\begin{Bmatrix} -0.125 \\ -0.375 \end{Bmatrix} + 2\begin{Bmatrix} -0.094 \\ -0.344 \end{Bmatrix} + \begin{Bmatrix} -0.172 \\ -0.203 \end{Bmatrix}\right]$$

therefore

$$\begin{Bmatrix} y \\ z \end{Bmatrix}_{0.5} = \begin{Bmatrix} 0.898 \\ -0.357 \end{Bmatrix}$$

$y(0.5) \simeq 0.898$

The exact solution in this case is given by $y=\cos x-\sin x+x$, hence $y(0.5)=0.898$.

Program 7.2. One-step methods for a system of first order equations
This program is an extension of Program 7.1 and allows us to obtain numerical solutions to a system of n equations of the form

$$\begin{aligned} y_1' &= f_1(x, y_1, y_2, \dots, y_n) & y_1(x_0) &= A_1 \\ y_2' &= f_2(x, y_1, y_2, \dots, y_n) & y_2(x_0) &= A_2 \\ &\vdots & &\vdots \\ y'_n &= f_n(x, y_1, y_2, \dots, y_n) & y_n(x_0) &= A_n \end{aligned}$$

```
      PROGRAM P72
C
C      PROGRAM 7.2  ONE-STEP METHODS FOR SYSTEMS OF EQUATIONS
C
C      ITYPE= 1 (EULER'S METHOD)    ITYPE= 2 (MODIFIED EULER'S METHOD)
C      ITYPE= 3 (MID-POINT RULE)    ITYPE= 4 (4TH ORDER RUNGE-KUTTA)
C
C      ALTER NEXT LINE TO CHANGE PROBLEM SIZE
C
      PARAMETER(MEQ=10)
C
      REAL Y(MEQ),Y0(MEQ),K0(MEQ),K1(MEQ),K2(MEQ),K3(MEQ)
C
      READ (5,*) ITYPE,N,NSTEPS,H,X
      READ (5,*) (Y(I),I=1,N)
      WRITE (6,*) ('******* SYSTEMS OF EQUATIONS **********')
      WRITE (6,*)
      GO TO (1,2,3,4) ITYPE
    1 WRITE (6,*) ('************ EULERS METHODS ***********')
      WRITE (6,*)
      WRITE (6,200) N
      DO 10 J = 0,NSTEPS
          WRITE (6,100) X, (Y(I),I=1,N)
          CALL FUNC(K0,Y,X)
          DO 11 I = 1,N
   11         Y(I) = Y(I) + K0(I)*H
   10     X = X + H
      GO TO 50
    2 WRITE (6,*) ('******* MODIFIED EULERS METHOD ********')
      WRITE (6,*)
      WRITE (6,200) N
      DO 20 J = 0,NSTEPS
          WRITE (6,100) X, (Y(I),I=1,N)
          CALL FUNC(K0,Y,X)
          DO 21 I = 1,N
              Y0(I) = Y(I)
   21         Y(I) = Y0(I) + K0(I)*H
          X = X + H
          CALL FUNC(K1,Y,X)
          DO 22 I = 1,N
   22         Y(I) = Y0(I) + .5*H* (K0(I)+K1(I))
   20     CONTINUE
      GO TO 50
    3 WRITE (6,*) ('*********** MID-POINT RULE ************')
      WRITE (6,*)
      WRITE (6,200) N
      DO 30 J = 0,NSTEPS
          WRITE (6,100) X, (Y(I),I=1,N)
          CALL FUNC(K0,Y,X)
          DO 31 I = 1,N
              Y0(I) = Y(I)
   31         Y(I) = Y0(I) + .5*H*K0(I)
          X = X + .5*H
          CALL FUNC(K1,Y,X)
          DO 32 I = 1,N
   32         Y(I) = Y0(I) + H*K1(I)
   30     X = X + .5*H
      GO TO 50
    4 WRITE (6,*) ('******** 4TH ORDER RUNGE-KUTTA ********')
      WRITE (6,*)
      WRITE (6,200) N
      DO 40 J = 0,NSTEPS
          WRITE (6,100) X, (Y(I),I=1,N)
          CALL FUNC(K0,Y,X)
          DO 42 I = 1,N
              Y0(I) = Y(I)
   42         Y(I) = Y0(I) + K0(I)*.5*H
```

```
            X = X + H*.5
            CALL FUNC(K1,Y,X)
            DO 43 I = 1,N
   43           Y(I) = Y0(I) + K1(I)*.5*H
            CALL FUNC(K2,Y,X)
            DO 44 I = 1,N
   44           Y(I) = Y0(I) + K2(I)*H
            X = X + H*.5
            CALL FUNC(K3,Y,X)
            DO 45 I = 1,N
   45           Y(I) = Y0(I) + (K0(I)+2.* (K1(I)+K2(I))+K3(I))/6.*H
   40       CONTINUE
   50 CONTINUE
  100 FORMAT (10E13.5)
  200 FORMAT ('      X              Y(I) , I = 1,',I2)
      STOP
      END
C
      SUBROUTINE FUNC(F,Y,X)
C
C      THIS SUBROUTINE PROVIDES THE VALUES OF F(X,Y) FOR EACH
C      EQUATION AND WILL VARY FROM ONE PROBLEM TO THE NEXT
C
      REAL F(*),Y(*)
C
      F(1) = 3.*x*y(2)+4.
      F(2) = x*y(1)-y(2)-exp(x)
      RETURN
      END
```

The same four methods available in Program 7.1 can be used, and the choice is left to the user through the input parameter ITYPE. The number of equations to be solved is read into the variable N. The actual form of the equations $f_i(x, y_1, y_2 \dots y_n)$ for $i=1, 2, \dots n$ is evaluated by subroutine FUNC which will be changed by the user from one problem to the next. It is readily seen that Program 7.1 is a special case of Program 7.2 with N=1.

With the exception of simple integer counters, the variables can be summarised thus

ITYPE	Defines method to be used (see main program)
N	Number of equations
NSTEPS	Number of steps
H	Step length
X	Initial and updated values of X
Y	1-d array holding initial and updated values of dependent variables
F	1-d array holding function values
Y_0, K_0, K_1, K_2, K_3	Working space for advancing the solution by one step

PARAMETER *restriction* MEQ ≥ N.

To illustrate use of the program, the following problem is to be solved using the fourth order Runge–Kutta method. Given that

$$\frac{dy}{dx}=3xz+4, \qquad y(0)=4$$

$$\frac{dz}{dx} = xy - z - e^x, \qquad z(0) = 1$$

estimate $y(0.5)$ and $z(0.5)$ using five steps of 0.1.

The input data and output from Program 7.2 are given in Figs 7.5(a) and (b) respectively. The functions representing dy/dx and dz/dx have been written into user-supplied subroutine FUNC at the foot of the main program.

As shown in Fig. 7.5(b) the computed results $y(0.5) \simeq 6.2494$ and $z(0.5) \simeq 0.6739$ are obtained.

One-step method to be used	ITYPE 4	
Number of equations	N 2	
Number and size of steps	NSTEPS 5	H 0.1
Initial value of x	X 0.0	
Initial values of dependent variables	Y(I), I=1, N 4.0	 1.0

Fig. 7.5(a) Data for Program 7.2

```
******* SYSTEMS OF EQUATIONS **********
******** 4TH ORDER RUNGE-KUTTA ********
      X             Y(I) , I = 1, 2
  .00000E+00    .40000E+01    .10000E+01
  .10000E+00    .44132E+01    .82536E+00
  .20000E+00    .48471E+01    .70295E+00
  .30000E+00    .52967E+01    .63519E+00
  .40000E+00    .57623E+01    .62459E+00
  .50000E+00    .62494E+01    .67386E+00
```

Fig. 7.5(b) Results from Program 7.2

7.3.1.8 θ-Methods for linear equations

All of the methods described so far for numerical solution of initial value problems have been suitable for both linear or nonlinear equations. If the differential equation is linear, a slightly different one-step approach is possible involving linear interpolation of derivatives between the beginning and end of each step.

Consider the first order linear equation

$$y' = f(x, y) = l(x)y + k(x) \quad \text{with } y(x_0) = y_0 \tag{7.51}$$

where $l(x)$ and $k(x)$ are functions of x.

Writing the differential equation at x_0 and x_1, distance h apart, and using an abbreviated notation whereby $l(x_i) = l_i$ and $k(x_i) = k_i$, we get

$$y_0' = l_0 y_0 + k_0 \tag{7.52}$$

$$y_1' = l_1 y_1 + k_1 \tag{7.53}$$

We now introduce the parameter θ which can be varied in the range

$$0 \leq \theta \leq 1 \tag{7.54}$$

and write our one-step method as follows:

$$y_1 = y_0 + h[(1-\theta)y_0' + \theta y_1'] \tag{7.55}$$

The parameter θ acts like a weighting coefficient on the gradients at the beginning and end of the step. When $\theta = 0$, the simple Euler method is obtained, where only the gradient at x_0 is included whereas when $\theta = 1$ only the gradient at x_1 is taken into account.

The most popular choice is $\theta = 0.5$, which gives equal weights to the gradients at x_0 and x_1, and is equivalent to the trapezium rule of integration. In the solution of time-dependent systems of differential equations, the use of $\theta = 0.5$ is sometimes referred to as the 'Crank–Nicolson' method. These θ methods are popular because they can be used even when extra coupling of the derivative terms means that the equations cannot easily be reduced to the 'standard form' of eqs 7.41.

As the differential equation is linear, eqs 7.52 and 7.53 can be substituted into 7.55 and rearranged to give

$$y_1 = \frac{y_0 + h[(1-\theta)(l_0 y_0 + k_0) + \theta k_1]}{1 - h\theta l_1} \tag{7.56}$$

which for a single equation is an explicit formula for y_1 in terms of y_0 and the functions l and k. For systems of equations the denominator can become a matrix in which case the system is 'implicit'.

Example 7.8

Given the differential equation

$y' = y + x$ with $y(0) = 0$

estimate $y(0.5)$ using eq. 7.56 with $\theta = 0.5$. Let $h = 0.25$.

Solution 7.8

$$y(0.25) = \frac{0 + 0.25[0.5(0+0) + 0.5(0.25)]}{1 - 0.25(0.5)(1)}$$

$$= 0.036$$

$$y(0.5) = \frac{0.036 + 0.25[0.5(0.036 + 0.25) + 0.5(0.5)]}{1 - 0.25(0.5)(1)}$$

$$= 0.148$$

(cf. exact solution 0.149)

Second order linear equations can also be solved by linear interpolation using the parameter θ.

Given the equation

$$y'' = m(x)y' + l(x)y + k(x) \tag{7.57}$$

with initial conditions $y(x_0) = y_0$, $y'(x_0) = y_0'$, the equation can be written at x_0 and x_1, distance h apart, to give

$$y_0'' = m_0 y_0' + l_0 y_0 + k_0 \tag{7.58}$$

$$y_1'' = m_1 y_1' + l_1 y_1 + k_1 \tag{7.59}$$

We now obtain the following expressions for y_1 and y_1' using θ to 'weight' the derivatives at x_0 and x_1, hence

$$y_1 = y_0 + h[(1-\theta)y_0' + \theta y_1'] \tag{7.60}$$

$$y_1' = y_0' + h[(1-\theta)y_0'' + \theta y_1''] \tag{7.61}$$

Elimination of y_0'', y_1' and y_1'' from eqs 7.58 to 7.61 leads to the following explicit formula for y_1:

$$y_1 = \frac{y_0(1-h\theta m_1) + hy_0'[1-h\theta m_1(1-\theta)] + h^2\theta[(1-\theta)(m_0 y_0' + l_0 y_0 + k_0) + \theta k_1]}{1 - h\theta m_1 - h^2\theta^2 l_1} \tag{7.62}$$

Having obtained y_1, an expression for the first derivative follows from eq. 7.60, hence

$$y_1' = \frac{y_1 - y_0}{h\theta} - \frac{(1-\theta)}{\theta} y_0' \tag{7.63}$$

A value of $\theta = \frac{1}{2}$ remains the most logical choice as it gives equal weight to derivatives at each end of the step. It may also be noted from eq. 7.63 that the method would fail for $\theta = 0$.

For a single second order equation, the fourth order Runge–Kutta method is considerably more accurate than the trapezium rule with $\theta = \frac{1}{2}$ described in this section. However for large engineering systems involving equations with coupled derivatives, the linear interpolation methods involving θ are still frequently used because of their simplicity.

7.3.2 Predictor–corrector methods

Predictor–corrector methods use information from several previous known points to compute the next as indicated in Fig. 7.6. A disadvantage of the methods is that they are not self-starting and may often need a one-step method to generate a few points in order to get started. The attraction of the methods is that more efficient use is made of existing information in order to advance to the next step. This is in contrast to the fourth order Runge–Kutta method where at each step, four function evaluations are required which are never used again.

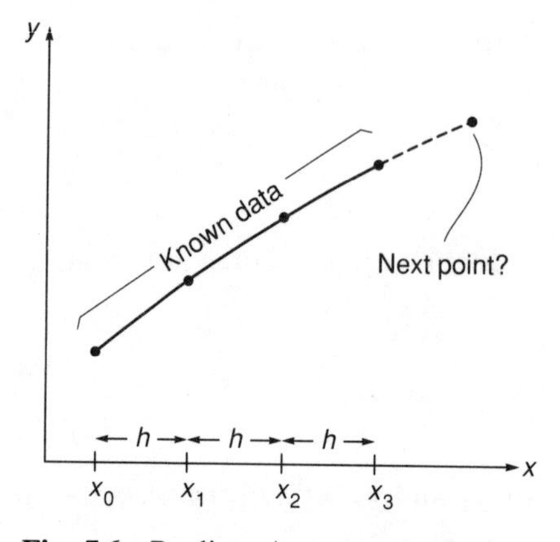

Fig. 7.6 Predictor/corrector methods

Predictor–corrector methods make use of two formulae; the predictor formula extrapolates existing data to estimate the next point and the corrector formula improves on this estimate. In some cases the corrector formula can be applied repeatedly until some convergence criterion is satisfied.

The flowchart in Fig. 7.7 describes a typical predictor–corrector algorithm.

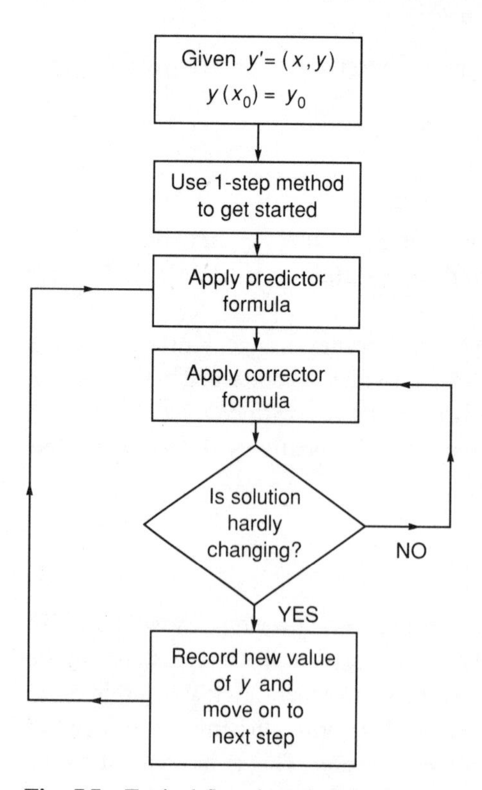

Fig. 7.7 Typical flowchart for predictor-corrector methods

Predictor formulae estimate the new value of y_{i+1} by integrating under the curve of y' vs. x using sampling points at x_i, x_{i-1}, x_{i-2} etc. Any numerical integration formula which does not require a prior estimate of y'_{i+1} is suitable for use as a predictor. Formulae of this type have already been discussed in Section 6.4.6, where the sampling points are outside the range of integration.

Corrector formulae improve on the predicted value of y_{i+1} by integrating under the curve of y' vs. x using more conventional sampling points at x_{i+1}, x_i, x_{i-1} etc. Note that the corrector formula is able to sample at x_{i+1} because a value of y'_{i+1} is now available from the predictor stage. Any numerical integration formula which requires a prior estimate of y'_{i+1} is suitable for use as a corrector.

7.3.2.1 Euler–trapezoidal method

This is the simplest predictor–corrector method, and is equivalent to the modified Euler method described under 'one-step methods'. The only difference is that the trapezoidal corrector formula can be applied several times in order to refine the solution.

Euler's method is the predictor, and is applied once, followed by several applications of the trapezoidal corrector, i.e.,

given $\quad y'=f(x, y) \quad$ with $y(x_0)=y_0$ (7.64)

Euler's method (predictor) $\quad y_1^{(0)}=y_0+hf(x_0, y_0)$ (7.65)

Trapezoidal method (corrector) $\quad y_1^{(1)}=y_0+\tfrac{1}{2}h[f(x_0, y_0)+f(x_1, y_1^{(0)})]$ (7.66)

$$y_1^{(2)}=y_0+\tfrac{1}{2}h[f(x_0, y_0)+f(x_1, y_1^{(1)})] \qquad (7.67)$$

etc.

where $\quad x_1=x_0+h$ (7.68)

Example 7.9

Given $\quad y'=2x^2+2y \quad$ with $y(0)=1$

estimate $y(0.2)$ using the Euler/trapezoidal predictor–corrector method. Let $h=0.2$.

Solution 7.9

Predictor $\quad y(0.2)\simeq 1+0.2(2)=1.4$

Corrector $\quad y(0.2)\simeq 1+0.1[2+2(0.2)^2+2(1.4)]=1.488$

$$y(0.2)\simeq 1+0.1[2+2(0.2)^2+2(1.488)]=1.506$$

$$y(0.2)\simeq 1+0.1[2+2(0.2)^2+2(1.506)]=1.509$$

$$y(0.2)\simeq 1+0.1[2+2(0.2)^2+2(1.509)]=1.510$$

Although 'convergence' has been achieved, the result has not converged on the exact solution of 1.498. A difficulty with repeated corrector applications, especially in a low order method such as this, is that the result tends to converge on the finite difference approximation to the differential equation rather than the actual differential equation.

7.3.2.2 Milne's method

This method uses a formula due to Milne as a predictor, and the familiar Simpson's rule as a corrector. The method is fourth order, i.e. the dominant error term in both the predictor and corrector includes h^5, and requires four initial values of y to get started. Note the smaller error term associated with the corrector formula as compared with the predictor.

Milne's formula (predictor)

$$y_{i+1}{}^{(0)} = y_{i-3} + \frac{4h}{3}[2f(x_i, y_i) - f(x_{i-1}, y_{i-1}) + 2f(x_{i-2}, y_{i-2})] + \frac{28}{90}h^5 y^{(v)}(\xi) \tag{7.69}$$

Simpson's rule (corrector)

$$y_{i+1}{}^{(1)} = y_{i-1} + \frac{h}{3}[f(x_{i-1}, y_{i-1}) + 4f(x_i, y_i) + f(x_{i+1}, y_{i+1}{}^{(0)})] - \frac{1}{90}h^5 y^{(v)}(\xi) \tag{7.70}$$

$$y_{i+1}{}^{(2)} = y_{i-1} + \frac{h}{3}[f(x_{i-1}, y_{i-1}) + 4f(x_i, y_i) + f(x_{i+1}, y_{i+1}{}^{(1)})] \text{ etc.} \tag{7.71}$$

Milne's predictor integrates under the curve of y' vs. x between limits of x_{i-3} and x_{i+1} using three centrally placed sampling points as shown in Fig. 7.8.

A danger in using Milne's method, or indeed any method which uses Simpson's rule

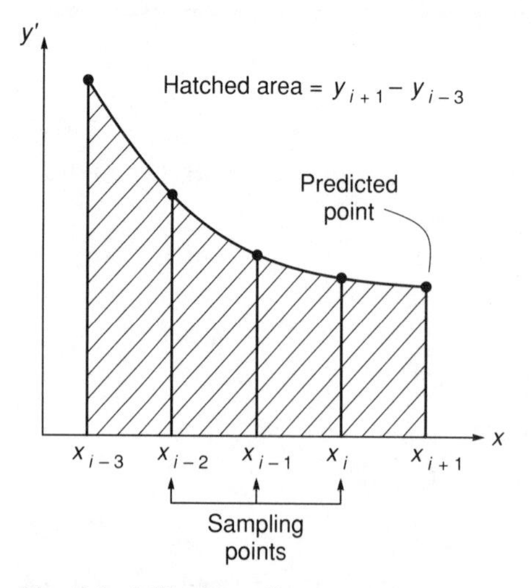

Fig. 7.8 Milne's predictor

as a corrector, is that errors generated at one stage of the calculation may subsequently grow in magnitude (see Section 7.3.4). For this reason other fourth order methods have tended to be more popular.

7.3.2.3 Adams–Moulton–Bashforth method

A more stable fourth order method, in which errors do not tend to grow so fast is based on the Adams–Bashforth predictor, together with an Adams–Moulton corrector, i.e.,

Adams–Bashforth predictor

$$y_{i+1}^{(0)}=y_i+\frac{h}{24}[55f(x_i, y_i)-59f(x_{i-1}, y_{i-1})+37f(x_{i-2}, y_{i-2})-9f(x_{i-3}, y_{i-3})]$$

$$+\frac{251}{720}h^5y^{(v)}(\xi) \tag{7.72}$$

Adams–Moulton corrector

$$y_{i+1}^{(1)}=y_i+\frac{h}{24}[f(x_{i-2}, y_{i-2})-5f(x_{i-1}, y_{i-1})+19f(x_i, y_i)+9f(x_{i+1}, y_{i+1}^{(0)})]$$

$$-\frac{19}{720}h^5y^{(v)}(\xi) \tag{7.73}$$

$$y_{i+1}^{(2)}=y_i+\frac{h}{24}[f(x_{i-2}, y_{i-2})-5f(x_{i-1}, y_{i-1})+19f(x_i, y_i)+9f(x_{i+1}, y_{i+1}^{(1)})] \tag{7.74}$$

etc.

The improved stability is obtained at a cost of somewhat larger error terms than in Milne's method, although the dominant terms still indicate that the corrector is considerably more accurate than the predictor.

The Adams–Bashforth predictor integrates under the curve of y' vs. x between limits of x_i and x_{i+1}, using sampling points shifted to the left of the range of integration as shown in Fig. 7.9.

Example 7.10

Repeat Example 7.9 using the Adams–Moulton–Bashforth method

Given $y'=2x^2+2y$

with $y(-0.6)=0.1918$

$y(-0.4)=0.4140$

$y(-0.2)=0.6655$

$y(0) \quad =1.0000$

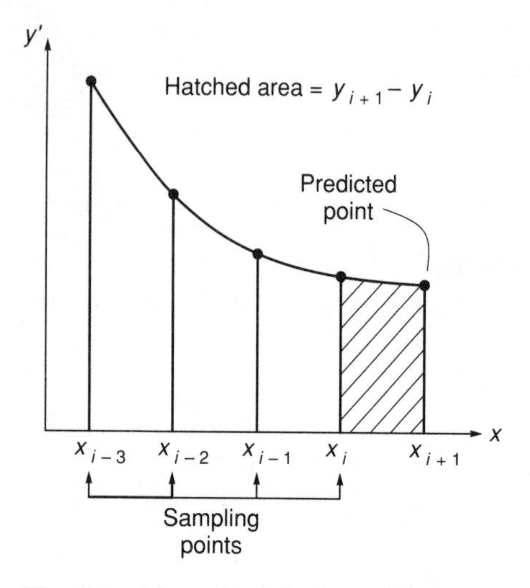

Fig. 7.9 Adams–Bashforth predictor

Solution 7.10

Predictor $y(0.2)=1+\frac{0.2}{24}\{55(2)-59[2(0.2)^2+2(0.6655)]$

$+37[2(0.4)^2+2(0.4140)]-9[2(0.6)^2+2(0.1918)]\}=1.4941$

Corrector $y(0.2)=1+\frac{0.2}{24}\{2(0.4)^2+2(0.4140)-5[2(0.2)^2+2(0.6655)]+19(2)$

$+9[2(0.2)^2+2(1.4941)]\}=1.4976$

$y(0.2)=1.4981$ (2nd iteration)

$y(0.2)=1.4982$ (3rd iteration)

(cf. exact solution 1.498)

7.3.2.4 Accuracy of predictor–corrector methods

An attractive feature of predictor–corrector methods is that they lead to a simple method for estimating the error in the computed solution after the first application of the corrector formula.

The estimate is based on the error terms associated with the predictor and corrector formulae. Let $y_{i+1}{}^{(0)}$ and $y_{i+1}{}^{(1)}$ represent the value of y_{i+1} computed by a predictor and the first application of a corrector respectively. If $y(x_{i+1})$ is the exact solution, then by Milne's method for example we can write

$$y(x_{i+1})-y_{i+1}{}^{(0)}=\tfrac{28}{90}h^5y^{(v)}(\xi_1) \tag{7.75}$$

and

$$y(x_{i+1}) - y_{i+1}^{(1)} = -\tfrac{1}{90} h^5 y^{(v)}(\xi_2) \tag{7.76}$$

Assuming that $\quad y^{(v)}(\xi_1) = y^{(v)}(\xi_2),$ (7.77)

then eqs 7.75 and 7.76 lead to

$$h^5 y^{(v)}(\xi) \simeq \tfrac{90}{29}(y_{i+1}^{(1)} - y_{i+1}^{(0)}) \tag{7.78}$$

and after substitution into 7.76 we get

$$y(x_{i+1}) - y_{i+1}^{(1)} \simeq -\tfrac{1}{29}(y_{i+1}^{(1)} - y_{i+1}^{(0)}) \tag{7.79}$$

Similar operation on the Adams–Moulton–Bashforth method leads to eq. 7.80

$$y(x_{i+1}) - y_{i+1}^{(1)} \simeq -\tfrac{1}{14}(y_{i+1}^{(1)} - y_{i+1}^{(0)}) \tag{7.80}$$

Equations 7.79 and 7.80 indicate that the error in the first corrected value of y_{i+1} is proportional to the difference between the predicted and corrected values. Subsequent applications of the corrector are therefore only of marginal value as they give no further information on the error size. It is recommended that when using Program 7.3, the tolerance is set to the required accuracy, and the step length reduced until convergence is achieved in one iteration.

The ability to estimate the error at each step of the solution process means that the step length can be adjusted to give the desired level of accuracy. If required, a program could take account of the error, and adjust the step size accordingly. If the error was too big, one strategy would be to keep halving the step size until the accuracy criterion was met. At a subsequent stage of the calculation, the previously obtained step length might become excessively small, in which case it could be increased systematically. It should be noted however, that if the step length is changed during the calculation, it will be necessary to recall some one-step starting procedure in order to generate enough points at the new step length for the predictor–corrector algorithm to proceed.

Program 7.3. Predictor–corrector using Adams–Moulton–Bashforth fourth order method
This program obtains a numerical solution to the initial value problem

$y' = f(x, y)$ with $y(x_0) = y_0$

using the fourth order Adams–Moulton–Bashforth predictor–corrector method. The method requires four initial values of x and y in order to proceed. The first piece of data read by the program is the integer NIC. If NIC=4 then the four data points should be read immediately. Alternatively, if NIC=1, only one initial condition should be provided, and the remaining three are generated automatically using the fourth order Runge–Kutta one-step method.

Once the four initial conditions have been either read or generated, the predictor and corrector formulae are applied. After the first application of the corrector formula an error

```
      PROGRAM P73
C
C      PROGRAM 7.3  PREDICTOR CORRECTOR METHOD USING
C                   4TH ORDER ADAMS/BASHFORTH/MOULTON
C
C      FOUR INITIAL VALUES OF X AND Y PROVIDED (NIC=4)
C
C      ONE INITIAL VALUE PROVIDED AND THE REMAINING THREE
C      GENERATED USING 4TH ORDER RUNGE-KUTTA (NIC=1)
C
      REAL X(5),Y(5)
C
      READ (5,*) NIC,NSTEPS,H,TOL
      WRITE (6,*) ('********* PREDICTOR CORRECTOR METHOD *********')
      WRITE (6,*)
      WRITE (6,*) ('********** ADAMS/BASHFORTH/MOULTON ***********')
      WRITE (6,*)
      IF (NIC.EQ.4) THEN
          WRITE (6,*) ('********** 4 INITIAL VALUES PROVIDED *********')
          WRITE (6,*)
          WRITE (6,*) (
     +       '     X              Y              ERR       ITERATIONS')
          DO 10 I = 1,4
              READ (5,*) X(I),Y(I)
   10     WRITE (6,100) X(I),Y(I)
      ELSE
          WRITE (6,*) ('********** 1 INITIAL VALUE PROVIDED **********')
          WRITE (6,*) ('*** REMAINING 3 FROM 4TH ORDER RUNGE-KUTTA ***')
          WRITE (6,*)
          WRITE (6,*) (
     +       '     X              Y              ERR       ITERATIONS')
          READ (5,*) X(1),Y(1)
          WRITE (6,100) X(1),Y(1)
          DO 20 I = 1,3
              RK0 = H*F(X(I),Y(I))
              RK1 = H*F(X(I)+0.5*H,Y(I)+0.5*RK0)
              RK2 = H*F(X(I)+0.5*H,Y(I)+0.5*RK1)
              RK3 = H*F(X(I)+H,Y(I)+RK2)
              X(I+1) = X(I) + H
              Y(I+1) = Y(I) + (RK0+2.*RK1+2.*RK2+RK3)/6.
   20     WRITE (6,100) X(I+1),Y(I+1)
      END IF
      DO 30 I = 1,NSTEPS
          Y(5) = Y(4) + H/24.* (55.*F(X(4),Y(4))-59.*F(X(3),Y(3))+
     +           37.*F(X(2),Y(2))-9.*F(X(1),Y(1)))
          YHOLD = Y(5)
          X(5) = X(4) + H
          ITERS = 0
   40     ITERS = ITERS + 1
          Y(5) = Y(4) + H/24.* (9.*F(X(5),Y(5))+19.*F(X(4),Y(4))-
     +           5.*F(X(3),Y(3))+F(X(2),Y(2)))
          IF (ITERS.EQ.1) ERR = -1./14.* (Y(5)-YHOLD)
          IF (ABS((Y(5)-YHOLD)/YHOLD).LE.TOL) GO TO 50
          YHOLD = Y(5)
          GO TO 40
   50     WRITE (6,100) X(5),Y(5),ERR,ITERS
          DO 60 J = 1,4
              Y(J) = Y(J+1)
   60     X(J) = X(J+1)
   30 CONTINUE
  100 FORMAT (3E13.5,I8)
      STOP
      END
C
      FUNCTION F(X,Y)
C
```

```
C        THIS FUNCTION PROVIDES THE VALUE OF F(X,Y))
C        AND WILL VARY FROM ONE PROBLEM TO THE NEXT
C
       F = X*X + Y*Y
       RETURN
       END
```

estimate is made from eq. 7.80. If required, the corrector is applied repeatedly until the change in y_{i+1} from one iteration to the next is less than a tolerance TOL read in as data. The convergence criterion is dimensionless, and can be stated as follows:

$$\text{if} \quad \left| \frac{y_{i+1}^{(\text{new})} - y_{i+1}^{(\text{old})}}{y_{i+1}^{(\text{old})}} \right| < \text{TOL}$$

then the value of y_{i+1} is printed, together with the number of iterations required to achieve convergence. Also printed, is the error estimate after the first application of the corrector. It is recommended that once a value of TOL is decided upon, the value of the step length H should be reduced until convergence is achieved with a single application of the corrector.

With the exception of simple integer counters, the variables can be summarised as follows:

NIC	Number of initial conditions provided
NSTEPS	Number of steps (predictor/corrector stage)
H	Step length
TOL	Convergence tolerance
RK0, RK1, RK2, RK3	Intermediate values of $hf(x, y)$ for fourth order Runge–Kutta
ITERS	Iteration counter
ERR	Error estimates after first corrector application
YHOLD	Old value of y_{i+1}
X, Y	1-d arrays holding 5 values of x and y from $i-3$ to $i+1$

The function $f(x, y)$ is evaluated by user-supplied function F which will be changed by the user from one problem to the next.

Number of initial conditions	NIC 1	
Number* and size of steps	NSTEPS 7	H 0.05
Convergence tolerance	TOL 0.0001	
Initial value(s) of x and y	X 0.0	Y 0.0

*Number of steps during predictor–corrector stage only

Fig. 7.10(a) Data for Program 7.3

```
********* PREDICTOR CORRECTOR METHOD *********
********** ADAMS/BASHFORTH/MOULTON ***********
********** 1 INITIAL VALUE PROVIDED **********
*** REMAINING 3 FROM 4TH ORDER RUNGE-KUTTA ***
        X              Y              ERR         ITERATIONS
   .00000E+00    .00000E+00
   .50000E-01    .41667E-04
   .10000E+00    .33333E-03
   .15000E+00    .11250E-02
   .20000E+00    .26669E-02   -.36421E-08           1
   .25000E+00    .52093E-02   -.78165E-08           1
   .30000E+00    .90035E-02   -.13770E-07           1
   .35000E+00    .14302E-01   -.21287E-07           1
   .40000E+00    .21359E-01   -.30734E-07           1
   .45000E+00    .30435E-01   -.42043E-07           1
   .50000E+00    .41791E-01   -.55613E-07           1
```

Fig. 7.10(b) Results from Program 7.3

To illustrate use of the program, the following problem is to be solved. Given $y' = x^2 + y^2$ with $y(0) = 0$, estimate $y(0.5)$ using the fourth order Adams–Moulton–Bashforth method with $h = 0.05$.

As only one initial value is provided, NIC is set to 1 and the three other points are generated using the Runge–Kutta method. The input data and output from Program 7.3 are given in Figs 7.10(a) and 7.10(b) respectively. The function $x^2 + y^2$ has been written into function F at the foot of the main program. As shown in Fig. 7.10(b), with a tolerance value of TOL = 0.0001 the result $y(0.5) = 0.04179$ was achieved. As only 1 iteration was required at each step, no iterations on the corrector were necessary, and the error estimate ERR applies directly to the computed result. In this case, the error estimate on $y(0.5)$ is -0.558×10^{-7}.

7.3.3 Stiff equations

Certain types of differential equations do not lend themselves to numerical solution by the techniques described so far in this chapter. Problems can occur if the solution to the system of equations contains components with widely different 'time scales'.

For example, the solution to a second order differential equation might be of the form

$$y(x) = C_1 e^{-x} + C_2 e^{-100x} \tag{7.81}$$

where the second term decays very much more rapidly than the first. Such a system of equation is said to be 'stiff' and solutions are unreliable when treated by traditional methods. Any stepping method used to tackle such a problem numerically must have a step length small enough to account for the 'fastest-changing' component of the solution, and this step size must be maintained even after the 'fast' component has died out. As has been discussed previously, very small step lengths can have disadvantages from efficiency and accuracy consideration.

7.3.4 Error propagation – numerical stability

All numerical methods are approximate in nature, and some of the sources of these errors were discussed in Chapter 1.

When solving differential equations by repetitive algorithms where errors introduced at one stage are carried on into subsequent calculations, the question arises as to whether these errors will propagate with increased magnitude or remain within acceptable limits.

An 'unstable' process in the context of numerical methods is one in which a small perturbation introduced into the calculation will grow spontaneously. Several sources of 'instability' can occur.

One source of 'instability' can be in the differential equation itself rather than the numerical method being used to solve it. For example, the second order equation

$$y''-3y'-10y=0 \tag{7.82}$$

with

$$y(0)=1 \quad \text{and} \quad y'(0)=-2$$

has the exact solution $y=e^{-2x}$, which decays as x increases.

If a small perturbation ϵ is introduced into one of the initial conditions, i.e.

$$y(0)=1+\epsilon \tag{7.83}$$

then the new solution is given by

$$y=e^{-2x}+\epsilon\left(\tfrac{5}{7}e^{-2x}+\tfrac{2}{7}e^{5x}\right) \tag{7.84}$$

which tends to infinity as x becomes very large.

The small perturbation has caused a huge change in the solution for large x, hence this differential equation is said to be unstable. When tackling 'unstable' problems such as this by numerical methods, any of the sources of error described in Chapter 1 could contribute to an initial perturbation of this type and lead to completely erroneous solutions.

Another source of instability in the solution of differential equations can come from the difference formula itself. As mentioned earlier in this chapter, Simpson's corrector formula (7.70) can lead to instability. It can be shown that this formula does not cope well with differential equations of the form

$$y'=cy \qquad \text{where } c \text{ is negative} \tag{7.85}$$

In these cases, a spurious solution causes the errors to grow exponentially. Spurious solutions can occur whenever the order of the difference formula is higher than that of the differential equation being solved.

In some methods the spurious solutions can cause problems, whereas in others their effects die away. For example, the Adams–Moulton–Bashforth family of predictor–corrector formulae do not suffer from instability problems.

The interested reader is referred to Section 7.6 for further reading on this topic.

7.3.5 Concluding remarks on initial value problems

No single method for obtaining numerical solutions to initial value problems can be recommended for all occasions. Fourth order methods are to be recommended on the grounds of accuracy without undue complexity. The question still remains as to whether to use one-step or predictor–corrector methods.

The following points should be considered before deciding on a particular method.

(a) Predictor–corrector methods are not self-starting, and must actually rely on a one-step method in order to generate enough points to get started. If a change in step size is made during the solution process, a temporary reversion to the one-step method is usually required. Changes in the step size are very easy to implement in a one-step method.

(b) One-step methods are of comparable accuracy to predictor–corrector methods of the same order. However, the predictor–corrector methods provide a simple error estimate at each step, enabling the step size to be adjusted for maximum accuracy and efficiency. No such estimate is usually available with a one-step method, hence the step size h is often made smaller than necessary to be on the 'safe side'.

(c) To advance one step using the fourth order Runge–Kutta method requires four evaluations of the function $f(x, y)$ which are used just once and then discarded. To advance one step using a fourth order predictor–corrector method usually requires just one or two function evaluations for convergence. In addition, function evaluations in a predictor–corrector method can be stored and used again, as only whole intervals of x are required.

The fourth order methods of Runge–Kutta and Adams–Moulton–Bashforth are the preferred one-step and predictor–corrector approaches that have been described in this chapter. The fourth order Runge–Kutta method has the great advantage of simplicity, and for this reason alone is probably to be recommended for most engineering applications.

Many methods exist for numerical solution of differential equations, including newer hybrid methods which take advantage of both the simplicity of the one-step methods and the simple error-estimates of the predictor–corrector methods. These refinements are only obtained at the cost of greater complexity and more function evaluations per step. The reader is referred to more advanced texts on numerical analysis to learn of such developments.

7.4 Boundary value problems

When we attempt to solve ordinary differential equations of second order or higher with information provided at different values of the independent variable, we must usually resort to different methods of solution from those described in the previous section.

The initial value problems covered previously often involved 'time' as the independent variable, and solution techniques required us to 'march' along in steps until the required solution was reached. The domain of the solution in such cases is not finite, because in principle we can step along indefinitely in either the positive or negative direction.

Boundary value problems usually involve a domain of solution which is finite, and we wish to find solutions to the equation within that domain. The independent variable in such problems is usually a coordinate measuring distance in space. A typical second order boundary value problem might be of the form

$$y''=f(x, y, y') \tag{7.86}$$

with $y(A)=y_A$ and $y(B)=y_B$.

The domain of the solution, assuming $B>A$, is given by values of x in the range $A \leqslant x \leqslant B$ and we are interested in finding the values of y corresponding to this range of x.

Numerical solution methods that will be considered in this chapter fall into three categories:

(a) Techniques that replace the original differential equation by its finite difference equivalent.

(b) 'Shooting methods', which attempt to replace the boundary value problem by an equivalent initial value problem.

(c) Methods of 'weighted residuals', where a trial solution satisfying the boundary conditions is 'guessed', and certain parameters within that solution adjusted in order to minimise errors.

7.4.1 Finite difference methods

The finite difference method replaces the derivative terms in the differential equation by their 'finite difference' equivalents. As will be shown, if the differential equation is linear this process leads to a system of linear simultaneous equations, which can be solved using the methods described in Chapter 2.

Initially, we need to define some finite difference approximations to regularly encountered derivatives. Consider the solution curve in Fig. 7.11 in which the x-axis is subdivided into regular grid points, distance h apart.

We may express the first derivative of y with respect to x evaluated at x_i, in any of the following ways:

$$y_i' \simeq \frac{y_{i+1}-y_i}{h} \quad \text{'forward' difference} \tag{7.87}$$

$$y_i' \simeq \frac{y_i-y_{i-1}}{h} \quad \text{'backward' difference} \tag{7.88}$$

$$y_i' \simeq \frac{y_{i+1}-y_{i-1}}{2h} \quad \text{'central' difference} \tag{7.89}$$

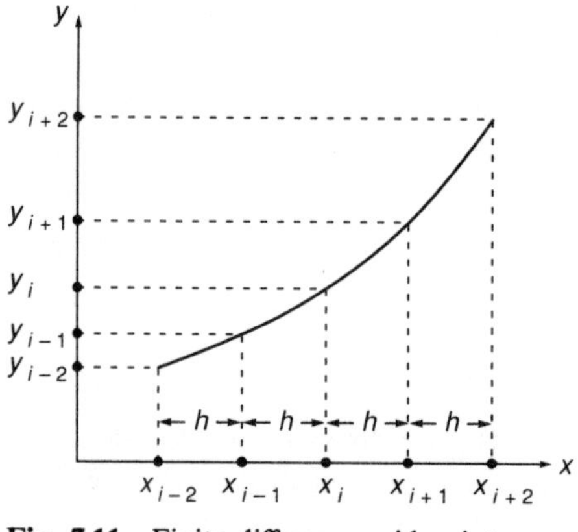

Fig. 7.11 Finite difference grid points

It appears that the required derivative is approximated by the slope of a straight line joining points in the vicinity of the function at x_i. For smooth functions, the central difference form is likely to be the most accurate, being the average of the forward and backward difference versions. The second derivative at x_i can be estimated by taking the difference of first derivatives, i.e.

$$y_i'' \simeq \frac{y_i' - y'_{i-1}}{h}$$

$$= \frac{\dfrac{y_{i+1} - y_i}{h} - \dfrac{y_i - y_{i-1}}{h}}{h} \tag{7.90}$$

$$y_i'' \simeq \frac{y_{i+1} - 2y_i + y_{i-1}}{h^2} \tag{7.91}$$

which is a central difference formula, being symmetrical about x_i.

Similarly a central difference formula for the third derivative at x_i is given by

$$y_i''' \simeq \frac{y''_{i+1} - y''_{i-1}}{2h}$$

$$= \frac{-y_{i-2} + 2y_{i-1} - 2y_{i+1} + y_{i+2}}{2h^3} \tag{7.92}$$

and for the fourth derivative at x_i by

$$y^{\text{iv}}{}_i \simeq \frac{y''_{i-1} - 2y''_i + y''_{i+1}}{h^2}$$

$$= \frac{y_{i-2} - 4y_{i-1} + 6y_i - 4y_{i+1} + y_{i+2}}{h^4} \tag{7.93}$$

Forward and backward difference versions of these higher derivatives are also readily obtained, and by including more points in the difference formulae, greater accuracy can be achieved. A summary of the coefficients for forward, central and backward difference formulae, was presented in Section 5.3.

Backward difference formulae are simply mirror images of the forward difference formulae, except for odd numbered derivatives (i.e., y', y''') where the signs must be reversed.

For example, a four-point forward difference formula for a first derivative at x_0 from Table 5.2 is given by

$$y'_0=\frac{1}{6h}(-11y_0+18y_1-9y_2+2y_3)-\tfrac{1}{4}h^3y^{iv} \tag{7.94}$$

and the corresponding backward difference formula from Table 5.4 by

$$y'_0=\frac{1}{6h}(11y_0-18y_{-1}+9y_{-2}-2y_{-3})+\tfrac{1}{4}h^3y^{iv} \tag{7.95}$$

The solution to a 'two-point' boundary value problem such as that given in eq. 7.86 involves splitting the range of x for which a solution is required into n equal parts, each of width h. Figure 7.12 gives an example of this with $n=4$.

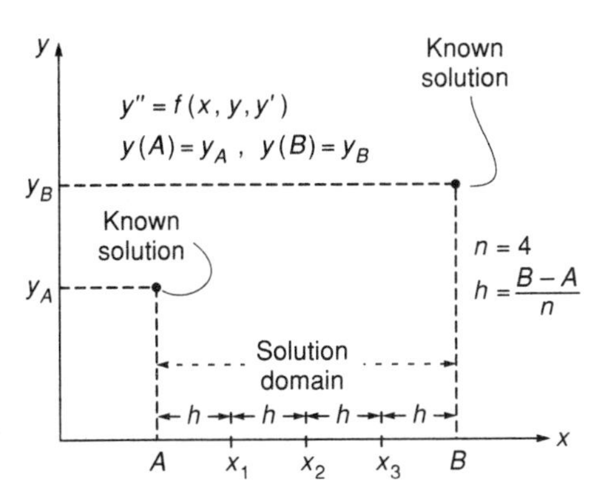

Fig. 7.12 Two-point boundary problem (n = 4)

Hence, if the boundary conditions are given at

$x=A$ and $x=B$,

let $x_i=x_0+ih$, $i=1,2,\ldots,n$

where $x_0=A$ and $x_n=B$.

The differential equation is then written in finite difference form at each of the internal points $i=1,2,\ldots,n-1$. If the differential equation is linear, this leads to $n-1$ simultaneous linear equations in the unknown values $y_1, y_2,\ldots, y_{n-1}$. The more subdivisions made in the solution domain, the more simultaneous equations to be solved, but the greater the detail and accuracy of the solution.

Example 7.11

Given $y''=3x+4y$ subject to boundary conditions $y(0)=0$, $y(1)=1$, solve the equation in the range $0 \leqslant x \leqslant 1$ by finite differences using $h=0.2$.

Solution 7.11

Firstly we write the differential equation in finite difference form. From Tables 5.2–5.4 we have a choice of formulae for the required second derivative. It is usual to use the lowest order central difference form, unless there is a particular reason to use the less accurate forward or backward difference forms, hence from eq. 7.91.

$$y_i'' \simeq \frac{1}{h^2}(y_{i-1}-2y_i+y_{i+1})$$

and the differential equation can be written as

$$\frac{1}{h^2}(y_{i-1}-2y_i+y_{i+1})=3x_i+4y_i.$$

The solution domain is split into five equal strips, each of width $h=0.2$ as shown in Fig. 7.13. The finite difference equation is then written at each of the grid points for which a solution is required. Known boundary conditions are introduced at this stage, with typical values as shown in Table 7.1.

These four equations in the four unknown values of y_1, y_2, y_3 and y_4 can be solved using one of the methods described in Chapter 2. A summary of the finite difference

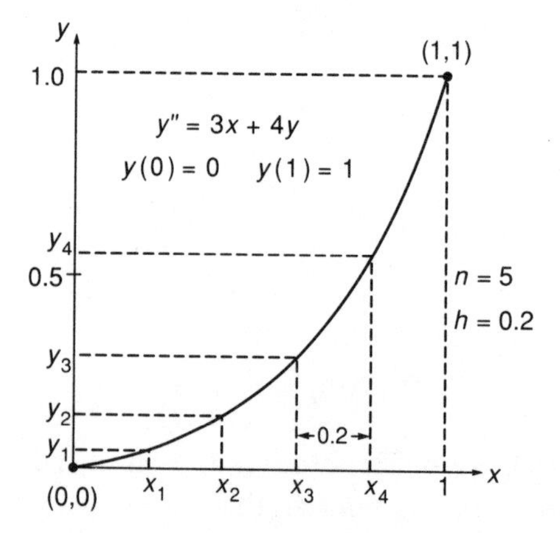

Fig. 7.13 Finite difference Example 7.11

Table 7.1. Difference equations in Example 7.11

i	x_i	Finite difference equation
1	0.2	$\frac{1}{0.04}(y_2-2y_1+0)=0.6+4y_1$
2	0.4	$\frac{1}{0.04}(y_3-2y_2+y_1)=1.2+4y_2$
3	0.6	$\frac{1}{0.04}(y_4-2y_3+y_2)=1.8+4y_3$
4	0.8	$\frac{1}{0.04}(1-2y_4+y_3)=2.4+4y_4$

solutions compared with the exact solution given by

$$y=\frac{7(e^{2x}-e^{-2x})}{4(e^2-e^{-2})}-\frac{3}{4}x$$

is presented in Table 7.2.

Table 7.2. Exact and finite difference solutions from Example 7.11

x	y_{exact}	y_{FD}
0.0	0.0	0.0
0.2	0.0482	0.0495
0.4	0.1285	0.1310
0.6	0.2783	0.2814
0.8	0.5472	0.5488
1.0	1.0	1.0

Clearly the accuracy of the finite difference solution could have been further improved by taking more subdivisions in the range $0 \leq x \leq 1$, at the expense of solving more equations. It should be noted that the equation coefficients in this class of problem have a narrow bandwidth (see Section 2.4).

Example 7.11 highlights a problem that would be encountered if one of the higher order finite difference representations of y'' had been used. For example, if we had used the five-point central different formula for y'' (see Table 5.3), values of y outside the solution domain would have been required in order to express y_1'' and y_4''. Further information would then be required in order to solve the system, because there would be more unknowns than equations.

The need to introduce points outside the solution domain, at least temporarily, is encountered quite frequently and is resolved by incorporating the appropriate boundary conditions.

Consider a boundary value problem where

$$y''=f(x, y, y') \qquad (7.96)$$

with $y(A)=(y_A$ and $y'(B)=y'_B$

In this case, $y(B)$ remains unknown, so the finite difference version of the differential equation will need to be written at $x=B$. Even when using the simplest central difference formula for y'' given by eq. 7.91, a value of y corresponding to $x=B+h$ will be introduced as shown in Fig. 7.14. In this case, we have 5 unknowns but only 4 equations.

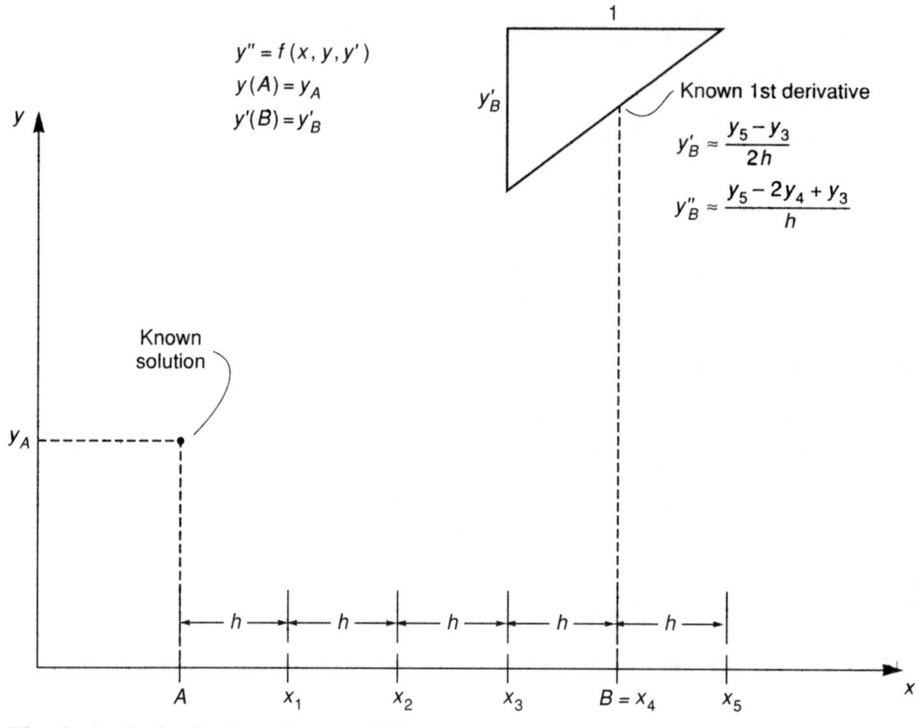

Fig. 7.14 Derivative boundary condition

The fifth equation comes from the derivative boundary condition which can also be written in finite difference form, i.e., using central differences

$$\frac{y_5-y_3}{2h}=y'_B. \qquad (7.97)$$

Example 7.12

Given $x^2y''+2xy'-2y=x^2$

$y(1)=1,\ y'(1.5)=-1$

solve for y in the range $1 \leqslant x \leqslant 1.5$ using finite differences.

Solution 7.12

Let $h=0.1$ and write the equation in finite difference form using central differences, hence

$$\frac{x_i^2}{h^2}(y_{i+1}-2y_i+y_{i-1})+\frac{x_i}{h}(y_{i+1}-y_{i-1})-2y_i=x_i^2$$

leading to the equations indicated in Table 7.3.

As shown in Fig. 7.15, $y_6=y(1.6)$ has been introduced into the system of equations. The derivative boundary condition is now used in order to provide the sixth equation,

Table 7.3. Difference equations in Example 7.12

i	x_i	Finite difference equations
1	1.1	$\frac{1.1^2}{0.1^2}(y_2-2y_1+1)+\frac{1.1}{0.1}(y_2-1)-2y_1=1.1^2$
2	1.2	$\frac{1.2^2}{0.1^2}(y_3-2y_2+y_1)+\frac{1.2}{0.1}(y_3-y_1)-2y_2=1.2^2$
3	1.3	$\frac{1.3^2}{0.1^2}(y_4-2y_3+y_2)+\frac{1.3}{0.1}(y_4-y_2)-2y_3=1.3^2$
4	1.4	$\frac{1.4^2}{0.1^2}(y_5-2y_4+y_3)+\frac{1.4}{0.1}(y_5-y_3)-2y_4=1.4^2$
5	1.5	$\frac{1.5^2}{0.1^2}(y_6-2y_5+y_4)+\frac{1.5}{0.1}(y_6-y_4)-2y_5=1.5^2$

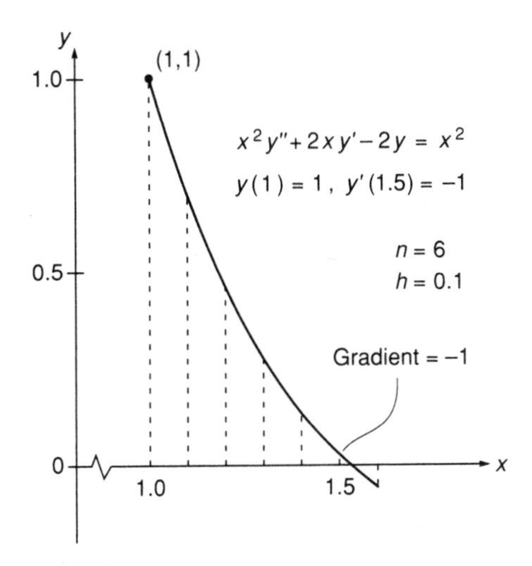

Fig. 7.15 Finite difference Example 7.12

hence

$$\frac{y_6 - y_4}{0.2} = -1$$

The solution to these six linear equations, together with the exact solution given by

$$y = \frac{135}{86x^2} - \frac{141x}{172} + \frac{x^2}{4}$$

is presented in Table 7.4.

Table 7.4. Exact and finite difference solution from Example 7.12

x	y_{exact}	y_{FD}
1.0	1.0	1.0
1.1	0.7981	0.7000
1.2	0.4774	0.4797
1.3	0.2857	0.2902
1.4	0.1432	0.1487
1.5	0.0305	0.0379
1.6	−0.0584	−0.0513

The finite difference solutions given in this section have performed quite well with relatively few grid points. This will not always be the case, especially if quite sudden changes in the derivatives occur within the solution domain. In cases where no exact solution is available for comparison, it is recommended that the problem should be solved using two or three different gradations of the solution domain. The sensitivity of the solutions to the grid size parameter h, will often indicate the accuracy of the solution. From the user's point of view, h should be made small enough to enable the desired accuracy to be achieved, but no smaller than necessary, as this would lead to excessively large systems of banded equations.

7.4.2 Shooting methods

Shooting methods attempt to solve boundary value problems as if they were initial value problems. Consider the following second order linear equation:

$$y'' = f_1(x)y' + f_2(x)y + f_3(x) \tag{7.98}$$

with $y(A) = y_A$ and $y(B) = y_B$.

Form the equivalent set of first order equations, i.e.

let $\quad y' = z \qquad (7.99)$

hence $\quad z' = f_1(x)z + f_2(x)y + f_3(x) \qquad (7.100)$

In order to solve this problem in the range $A \leq x \leq B$ as an initial value problem, we need to make assumptions about the gradient at the initial point $x=A$.

Hence let $z(A)=y'(A)=a_0$ where a_0 is a 'guess' as to the gradient at $x=A$. Sometimes, a_0 can be estimated from graphical or physical interpretations of the solution.

Knowing the initial conditions, we can solve for y and z by any one-step method using steps of h, until the solution at $x=B$ is reached. Let this 'solution' for y be given by $y_0(x)$.

We now repeat the procedure using a different initial gradient, $z(A)=y'(A)=a_1$, leading to a new 'solution' for y, namely $y_1(x)$.

If $y_0(B)=b_0$ and $y_1(B)=b_1$ then provided $b_0 \neq b_1$, the solution to the boundary value problem can be written as follows:

$$y(x)=\frac{1}{b_0-b_1}[(y_B-b_1)y_0(x)+(b_0-y_B)y_1(x)] \tag{7.101}$$

This approximate solution is a linear combination of the two 'incorrect' solutions $y_0(x)$ and $y_1(x)$ ensuring that both boundary conditions are satisfied.

Program 7.4. Shooting method for a second order linear boundary value problem
This program obtains a numerical solution to the linear second order differential equation

$$y''=f_1(x)y'+f_2(x)y+f_3(x)$$

subject to boundary conditions

$$y(A)=y_A \quad \text{and} \quad y(B)=y_B.$$

The shooting method described previously is used, leading to a linear combination of the solution to two initial value problems.

Initially, the second order equation must be broken down into two first order equations, i.e.

$$y'=z$$
$$z'=f_1(x)z+f_2(x)y+f_3(x)$$

which are computed by user-supplied subroutine FUNC. A PARAMETER statement defines the constant MAX, which must always be greater than the required number of steps in the solution domain, hence MAX $\geq$ NSTEPS + 1.

```
      PROGRAM P74
C
C      PROGRAM 7.4   SHOOTING METHODS FOR LINEAR 2ND ORDER
C                    BOUNDARY VALUE PROBLEMS
C
C      ALTER NEXT LINE TO CHANGE PROBLEM SIZE
C
      PARAMETER (MAX=25)
C
      REAL Y(2),Y0(MAX),Y1(MAX),YS(MAX)
C
      READ (5,*) NSTEPS,H,XA,YA,XB,YB,A0,A1
```

```
      WRITE (6,*) ('******** SHOOTING METHOD FOR **********')
      WRITE (6,*) ('******** LINEAR BVP"S USING ***********')
      WRITE (6,*) ('******* 2 TRIAL INITIAL GRADIENTS *****')
      WRITE (6,*)
      X = XA
      Y(1) = YA
      Y(2) = AO
      YO(1) = YA
      DO 10 I = 2,NSTEPS + 1
          CALL RUNG4(Y,X,H)
   10 YO(I) = Y(1)
      BO = YO(NSTEPS+1)
      X = XA
      Y(1) = YA
      Y(2) = A1
      Y1(1) = YA
      DO 20 I = 2,NSTEPS + 1
          CALL RUNG4(Y,X,H)
   20 Y1(I) = Y(1)
      B1 = Y1(NSTEPS+1)
      WRITE (6,*) ('      X              Y')
      X = XA
      DO 30 I = 1,NSTEPS + 1
          YS(I) = ((YB-B1)*YO(I)+ (BO-YB)*Y1(I))/ (BO-B1)
          WRITE (6,100) X,YS(I)
   30 X = X + H
  100 FORMAT (10E13.5)
      STOP
      END
C
      SUBROUTINE RUNG4(Y,X,H)
C
C      ADVANCES TWO 1ST ORDER EQUATIONS
C      USING THE 4TH ORDER RUNGE-KUTTA METHOD
C
      REAL Y(*),KO(2),K1(2),K2(2),K3(2),YO(2)
C
      CALL FUNC(KO,Y,X)
      DO 2 I = 1,2
          YO(I) = Y(I)
    2 Y(I) = YO(I) + KO(I)*.5*H
      X = X + H*.5
      CALL FUNC(K1,Y,X)
      DO 3 I = 1,2
    3 Y(I) = YO(I) + K1(I)*.5*H
      CALL FUNC(K2,Y,X)
      DO 4 I = 1,2
    4 Y(I) = YO(I) + K2(I)*H
      X = X + H*.5
      CALL FUNC(K3,Y,X)
      DO 5 I = 1,2
    5 Y(I) = YO(I) + (KO(I)+2.* (K1(I)+K2(I))+K3(I))/6.*H
      RETURN
      END
C
      SUBROUTINE FUNC(F,Y,X)
C
C      THIS SUBROUTINE PROVIDES THE VALUES OF F(X,Y) FOR EACH
C      EQUATION AND WILL VARY FROM ONE PROBLEM TO THE NEXT
C
      REAL F(2),Y(2)
C
      F(1) = Y(2)
      F(2) = 3.*X + 4.*Y(1)
      RETURN
      END
```

With the exception of simple integer counters, the variables can be summarised as follows:

NSTEPS	Number of steps
H	Step length
XA, YA	Boundary value (A, y_A)
XB, YB	Boundary value (B, y_B)
A0, A1	Trial gradients a_0 and a_1 at $x=A$
X	x-coordinate
Y	1-d array holding initial and updated dependent variables
F	1-d array holding function values
Y0	1-d array holding first solution
Y1	1-d array holding second solution
B0	Value of Y0 at $x=B$
B1	Value of Y1 at $x=B$
YS	1-d array holding final solution

The fourth order Runge–Kutta method for two first order equations is given in subroutine RUNG 4.

To illustrate use of the program, the following problem is to be solved. Given that $y''=3x+4y$ with $y(0)=0$, $y(1)=1$ solve for y in the range $0\leqslant x\leqslant 1$ using a shooting method.

The second order equation has been reduced to two first order equations, leading to the following two initial value problems.

$$y'=z \qquad y(0)=0$$
$$z'=3x+4y \qquad z(0)=0$$

$$y'=z \qquad y(0)=0$$
$$z'=3x+4y \qquad z(0)=1$$

The initial gradients input as A0 and A1, corresponding to $z(0)$, have arbitrarily been given the values 0 and 1 respectively. Input and output to the program are shown in Figs 7.16(a) and (b)

Number and size of steps	NSTEPS 5	H 0.2		
Boundary values	XA 0.0	YA 0.0	XB 1.0	YB 1.0
Trial gradients	A0 0.0	A1 1.0		

Fig. 7.16(a) Data for Program 7.4

```
******** SHOOTING METHOD FOR **********
******** LINEAR BVP"S USING ***********
******* 2 TRIAL INITIAL GRADIENTS *****

      X              Y
  .00000E+00     .00000E+00
  .20000E+00     .48215E-01
  .40000E+00     .12857E+00
  .60000E+00     .27839E+00
  .80000E+00     .54629E+00
  .10000E+01     .10000E+01
```

Fig. 7.16(b) Results from Program 7.4

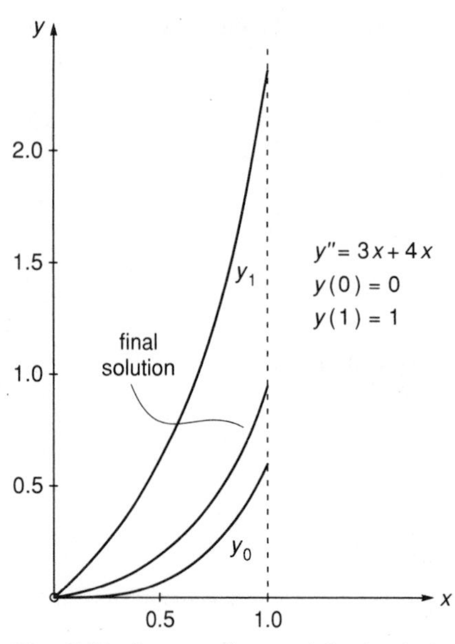

Fig. 7.17 Intermediate and final solutions by shooting method

respectively. The final solution which is printed in steps of $h=0.2$ is in good agreement with the analytical solution (see Example 7.11). The intermediate solutions Y1 and Y2, together with the final solution are shown graphically in Fig. 7.17.

The shooting method can also be used to obtain solutions to nonlinear differential equations, but we will not be able to use linear combinations of two solutions as we did previously. An iterative approach will be required to tackle the nonlinear problem, and extra care must be taken as multiple solutions may be present.

Consider the nonlinear second order differential equation

$$y''=f(y', y, x) \tag{7.102}$$

with boundary condition $y(A)=y_A$, $y(B)=y_B$.

We will solve a sequence of initial value problems of the form

$$y''=f(y', y, x) \text{ with } y(A)=y_A,\ y'(A)=a_i$$

By varying the initial gradient a_i in a methodical way, we can eventually reach a solution at $x=B$ that is sufficiently close to the required boundary value y_B. There are several different strategies for 'homing' in on the required solution, and these are similar to methods for finding roots of nonlinear algebraic equations (Chapter 3).

Once we have two initial gradients, say a_0 and a_1, which give values of y at $x=B$ which straddle the correct solution y_B, linear interpolation can then be used to obtain an 'improved' value, as was done in the 'false position' method (see Section 3.3.2).

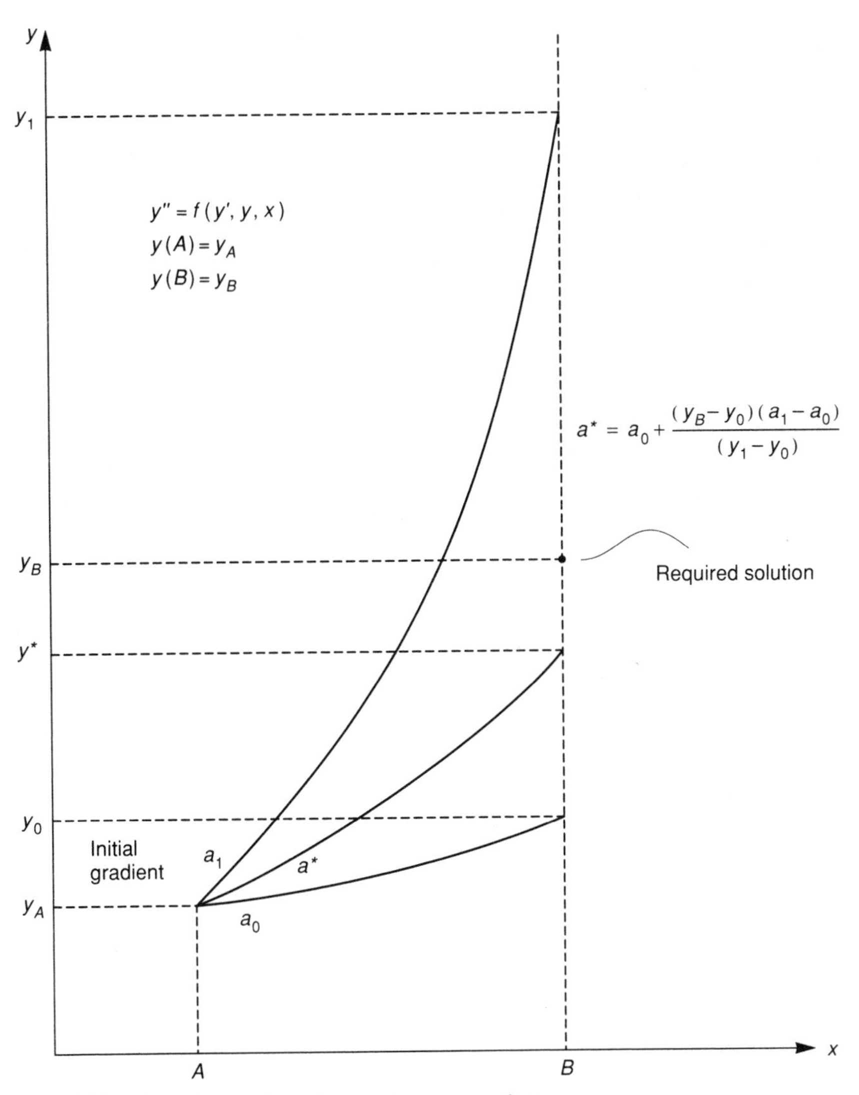

Fig. 7.18 Shooting method for nonlinear equation

Figure 7.18 shows an example where

gradient a_0 gives $y(B) = y_0$ and
gradient a_1 gives $y(B) = y_1$
where $y_0 < y_B < y_1$.

By linear interpolation, an improved value of the gradient is given by

$$a^* = a_0 + \frac{(y_B - y_0)(a_1 - a_0)}{y_1 - y_0} \tag{7.103}$$

which gives $y(B) = y^*$.

If $y^* > y_B$, a^* replaces a_1 and y^* replaces y_1, else
if $y^* < y_B$, a^* replaces a_0 and y^* replaces y_0.

Linear interpolation then proceeds as before, ensuring that the two solutions operating at any time, straddle the true solution. Iterations stop when successive values of $y(B)$ became sufficiently close to the required value y_B.

Program 7.5. Shooting method for a second order nonlinear boundary value problem
The program obtains a numerical solution to the second order nonlinear differential equation

$$y'' = f(y', y, x)$$

subject to boundary conditions $y(A) = y_A$ and $y(B) = y_B$. A sequence of initial value problems are solved by systematically altering the initial gradient until both boundary conditions are satisfied.

Initially, the second order equation must be broken down into two first order equations, i.e.

$$y' = z$$

$$z' = f(z, y, x)$$

```
      PROGRAM P75
C
C      PROGRAM 7.5   SHOOTING METHODS FOR NONLINEAR 2ND ORDER
C                    BOUNDARY VALUE PROBLEMS
C
C      ALTER NEXT LINE TO CHANGE PROBLEM SIZE
C
      PARAMETER (MAX=25)
C
      REAL Y(2),Y0(MAX),Y1(MAX),YS(MAX)
C
      READ (5,*) NSTEPS,H,XA,YA,XB,YB,A0,A1,TOL,ITS
      WRITE (6,*) ('******** SHOOTING METHODS FOR *********')
      WRITE (6,*) ('******** NONLINEAR BVP"S USING ********')
      WRITE (6,*) ('******** LINEAR INTERPOLATION *********')
      WRITE (6,*)
      X = XA
      Y(1) = YA
      Y(2) = A0
      Y0(1) = YA
      DO 10 I = 2,NSTEPS + 1
         CALL RUNG4(Y,X,H)
   10 Y0(I) = Y(1)
      B0 = Y0(NSTEPS+1)
      X = XA
      Y(1) = YA
      Y(2) = A1
      Y1(1) = YA
      DO 20 I = 2,NSTEPS + 1
         CALL RUNG4(Y,X,H)
   20 Y1(I) = Y(1)
      B1 = Y1(NSTEPS+1)
      IF ((B0-YB)*(B1-YB).GT.0.) THEN
         WRITE (6,*) ('TRY NEW GRADIENTS')
         STOP
      END IF
      ITERS = 0
   30 ITERS = ITERS + 1
      AS = A0 + (YB-B0)*(A1-A0)/(B1-B0)
      X = XA
      Y(1) = YA
```

```
      Y(2) = AS
      YS(1) = YA
      DO 40 I = 2,NSTEPS + 1
          CALL RUNG4(Y,X,H)
   40 YS(I) = Y(1)
      BS = YS(NSTEPS+1)
      IF ((BS-YB)* (B0-YB).GT.0.) THEN
          A0 = AS
          DO 50 I = 2,NSTEPS + 1
   50     Y0(I) = YS(I)
          B0 = BS
      ELSE
          A1 = AS
          DO 60 I = 2,NSTEPS + 1
   60     Y1(I) = YS(I)
          B1 = BS
      END IF
      IF (ABS((BS-YB)/YB).GT.TOL .AND. ITERS.LT.ITS) GO TO 30
      WRITE (6,*) ('     X            Y')
      X = XA
      WRITE (6,100) X,YS(1)
      DO 70 I = 2,NSTEPS + 1
          X = X + H
   70 WRITE (6,100) X,YS(I)
      WRITE (6,*)
      WRITE (6,*) 'AFTER',ITERS,'ITERATIONS'
  100 FORMAT (10E13.5)
      STOP
      END
C
      SUBROUTINE RUNG4(Y,X,H)
C
C      ADVANCES TWO 1ST ORDER EQUATIONS
C      USING THE 4TH ORDER RUNGE-KUTTA METHOD
C
      REAL Y(*),K0(2),K1(2),K2(2),K3(2),Y0(2)
C
      CALL FUNC(K0,Y,X)
      DO 2 I = 1,2
          Y0(I) = Y(I)
    2 Y(I) = Y0(I) + K0(I)*.5*H
      X = X + H*.5
      CALL FUNC(K1,Y,X)
      DO 3 I = 1,2
    3 Y(I) = Y0(I) + K1(I)*.5*H
      CALL FUNC(K2,Y,X)
      DO 4 I = 1,2
    4 Y(I) = Y0(I) + K2(I)*H
      X = X + H*.5
      CALL FUNC(K3,Y,X)
      DO 5 I = 1,2
    5 Y(I) = Y0(I) + (K0(I)+2.* (K1(I)+K2(I))+K3(I))/6.*H
      RETURN
      END
C
      SUBROUTINE FUNC(F,Y,X)
C
C      THIS SUBROUTINE PROVIDES THE VALUES OF F(X,Y) FOR EACH
C      EQUATION AND WILL VARY FROM ONE PROBLEM TO THE NEXT
C
      REAL F(2),Y(2)

      F(1) = Y(2)
      F(2) = -2*Y(2)**2/Y(1)
      RETURN
      END
```

which are computed by user-supplied subroutine FUNC. A PARAMETER statement defines the constant MAX, which must be always greater than or equal to the required number of steps in the solution domain, hence

$$\text{MAX} \geqslant \text{NSTEPS} + 1$$

The input data are very similar to those required for Program 7.4 except for the convergence tolerance TOL. Computation will stop, and a solution will be written to the output channel if the following condition is satisfied:

$$\left| \frac{y^* - y_B}{y_B} \right| < \text{TOL}$$

or alternatively if a maximum of ITS iteration is reached. Clearly, if a very strict tolerance is required, more iterations will be needed.

With the exception of simple integer counters, the variables can be summarised as follows:

NSTEPS	Number of steps
H	Step length
XA, YA	Boundary value (A, y_A)
XB, YB	Boundary value (B, y_B)
A0, A1	Initial trial gradients a_0 and a_1 at $x=A$
TOL	Convergence tolerance
ITS	Maximum number of iterations
AS	Interpolated gradient a^*
X	x-coordinate
Y	1-d array holding initial and updated dependent variables
F	1-d array holding function values
Y0	1-d array holding 'one' solution
Y1	1-d array holding 'the other' solution
YS	1-d array holding interpolated (final) solution
B0	Value of Y0 at $x=B$
B1	Value of Y1 at $x=B$
BS	Value of YS at $x=B$

The fourth order Runge–Kutta method is used to solve the initial value problems using subroutine RUNG4 listed at the end of the main program. If the initial gradients A0 and A1 chosen as data, lead to values of $y(B)$ which do not lie on either side of the correct boundary value, the program stops and new values should be tried. Unless information is available to suggest suitable starting values for $y'(A)$, the selection of A0 and A1 is a trial and error process.

To illustrate use of the program, the following problem is to be solved: given that $y'' = -2(y')^2/y$ with $y(0)=1$, $y(1)=2$, solve for y in the range $0 \leq x \leq 1$ using a shooting method.

The second order equation has been reduced to two first order equations leading to the following initial value problems:

$$y' = z \qquad y(0) = 1$$

$$z' = \frac{-2z^2}{y} \qquad z(0) = 1$$

$$y' = z \qquad y(0) = 1$$

$$z' = \frac{-2z^2}{y} \qquad z(0) = 3$$

The two initial gradients input as A0 and A1, corresponding to $z(0)$, have arbitrarily been given the values 1 and 3 respectively. Input and output to the program are shown in Figs 7.19(a) and (b) respectively. The converged solution was obtained after 5 iterations with a tolerance of 0.0001. A step length of $h = 0.1$ was used to obtain the solutions which agree well with the exact solution given by

$$y = (7x + 1)^{1/3}.$$

Number and size of steps	NSTEPS 10		H 0.1	
Boundary values	XA 0.0	YA 1.0	XB 1.0	YB 2.0
Trial gradients	A0 1.0	A1 3.0		
Convergence Tolerance	TOL 0.0001	ITS 25		

Fig. 7.19(a) Data for Program 7.5

```
******** SHOOTING METHODS FOR *********
******** NONLINEAR BVP"S USING ********
******** LINEAR INTERPOLATION *********

       X              Y
  .00000E+00    .10000E+01
  .10000E+00    .11932E+01
  .20000E+00    .13387E+01
  .30000E+00    .14579E+01
  .40000E+00    .15604E+01
  .50000E+00    .16509E+01
  .60000E+00    .17324E+01
  .70000E+00    .18070E+01
  .80000E+00    .18758E+01
  .90000E+00    .19399E+01
  .10000E+01    .20000E+01

AFTER 5 ITERATIONS
```

Fig. 7.19(b) Results from Program 7.5

7.4.3 Weighted residual methods

These methods have been the subject of a great deal of research in recent years. The methods can form the basis of the finite element method which is now the most widely used numerical method for the solution of large boundary value problems.

The starting point for weighted residual methods is to 'guess' a solution to the differential equation which satisfies the boundary conditions. This 'trial solution' will contain certain parameters which can be adjusted to minimise the errors. Several different methods are available for minimising the error or 'residual', such that the trial solution is as close to the exact solution as possible.

Consider the second order boundary value problem

$$y''=f(y', y, x) \tag{7.104}$$

with $\quad y(A)=y_A \quad$ and $\quad y(B)=y_B$

The differential equation can be rearranged as

$$R=y''-f(y', y, x) \tag{7.105}$$

where R is the 'residual' of the equation. Only the exact solution $y(x)$ will satisfy the boundary conditions and cause R to equal zero for all x.

Consider a trial solution $\phi(x)$ of the form

$$\phi(x)=F(x)+C_1\psi_1(x)+C_2\psi_2(x)+\cdots \tag{7.106}$$

where $\quad \phi(A)=y_A, \quad \phi(B)=y_B \quad$ and $F(x)$ can be zero if required.

The trial solution, which is made up of trial functions ψ_1, ψ_2, etc., satisfies the boundary conditions of the differential equation, and contains a number of 'undetermined parameters' C_1, C_2, etc. The trial solution (7.106) is differentiated twice, and substituted into (7.105) to give

$$R=\phi''-f(\phi', \phi, x) \tag{7.107}$$

We wish to find the values of C_1, C_2, etc., which minimise R in the solution domain $A \leq x \leq B$. Various methods for minimising R will be described.

Clearly the choice of a suitable trial solution is crucial to the whole process. Frequently the trial functions ψ_1, ψ_2, etc., are simple polynomials, although transcendental functions such as sine, cosine, $\log_e$ etc., could also be used. Sometimes the physics of the problem points to a 'sensible' choice of trial functions.

There is no limit to the number of undetermined parameters that may be included in the trial solution. The more undetermined parameters, the more accurate the final solution, but the more work that must be done in order to find their values.

Example 7.13

Given the boundary value problem

$y''=3x+4y$ subject to boundary conditions
$y(0)=0,\ y(1)=1$

obtain an expression for the residual using a trial solution with one undetermined parameter.

Solution 7.13

In order to find a trial solution which satisfies the boundary condition, we could derive a Lagrangian polynomial (see Section 5.2.1) which passes through the points

x	y
0	0
0.5	a
1	1

The value of the polynomial at $x=0.5$ remains variable at this stage and will be related to our single undetermined parameter. The resulting polynomial is of second order, and given by

$$\phi(x)=x^2(2-4a)-x(1-4a)$$

which can be rearranged in the general form given by eq. 7.106 to give

$$\phi(x)=F(x)+C_1\psi_1(x)$$

where $F(x)=2x^2-x$

$C_1=-4a$

$\psi_1(x)=x^2-x$

In order to find the residual of the equation, rearrange it thus

$$R=y''-3x-4y$$

and differentiate the trial solution twice to give

$$\phi'=2x(2-4a)-(1-4a)$$

i.e.

$$\phi''=2(2-4a)$$

The trial solution and its derivatives are now substituted into the differential equation replacing y, y' and y'' by ϕ, ϕ' and ϕ''. In this example there is no first derivative, but after rearrangement and substitution for C_1 we get

$$R=-4x^2(2+C_1)+x(1+4C_1)+2(2+C_1)$$

Example 7.14

Obtain a trial solution to the problem of Example 7.13 involving two undetermined parameters.

Solution 7.14

Derive a Lagrangian polynomial which passes through the following points

x	y
0	0
$\frac{1}{3}$	a
$\frac{2}{3}$	b
1	1

The values of the polynomial at $x=\frac{1}{3}$ and $x=\frac{2}{3}$ remain variable at this stage and will be related to our two undetermined parameters.

The resulting polynomial is cubic, and given by

$$\phi(x)=\tfrac{1}{2}\left[x^3(27a-27b+9)-x^2(45a-36b+9)+x(18a-9b+2)\right]$$

which can be rearranged in the general form given by eq. 7.106, i.e.

$$\phi(x)=F(x)+C_1\psi_1(x)+C_2\psi_2(x)$$

where

$$\begin{aligned} F(x) &= \tfrac{1}{2}(9x^3-9x^2+2x) \\ C_1 &= \tfrac{27}{2}(a-b) \\ \psi_1(x) &= x^3-x^2 \\ C_2 &= -\tfrac{9}{2}(2a-b) \\ \psi_2(x) &= x^2-x \end{aligned}$$

Four of the best-known methods for minimising the residual R are now covered briefly. In each case, the trial solution and residual from Example 7.13, i.e.

$$\phi(x)=(2x^2-x)+C_1(x^2-x) \tag{7.108}$$

$$R=-4x^2(2+C_1)+x(1+4C_1)+2(2+C_1) \tag{7.109}$$

involving one undetermined parameter are operated on.

7.4.3.1 Collocation

In this method, the residual is made equal to zero at as many points in the solution domain as there are unknown C_i's. Hence, if we have n undetermined parameters, we write

$$R(x_i)=0, \quad i=1,2,\ldots,n \tag{7.110}$$

where $x_1, x_2, \ldots, x_n$ are n points within the solution domain. This leads to n simultaneous equations in the unknown C_i's.

Example 7.15

Use collocation to find a solution to the differential equation of Example 7.13.

Solution 7.15

Using the trial solution and residual from Solution 7.13 we have one undetermined parameter C_1. It seems logical to collocate at the midpoint of the solution domain, $x=0.5$, hence

$$R(0.5)=-4(0.5)^2(2+C_1)+0.5(1+4C_1)+2(2+C_1)=0$$

which gives $C_1=-\frac{5}{6}$.

The trial solution is therefore given by

$$\phi(x)=(2x^2-x)-\tfrac{5}{6}(x^2-x)$$

hence

$$\phi(x)=\frac{x}{6}(7x-1)$$

Note: If our trial solution had two undetermined parameters such as that given in Example 7.14, we could have collocated at $x=\frac{1}{3}$ and $x=\frac{2}{3}$ to give two equations in the unknown C_1 and C_2.

7.4.3.2 Subdomain

In this method, the solution domain is split into as many 'subdomains' or parts of the domain as there are unknown C_i's. We then integrate the residual over each subdomain and set the result to zero. Hence, for n undetermined parameters we have

$$\int_{x_0}^{x_1} R\,\mathrm{d}x=0, \qquad \int_{x_1}^{x_2} R\,\mathrm{d}x=0,\ldots, \qquad \int_{x_{n-1}}^{x_n} R\,\mathrm{d}x=0 \tag{7.111}$$

where the solution domain is in the range

$$x_0 \leq x \leq x_n$$

and $x_1\ x_2 \ldots x_{n-1}$ are points within that range.

This leads to n simultaneous equations in the unknown C_i's.

Example 7.16

Use the subdomain method to find a solution to the differential equation of Example 7.13.

Solution 7.16

Using the trial solution and residual from Solution 7.13 we have one undetermined parameter C_1. In this case we require only one 'subdomain' which is the solution domain itself, hence

$$\int_0^1 R\,dx = [-\tfrac{4}{3}x^3(2+C_1)+\tfrac{1}{2}x^2(1+4C_1)+2x(2+C_1)]_0^1 = 0$$

which gives $C_1 = -\frac{11}{16}$

The trial solution is therefore given by

$$\phi(x) = (2x^2 - x) - \tfrac{11}{16}(x^2 - x)$$

hence $\quad \phi(x) = \dfrac{x}{16}(21x - 5)$

Note: If our trial solution had two undetermined parameters such as that given in Example 7.14, we would have needed to integrate over two subdomains, e.g.

$$\int_0^{0.5} R\,dx = 0 \qquad \text{and} \qquad \int_{0.5}^{1} R\,dx = 0$$

to give two equations in the unknown C_1 and C_2.

7.4.3.3 Least squares

In this method we integrate the square of the residual over the full solution domain. We then differentiate with respect to each of the undetermined parameters in turn, and set the result to zero. This has the effect of minimising the integral of the square of the residual.

In general we write

$$\frac{\partial}{\partial C_i}\int_A^B R^2\,dx \equiv 2\int_A^B R\frac{\partial R}{\partial C_i}\,dx = 0 \quad \text{for } i = 1, 2, \ldots, n \tag{7.112}$$

where the solution domain is given by $A \leq x \leq B$.

Hence for n undetermined parameters, we have

$$\int_A^B R\frac{\partial R}{\partial C_1}\,dx = 0, \qquad \int_A^B R\frac{\partial R}{\partial C_2}\,dx = 0, \ldots, \qquad \int_A^B R\frac{\partial R}{\partial C_n}\,dx = 0 \tag{7.113}$$

which leads to n simultaneous equations in the unknown C_i's.

Example 7.17

Use the least squares method to find a solution to the differential equation of Example 7.13.

Solution 7.17

Using the trial solution and residual from Solution 7.13 we have one undetermined parameter C_1. Hence set

$$\int_0^1 R\frac{\partial R}{\partial C_1}\,\mathrm{d}x=0$$

i.e. $$\int_0^1 [-4x^2(2+C_1)+x(1+4C_1)+2(2+C_1)](-4x^2+4x+2)\,\mathrm{d}x=0$$

which gives $C_1=\frac{-76}{108}$

The trial solution is therefore given by

$\phi(x)=(2x^2-x)-\frac{76}{108}(x^2-x)$

hence, $\phi(x)=\dfrac{x}{27}(35x-8)$

Note: If our trial solution had two undetermined parameters such as that given in Example 7.14, we would have written

$$\int_0^1 R\frac{\partial R}{\partial C_1}\,\mathrm{d}x=0 \quad \text{and} \quad \int_0^1 R\frac{\partial R}{\partial C_2}\,\mathrm{d}x=0$$

to give two equations in the unknown C_1 and C_2.

7.4.3.4 Galerkin's method

In this method, which is the most popular in finite element applications we 'weight' the residual by the trial functions $\psi_1(x), \psi_2(x), \ldots, \psi_n(x)$, and set the integrals to zero. Hence for n undetermined parameters, we have

$$\int_A^B R\psi_1(x)\,\mathrm{d}x=0, \qquad \int_A^B R\psi_2(x)\,\mathrm{d}x=0, \ldots, \qquad \int_A^B R\psi_n(x)\,\mathrm{d}x=0 \tag{7.114}$$

where the solution domain is given by $A\leq x\leq B$. This leads to n simultaneous equations in the unknown C_i's.

Example 7.18

Use Galerkin's method to find a solution to the differential equation of Example 7.13.

Solution 7.18

Using the trial solution and residual from Solution 7.13 we have one undetermined parameter C_1. Hence set

$$\int_0^1 R\psi_1(x)\,dx=0$$

i.e.

$$\int_0^1 [-4x^2(2+C_1)+x(1+4C_1)+2(2+C_1)](x^2-x)\,dx=0$$

which gives $C_1=-\frac{3}{4}$

The trial solution is therefore given by

$$\phi(x)=(2x^2-x)-\tfrac{3}{4}(x^2-x)$$

$$\phi(x)=\frac{x}{4}(5x-1)$$

Note: If our trial solution had two undetermined parameters, such as that given in Example 7.14, we would have written

$$\int_0^1 R\psi_1(x)\,dx=0 \qquad \text{and} \qquad \int_0^1 R\psi_2(x)\,dx=0$$

to give two equations in the unknown C_1 and C_2.

7.4.3.5 Concluding remarks on weighted residual procedures

The problem considered in the preceding sections was

$$y''=3x+4y \quad \text{with} \quad y(0)=0 \quad \text{and} \quad y(1)=1 \tag{7.115}$$

and a summary of the trial solutions using one undetermined parameter is as follows:

$$\begin{aligned}
&\text{Collocation} && y=\frac{x}{6}(7x-1)\\
&\text{Subdomain} && y=\frac{x}{16}(21x-5)\\
&\text{Least squares} && y=\frac{x}{27}(35x-8)\\
&\text{Galerkin} && y=\frac{x}{4}(5x-1)\\
&\text{Exact} && y=\frac{7(e^{2x}-e^{-2x})}{4(e^2-e^{-2})}-\frac{3}{4}x
\end{aligned} \tag{7.116}$$

The error associated with each of the trial solutions is shown in Fig. 7.20. It is clear that for this example, no single method emerges as the 'best'. Indeed each of the methods is the 'most accurate' at some point within the solution domain $0\leq x\leq 1$.

For simple problems such as the one demonstrated here, the collocation method is the easiest to apply, because it does not require any integration. For finite element

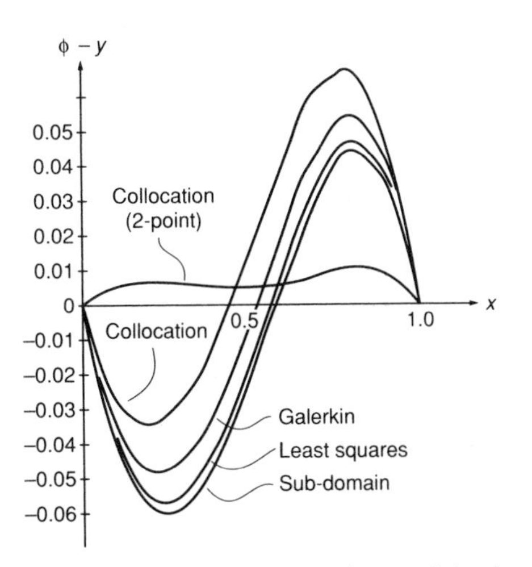

Fig. 7.20 Error due to various weighted residual methods

methods however, the Galerkin method is most often employed, because it can be shown that use of the trial functions (or shape functions) as 'weighting' terms leads to desirable properties such as symmetry of simultaneous equation coefficient matrices.

For comparison, the trial solution with two undetermined parameters developed in Example 7.14, has been derived using two-point collocation. It is left to the reader to show that the resulting trial solution becomes

$$\phi(x)=\tfrac{1}{2}(9x^3-9x^2+2x)-3.48388(x^3-x^2)+0.70348(x^2-x) \tag{7.117}$$

leading to considerably reduced errors as shown in Fig. 7.20.

All the four methods described in this section for optimising the accuracy of a trial solution employ the philosophy that weighted averages of the residual should vanish. The difference between methods therefore, lies only in the way the residual is weighted. In general, if we have n undetermined parameters, we set

$$\int_A^B W_i(x)R\,\mathrm{d}x=0 \qquad i=1,2,\ldots,n \tag{7.118}$$

where the solution domain is $A\leq x\leq B$. In the previous examples with one undetermined parameter, the weighting term W_i was of the following form:

$$\begin{array}{ll} \text{Collocation} & \begin{cases} W_1(x)=0 & x\neq\tfrac{1}{2} \\ W_1(x)=1 & x=\tfrac{1}{2} \end{cases} \\ \text{Subdomain} & W_1(x)=1 \\ \text{Least squares} & W_1(x)=-4x^2+4x+2 \\ \text{Galerkin} & W_1(x)=x^2-x \end{array} \tag{7.119}$$

where the solution domain was in the range $0\leq x\leq 1$.

These weighting functions are summarised in Fig. 7.21.

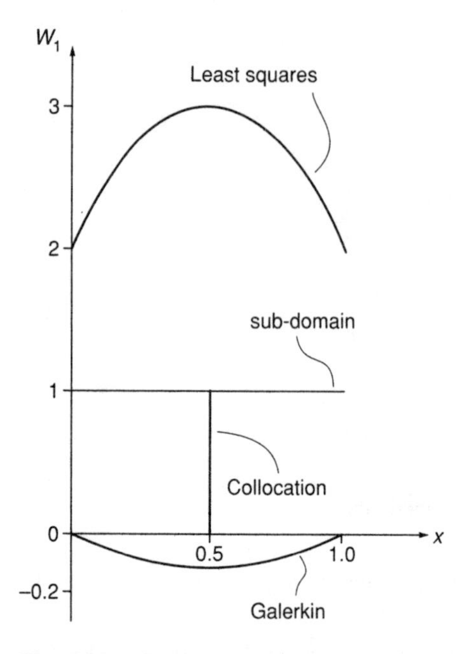

Fig. 7.21 Weighting functions

7.5 Exercises

1 Given $y' = xy \sin y$ with $y(1) = 1$; estimate $y(1.5)$ using Euler's method. Let (a) $h = 0.25$, (b) $h = 0.1$.
Answer: (a) 1.5643, (b) 1.6963

2 Given $y' = (x^2 + y^2)/y$ with $y(1) = 1$ estimate $y(0.5)$ using the modified Euler method with (a) $h = -0.1$ and (b) $h = -0.05$.
Answer: (a) 0.1869 (b) 0.1923

3 Rework Exercise 2 using the midpoint rule.
Answer: (a) 0.1952 (b) 0.1943

4 Rework Exercise 2 using the fourth order Runge–Kutta method with two steps of -0.25.
Answer: 0.1941

5 Use Euler's method to solve numerically each of the following:
(a) $y' = y$, $y(0) = 1$ find $y(0.2)$ using $h = 0.05$
(b) $y' = x + y$, $y(0) = 0$ find $y(0.5)$ using $h = 0.1$
(c) $y' = 3x - y$, $y(1) = 0$ find $y(0.5)$ using $h = -0.1$
Answer: (a) 1.2155, (b) 0.1105, (c) -1.5
Exact 1.2214, 0.1487 -1.5

6 Since the equations in Exercise 5 are linear, rework them using a 'θ-method' with $\theta = \frac{1}{2}$.
Answer: (a) 1.2215, (b) 0.1494, (c) -1.5

7 Given $y' = (x + y)^2$ with $y(0) = 1$, estimate $y(0.2)$ using a suitable numerical method.

Answer: 1.3085 with fourth order Runge–Kutta ($h=0.2$)
Exact 1.3085

8 Given the equation

$$y'=x+y+xy+1 \quad \text{with} \quad y(0)=2$$

estimate $y(0.4)$ using the modified Euler method with step lengths 0.1 and 0.2, and use these to make a further, improved estimate.
Answer: 3.8428, 3.8277, 3.8478

9 Given the equation

$$\frac{y''}{[1+(y')^2]^{3/2}}=1-x$$

with $y(0)=y'(0)=0$, estimate $y(0.5)$ using the midpoint rule with $h=0.25$ and $h=0.05$.
Answer: 0.1124, 0.1087

10 Given the equation

$$y''+y+(y')^2=0$$

subject to initial conditions $y(0)=-1$, $y'(0)=1$ estimate $y(0.5)$ and $y'(0.5)$ using the fourth order Runge–Kutta method with a step length of $h=0.25$.
Answer: -0.5164, 0.9090

11 Given the equation

$$y''-2y'+y=2x \quad \text{with } y(1)=0,\ y'(1)=0$$

find $y(0.4)$ using a suitable numerical method.
Answer: 0.1900, fourth order Runge–Kutta ($h=-0.15$)

12 The angular acceleration of a pendulum is given by

$$\ddot{\theta}=-\frac{g}{L}\sin\theta$$

with initial condition $\theta(0)=\pi/4$, $\ddot{\theta}(0)=0$. If $L=10$ m use the fourth order Runge–Kutta method to find the angular position of the rod after 0.5 s.
Answer: 0.7 ($h=0.25$)

13 Given the equation

$$y''+y=x \quad \text{with } y(0)=1 \quad \text{and} \quad y'(0)=0,$$

find $y(0.5)$ using a suitable numerical method.
Answer: 0.8982, fourth order Runge–Kutta ($h=0.25$)

14 Use the modified Euler method to estimate $y'(0.2)$ for the equation.

$$y''=-x/y' \quad \text{with } y(0)=y'(0)=1.$$

Answer: 0.9824 ($h=0.05$)

15 Given the equation

$$y''-2xy'+(3+x^2)y=0$$

with $y(0)=2$, $y'(0)=0$, estimate $y(0.1)$ using one step of the fourth order Runge–Kutta method.
Answer: 0.1970

16 The differential equation

$y'=\sec x+y\tan x$ has the following solutions in the range $0\leq x\leq 0.5$.

x	y
0	0
0.1	0.1005
0.2	0.2041
0.3	0.3140
0.4	0.4343
0.5	0.5697

continue the solution to estimate $y(0.7)$ using the Adams–Bashforth–Moulton predictor–corrector method.
Answer: $y(0.6)=0.7270$, $y(0.7)=0.9153$ with 10^{-4} tolerance

17 Rework Exercise 16 using the Milne–Simpson predictor–corrector method.
Answer: $y(0.6)=0.7270$, $y(0.7)=0.9152$ with 10^{-4} tolerance

18 Use polynomial substitution to derive the weights in the predictor formula

$$y_{n+1}=y_{n-3}+\sum_{i=n-2}^{n} w_i y_i'$$

where the sampling points are equally spaced, distance h apart.
Use the formula to estimate $y(0.4)$ given the differential equation:

$y'=\tan(xy)$

and the initial values

x	y
0	0.1
0.1	0.1005
0.2	0.1020
0.3	0.1046

Answer: Adams–Bashforth predictor $y(0.4)\simeq 0.1083$

19 Given the boundary value problem

$$y''=\frac{2}{x}y'+\frac{2}{x^2}y-\sin x$$

with $y(1)=1$, $y(2)=2$, use a shooting method to estimate $y(1.5)$.
Answer: 1.3092

20 Given the nonlinear boundary value problem

$$2yy''-(y')^2+4y^2=0$$

with $y(\pi/6)=0.75$, $y(5\pi/12)=0.0670$, take $a_0=-0.5$ and $a_1=-0.9$ as the initial estimates of $y'(\pi/6)$ and hence estimate $y(\pi/3)$.
Answer: 0.2514 (exact solution $y=\cos^2 x$)

21 Solve the following problems using a shooting method:

(a) $y''=6y^2$, $y(1)=1$, $y(2)=0.25$
estimate $y(1.5)$.
Try initial gradients $a_0=-1.8$, $a_1=-2.1$.

(b) $y''=-e^{-2y}$, $y(1)=0$, $y(1.5)=0.4055$
estimate $y(1.25)$
Try initial gradients $a_0=0.8$, $a_1=1.2$.

Answer: (a) 0.4445, (b) 0.2232

22 A cantilever of unit length and flexural rigidity supports a unit load at its free end. Use a finite difference scheme with a step length of (a) 0.25 and (b) 0.125 to estimate the free end deflection.

The governing equation in this case is

$$y''=M(x)$$

where $M(x)$ is the bending moment and y is the beam deflection.
Note: Use a program from Chapter 2 to solve the equations.
Answer: (a) 0.344, (b) 0.336

23 A beam of length L and flexural rigidity EI rests on an elastic foundation of modulus k. The beam supports a uniform load of w. If y is the deflection of the beam, the governing equation in dimensionless form is given by

$$\frac{d^4\phi}{dz^4}+k\phi=1 \qquad \begin{array}{l}\phi(0)=\phi(1)=0\\ \phi''(0)=\phi''(1)=0\end{array}$$

where $\phi=\dfrac{EIy}{wL^4}$, $z=\dfrac{x}{L}$, $K=\dfrac{kL^4}{EI}$

Using a finite difference scheme with a step length of $h=\frac{1}{2}$, estimate the midpoint deflection as a function of k, w, L, E and I. By putting $k=0$, what is the percentage error of the finite difference solution if the exact solution at the centre is given by

$$y=\frac{5}{384}\frac{wL^4}{EI}?$$

Answer: 20%

24 Rework Exercise 23 with a smaller step length of $h=\frac{1}{4}$.
Answer: 5%

25 Given the differential equation

$$y''=-y+2$$

subject to boundary conditions $y(0)=1$
$y'(1)+y(1)=3.682941$

estimate y at intervals of 0.2 in the range $0 \le x \le 1$ using a finite difference scheme.

Note: Use a program from Chapter 2 to solve the equations.
Answer: $y(0.2)=1.220$, $y(0.4)=1.471$, $y(0.6)=1.743$, $y(0.8)=2.025$, $y(1.0)=2.307$

26 Estimate the value of the cantilever tip deflection from Exercise 22 using a method of weighted residuals with the trial solution

$$y=C_1x^2+C_2x^3.$$

Answer: All methods give exact solution 0.333.

27 Given the equation

$$y''+\pi^2y=x$$

subject to boundary condition $y(0)=1$, $y(1)=-0.8987$, estimate $y(0.5)$ using the trial solution

$$y=1-1.8987x+\sum_{i=1}^{n} C_i x^i(1-x) \text{ with } n=2$$

Use (a) collocation and (b) Galerkin.
Answer: (a) 0.0508, (b) 0.0495

28 Apply the Galerkin method to

$$y''+\lambda y=0 \quad \text{with } y(0)=y(1)=0$$

using the trial solution

$$y=C_1x(1-x)+C_2x^2(1-x)^2$$

and hence estimate the smallest eigenvalue λ.
Answer: 9.86975

29 Rework Exercise 28 using finite differences with a step length of

(a) $\frac{1}{3}$ and (b) $\frac{1}{4}$.

It may be assumed that y will be symmetrical about the point $x=\frac{1}{2}$.
Answer: (a) 9, (b) 9.37

7.6 Further reading

Burden, R.L. and Faires, J.D. (1985). *Numerical Analysis*, 3rd edn, Prindle, Weber and Schmidt, Boston, Mass.
Collatz, L. (1966). *The Numerical Treatment of Differential Equations*, 3rd edn, Springer-Verlag, New York.
Conte, S.D. and de Boor, C. (1980). *Elementary Numerical Analysis*, McGraw-Hill, New York.
Crandall, S.H. (1956). *Engineering Analysis*, McGraw-Hill, New York.
Dahlquist, G. and Bjorck, A. (1974). *Numerical Methods*, Prentice-Hall, Englewood Cliffs, New Jersey.
Gear, C.W. (1971). *Numerical Initial Value Problems in Ordinary Differential Equations*, Prentice-Hall, Englewood Cliffs, New Jersey.
Gladwell, I. and Sayers, D. (eds) (1980). *Computational Techniques for Ordinary Differential Equations*, Academic Press, New York.
Grove, W.E. (1966). *Brief Numerical Methods*, Prentice-Hall, Englewood Cliffs, New Jersey.
Hall, G. and Watt, J. (eds) (1976). *Modern Numerical Methods for Ordinary Differential Equations*, Oxford University Press.
Hamming, R.W. (1959). Stable predictor–corrector methods for ordinary differential equations, *Journal of the Association of Computer Machinery*, Vol. 6, No. 1, pp. 37–47.

Johnson, L.W. and Riess, R.D. (1982). *Numerical Analysis*, 3rd edn, Addison-Wesley, Reading, Mass.
Lapidus, L. and Seinfeld, J. (1971). *Numerical Solution of Ordinary Differential Equations*, Academic Press, New York.
Ralston, A. (1965). *A First Course in Numerical Analysis*, McGraw-Hill, New York.
Shoup, T.E. (1979). *A Practical Guide to Computer Methods for Engineers*, Prentice-Hall, Englewood Cliffs, New Jersey.
Strang, G. and Fix, G. (1973). *An Analysis of the Finite Element Method*, Prentice-Hall, Englewood Cliffs.

8

Introduction to Partial Differential Equations

8.1 Introduction

A partial differential equation (PDE) is a differential equation containing derivatives involving two or more independent variables. This is in contrast to the ordinary differential equations in the previous chapter which only involved one independent variable.

Many phenomena in engineering science are described by partial differential equations. For example a dependent variable such as a displacement or temperature may vary as a function of time (t) and space (x, y, z).

The solution of partial differential equations for realistic problems is beyond the scope of this book, but the subject is covered in a large number of more advanced texts, some of which are mentioned at the end of this chapter. Some of these texts describe the 'finite element method' which is undoubtedly the most widely used method of solution for partial differential equations at the present time.

The finite element method is not covered in this book, since the authors have already devoted a text to it (Smith and Griffiths 1988). However, some solutions involving the 'finite difference method', which is much simpler to describe, will be presented for a selection of rather elementary examples.

All that is attempted here is an introductory treatment of the subject. The aim is to enable the student to recognise certain types of partial differential equation and obtain some insight into the types of physical phenomena they describe.

8.2 Some definitions

Consider the following second order PDE in two independent variables

$$a\frac{\partial^2 u}{\partial x^2}+b\frac{\partial^2 u}{\partial x \partial y}+c\frac{\partial^2 u}{\partial y^2}+d\frac{\partial u}{\partial x}+e\frac{\partial u}{\partial y}+fu+g=0 \tag{8.1}$$

note the presence of 'mixed' derivatives such as that associated with the coefficient b.

If $a, b, c, \ldots, g$ are functions of x and y only, the equation is 'linear', but if these coefficients contain u or its derivatives, the equation is 'nonlinear'.

The degree of a PDE is the power to which the highest derivative is raised, thus eq. 8.1 is first degree. Only linear (first degree) equations will be considered in this chapter.

A regularly encountered shorthand in the study of second order PDEs is the 'Laplace operator' ∇^2 where

$$\nabla^2 f \equiv \frac{\partial^2 f}{\partial x^2} + \frac{\partial^2 f}{\partial y^2} \qquad \text{in 2 dimensions}$$

and

$$\nabla^2 f \equiv \frac{\partial^2 f}{\partial x^2} + \frac{\partial^2 f}{\partial y^2} + \frac{\partial^2 f}{\partial z^2} \quad \text{in 3 dimensions}$$

Another abbreviation is the 'biharmonic operator' ∇^4 where

$$\nabla^4 f \equiv \frac{\partial^4 f}{\partial x^4} + \frac{2\partial^4 f}{\partial x^2 \partial y^2} + \frac{\partial^4 f}{\partial y^4}$$

Table 8.1 summarises some commonly encountered PDEs and the type of physical situation in which they might arise.

Table 8.1. Common types of PDEs in engineering analysis

Type	Equation	Application
Laplace's eq.	$\nabla^2 f = 0$	Steady flow of heat and fluids
Poisson's eq.	$\nabla^2 f = f(x, y)$	Heat/fluid flow with sources or sinks, torsional problems
Diffusion eq.	$\nabla^2 f = \frac{1}{h^2}\frac{\partial f}{\partial t}$	Transient flow in heat conduction/consolidation
Wave eq.	$\nabla^2 f = \frac{1}{c^2}\frac{\partial^2 f}{\partial t^2}$	Vibration problems, wave propagation
Biharmonic eq.	$\nabla^4 f = f(x, y)$	Deformation of thin plates

8.3 First order equations

Although the majority of engineering applications involve second order PDEs, we start with a consideration of first order equations, as this will lead to a convenient introduction to the 'method of characteristics'.

Consider the first order equation

$$a\frac{\partial u}{\partial x} + b\frac{\partial u}{\partial y} = c \tag{8.2}$$

where a, b and c may be functions of x, y and u, but not derivatives of u.

The following substitutions can be made to simplify the algebra:

$$p = \frac{\partial u}{\partial x}, \quad q = \frac{\partial u}{\partial y}$$

hence

$$ap+bq=c \tag{8.3}$$

A 'solution' to eq. 8.2 will be an estimate of the value of u at any point within the x, y-plane. This plane is known as the 'solution domain', and in order to find values of u in this range we require some 'initial conditions' in order to get started.

As shown in Fig. 8.1, let values of u be known along the line I in the solution domain. We now consider an arbitrary line C which intersects our initial line I as shown in Fig. 8.2.

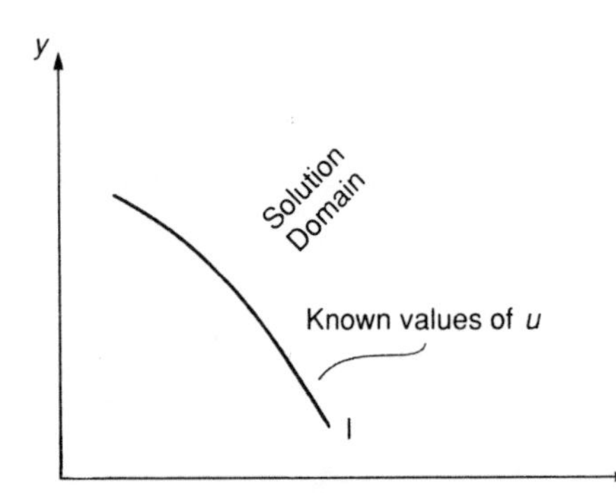

Fig. 8.1 Solution domain and 'initial conditions' line I

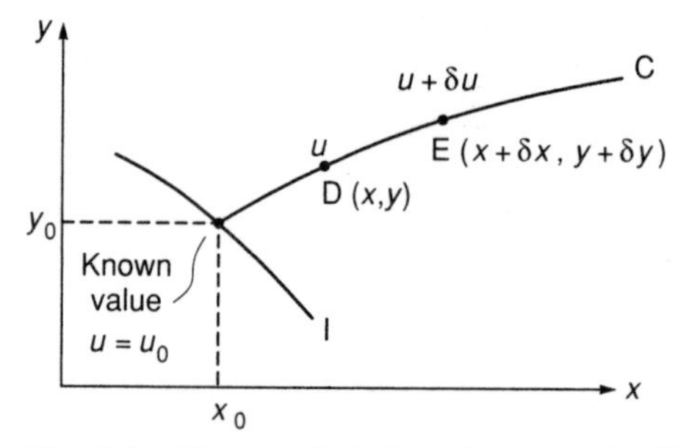

Fig. 8.2 Characteristic line C intersecting line I

Consider a small change in u along the line C between points D and E. This leads to the equation

$$\delta u=\frac{\partial u}{\partial x}\delta x+\frac{\partial u}{\partial y}\delta y \tag{8.4}$$

which in the limit as $\delta x\rightarrow 0$ and $\delta y\rightarrow 0$ becomes

$$\mathrm{d}u=\frac{\partial u}{\partial x}\mathrm{d}x+\frac{\partial u}{\partial y}\mathrm{d}y \tag{8.5}$$

or $\quad \mathrm{d}u=p\,\mathrm{d}x+q\,\mathrm{d}y \qquad (8.6)$

Eliminating p between eqs 8.3 and 8.6 and rearranging gives

$$q(a\,\mathrm{d}y - b\,\mathrm{d}x) + (c\,\mathrm{d}x - a\,\mathrm{d}u) = 0 \tag{8.7}$$

This statement would be true for any line C, but if we choose C so that at all points along its length, the following condition is satisfied

$$a\,\mathrm{d}y - b\,\mathrm{d}x = 0 \tag{8.8}$$

or $$\frac{\mathrm{d}y}{\mathrm{d}x} = \frac{b}{a}$$

then q can also be eliminated from eq. 8.7 which becomes

$$c\,\mathrm{d}x - a\,\mathrm{d}u = 0 \tag{8.9}$$

Combining this with eq. 8.8 gives the final relationship along this special line C as

$$\frac{\mathrm{d}x}{a} = \frac{\mathrm{d}y}{b} = \frac{\mathrm{d}u}{c} \tag{8.10}$$

Thus the original PDE given by eq. 8.2 has been reduced to an ordinary differential equation in x (or y).

This special C-line is called a 'characteristic' and it can be shown that within the solution domain, a family of similar lines exists along which the PDE reduces to an ordinary differential equation. Along these 'characteristics' ordinary differential equation solution techniques may be employed such as those described in Chapter 7.

Rearrangement of eq. 8.10 gives the first order ordinary differential equations

$$\frac{\mathrm{d}y}{\mathrm{d}x} = \frac{b}{a} \qquad y(x_0) = y_0$$

$$\frac{\mathrm{d}u}{\mathrm{d}x} = \frac{c}{a} \qquad u(x_0) = u_0 \tag{8.11}$$

where the initial conditions (x_0, y_0) and (x_0, u_0) correspond to values of x, y and u at the point of intersection of the characteristic line C with the initial conditions line I as shown in Fig. 8.2.

Thus, if we know the value of u_0 at any initial point (x_0, y_0), solution of the ordinary differential equations given by eq. 8.11 will lead to values of u along the characteristic line passing through (x_0, y_0).

Example 8.1

Given the first order PDE

$$\frac{\partial u}{\partial x} + 3x^2\frac{\partial u}{\partial y} = x + y$$

find u (3, 19) given the initial condition $u(x, 0) = x^2$.

Solution 8.1

The equation is linear so an analytical solution is possible in this case. Along the characteristics

$$\frac{dx}{1}=\frac{dy}{3x^2}=\frac{du}{x+y}$$

hence $$\frac{dy}{dx}=3x^2 \qquad \text{(A)}$$

$$\frac{du}{dx}=x+y \qquad \text{(B)}$$

From (A), $y=x^3+k$ which is the equation of the family of characteristic lines. We are interested in the characteristic passing through the point (3, 19), thus

$$19=27+k$$

$\therefore \quad k=-8$

$\therefore \quad y=x^3-8$ which is the equation of line C.

The initial value line I, is the x-axis as shown in Fig. 8.3. The characteristic line of interest intersects the x-axis at (2, 0), at which point the value of $u=4$.

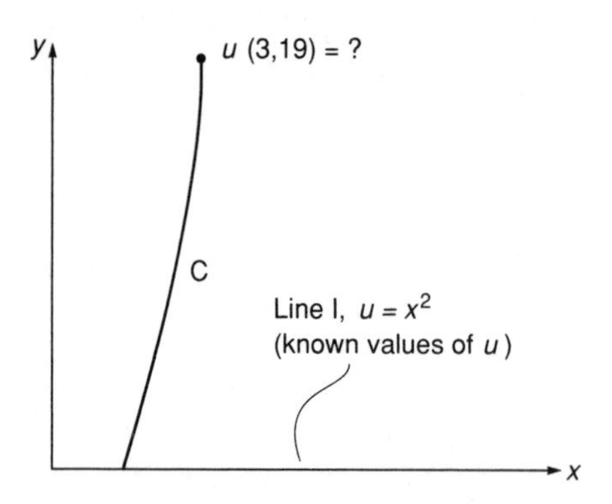

Fig. 8.3 Lines C and I in Example 8.1

Substituting for y in equation (B) gives

$$\frac{du}{dx}=x+x^3-8$$

$\therefore \quad u=\frac{1}{2}x^2+\frac{1}{4}x^4-8x+K.$

However, it is known that when $x=2, u=4$, hence $K=14$. Thus along the characteristic line given by

$$y=x^3-8$$

we have the solution

$$u = \tfrac{1}{4}x^4 + \tfrac{1}{2}x^2 - 8x + 14.$$

Hence when $x=3$, $u=14.75$.

Example 8.2

Given the equation

$$\sqrt{x}\frac{\partial u}{\partial x} + u\frac{\partial u}{\partial y} = -u^2$$

Find the value of u when $x=1.1$ on the characteristic passing through the point $(1, 0)$ given the initial condition $u(x, 0)=1$.

Solution 8.2

Write the equations along a characteristic as

$$\frac{\mathrm{d}x}{\sqrt{x}} = \frac{\mathrm{d}y}{u} = \frac{\mathrm{d}u}{-u^2}$$

$$\therefore \qquad \frac{\mathrm{d}u}{\mathrm{d}x} = \frac{-u^2}{\sqrt{x}} \qquad u(1)=1$$

$$\frac{\mathrm{d}y}{\mathrm{d}x} = \frac{u}{\sqrt{x}} \qquad y(1)=0$$

For nonlinear equations such as this, a numerical solution will usually be more convenient.

We require $u(1.1)$ along this characteristic and may use any of the methods described in Chapter 7.

Using Program 7.2 and the fourth order Runge–Kutta method with a step length of 0.1 we get

$u(1.1)=0.9111$

$y(1.1)=0.0931$

8.4 Second order equations

The majority of partial differential equations that are likely to be encountered in engineering analysis are of second order. The concept of characteristic lines introduced in the preceding section for first order equations is used again here, because it leads to an important means of classifying second order equations.

Consider the general eq. 8.1, but exclude the first derivative terms and the term containing u, i.e.,

$$a\frac{\partial^2 u}{\partial x^2}+b\frac{\partial^2 u}{\partial x \partial y}+c\frac{\partial^2 u}{\partial y^2}+g=0 \tag{8.12}$$

We assume for now that the equation is linear, that is the terms **a, b, c and g are** functions of x and y only and not of u or its derivatives.

The following substitutions can be made to simplify the algebra:

$$p=\frac{\partial u}{\partial x}, \quad q=\frac{\partial u}{\partial y}, \quad r=\frac{\partial^2 u}{\partial x^2}, \quad s=\frac{\partial^2 u}{\partial x \partial y}, \quad t=\frac{\partial^2 u}{\partial y^2}$$

hence, $$ar+bs+ct+g=0 \tag{8.13}$$

Considering small changes in p and q with respect to x and y, we can write

$$\begin{aligned} \mathrm{d}p&=\frac{\partial p}{\partial x}\mathrm{d}x+\frac{\partial p}{\partial y}\mathrm{d}y=r\,\mathrm{d}x+s\,\mathrm{d}y \\ \mathrm{d}q&=\frac{\partial q}{\partial x}\mathrm{d}x+\frac{\partial q}{\partial y}\mathrm{d}y=s\,\mathrm{d}x+t\,\mathrm{d}y \end{aligned} \tag{8.14}$$

Eliminating r and t from eqs 8.13 and 8.14 gives

$$\frac{a}{\mathrm{d}x}(\mathrm{d}p-s\,\mathrm{d}y)+bs+\frac{c}{\mathrm{d}y}(\mathrm{d}q-s\,\mathrm{d}x)+g=0 \tag{8.15}$$

which after multiplication by $\mathrm{d}y/\mathrm{d}x$ leads to

$$s\left[a\left(\frac{\mathrm{d}y}{\mathrm{d}x}\right)^2-b\left(\frac{\mathrm{d}y}{\mathrm{d}x}\right)+c\right]-\left[a\frac{\mathrm{d}p}{\mathrm{d}x}\frac{\mathrm{d}y}{\mathrm{d}x}+c\frac{\mathrm{d}q}{\mathrm{d}x}+g\frac{\mathrm{d}y}{\mathrm{d}x}\right]=0 \tag{8.16}$$

The only remaining partial derivative s can be eliminated by choosing curves or characteristic lines in the solution domain satisfying

$$a\left(\frac{\mathrm{d}y}{\mathrm{d}x}\right)^2-b\left(\frac{\mathrm{d}y}{\mathrm{d}x}\right)+c=0 \tag{8.17}$$

Depending on the roots of 8.17 we can classify three different types of PDE.

(a) $b^2-4ac<0$ Equation is 'ELLIPTIC'
No real characteristic lines exist.

e.g. Laplace's eq. $\dfrac{\partial^2 \phi}{\partial x^2}+\dfrac{\partial^2 \phi}{\partial y^2}=0$

Poisson's eq. $\dfrac{\partial^2 \phi}{\partial x^2}+\dfrac{\partial^2 \phi}{\partial y^2}=f(x, y)$

In both cases $b=0$, $a=c=1$, $b^2-4ac=-4$

(b) $b^2-4ac=0$ Equation is 'PARABOLIC'
Characteristics are coincident

e.g. consolidation eq. $c_v\dfrac{\partial^2 u}{\partial z^2}-\dfrac{\partial u}{\partial t}=0$

In this case $a=c_v$, $b=c=0$, $b^2-4ac=0$

(c) $b^2-4ac>0$ Equation is 'HYPERBOLIC'
Two families of real characteristics exist

e.g. wave eq. $\dfrac{\partial^2 u}{\partial x^2}-\dfrac{1}{k^2}\dfrac{\partial^2 u}{\partial t^2}=0$

In this case $a=1$, $b=0$, $c=-1/k^2$, $b^2-4ac=4/k^2$

The characteristics method is most useful for hyperbolic partial differential equations, where essentially the same techniques of 'integration along the characteristics' as was used for first order equations can be employed again.

The most popular methods for numerical solution of all PDEs, irrespective of their classification, are the finite element method and the finite difference method.

The finite element method is the most versatile approach and is dealt with in numerous texts. The finite difference method is more appropriate to an introductory text and has already been applied to ordinary differential equations in Chapter 7. The remainder of this chapter is devoted to application of the finite difference method to some simple PDEs. Initially, a review of the finite difference method is presented for more than one independent variable.

8.5 Finite differences in two dimensions

In the finite difference method, derivatives that occur in the governing PDE are replaced by their finite difference equivalents. These derivatives are approximated by various combinations of the unknown variable occurring at 'grid points' surrounding the location at which a derivative is required.

Consider in Fig. 8.4 a two-dimensional Cartesian solution domain, split into a regular rectangular grid of width h and height k. The dependent variable u is a function of the two independent variables x and y. The point at which a derivative is required is given the subscript i, j, where i increases in the x-direction and j increases in the y-direction.

Any partial derivative may now be approximated using finite differences and all the expressions given in Tables 5.2, 5.3 and 5.4 are valid.

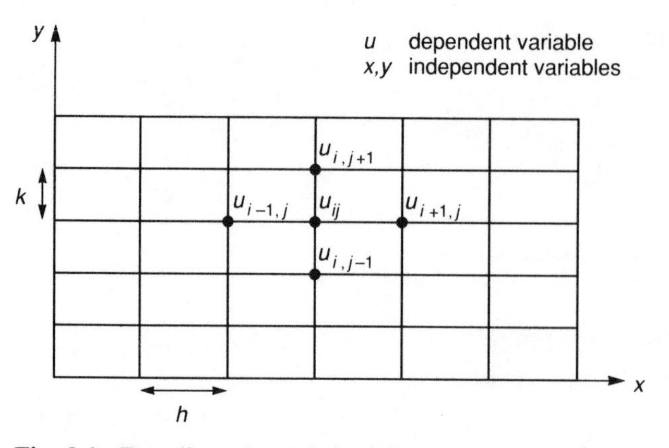

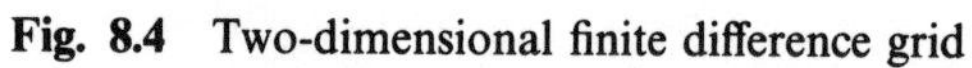
Fig. 8.4 Two-dimensional finite difference grid

For example,

$$\left(\frac{\partial u}{\partial x}\right)_{ij} \simeq \frac{1}{2h}(u_{i+1,j} - u_{i-1,j})$$

$$\left(\frac{\partial u}{\partial y}\right)_{ij} \simeq \frac{1}{2k}(u_{i,j+1} - u_{i,j-1}) \tag{8.18}$$

are central difference formulae for first derivatives.

These are conveniently expressed as computational 'molecules' where,

$$\left(\frac{\partial u}{\partial x}\right)_{ij} \simeq \frac{1}{2h}\left[\textcircled{-1} \!-\! \textcircled{0}_{ij} \!-\! \textcircled{1} \right]$$

$$\left(\frac{\partial u}{\partial y}\right)_{ij} \simeq \frac{1}{2k}\begin{bmatrix} \textcircled{1} \\ | \\ \textcircled{0}_{ij} \\ | \\ \textcircled{-1} \end{bmatrix} \tag{8.19}$$

Combining these two first derivatives, and letting $h=k$, which will always be assumed unless stated otherwise, we get

$$\left(\frac{\partial u}{\partial x} + \frac{\partial u}{\partial y}\right)_{ij} \simeq \frac{1}{2h}\begin{bmatrix} & \textcircled{1} & \\ & | & \\ \textcircled{-1} \!-\! & \textcircled{0}_{ij} & \!-\! \textcircled{1} \\ & | & \\ & \textcircled{-1} & \end{bmatrix} \tag{8.20}$$

Similarly, for second derivatives

$$\left(\frac{\partial^2 u}{\partial x^2}\right)_{ij} \simeq \frac{1}{h^2}\left[\textcircled{1} \!-\! \textcircled{-2}_{ij} \!-\! \textcircled{1} \right], \quad \left(\frac{\partial^2 u}{\partial y^2}\right)_{ij} \simeq \frac{1}{h^2}\begin{bmatrix} \textcircled{1} \\ | \\ \textcircled{-2}_{ij} \\ | \\ \textcircled{1} \end{bmatrix} \tag{8.21}$$

leading to the Laplacian 'molecule'

$$\nabla^2 u_{ij} \simeq \frac{1}{h^2}\left[\begin{array}{ccc} & 1 & \\ 1 & -4_{ij} & 1 \\ & 1 & \end{array}\right] \qquad (8.22)$$

Central difference 'molecules' for third and fourth derivatives can be written in the form

$$\left(\frac{\partial^3 u}{\partial x^3}\right)_{ij} \simeq \frac{1}{2h^3}\left[\begin{array}{ccccc} -1 & 2 & 0_{ij} & -2 & 1 \end{array}\right] \qquad (8.23)$$

$$\left(\frac{\partial^4 u}{\partial x^4}\right)_{ij} \simeq \frac{1}{h^4}\left[\begin{array}{ccccc} 1 & -4 & 6_{ij} & -4 & 1 \end{array}\right] \qquad (8.24)$$

and 'mixed' derivatives as

$$\frac{\partial}{\partial x}\left(\frac{\partial u}{\partial y}\right)_{ij} = \left(\frac{\partial^2 u}{\partial x \partial y}\right)_{ij} \simeq \frac{1}{4h^2}\left[\begin{array}{ccc} -1 & 0 & 1 \\ 0 & 0_{ij} & 0 \\ 1 & 0 & -1 \end{array}\right] \qquad (8.25)$$

The biharmonic operator can be given as

$$\nabla^4 u_{ij} \simeq \frac{1}{h^4}\left[\begin{array}{ccccc} & & 1 & & \\ & 2 & -8 & 2 & \\ 1 & -8 & 20_{ij} & -8 & 1 \\ & 2 & -8 & 2 & \\ & & 1 & & \end{array}\right] \qquad (8.26)$$

All these examples used the lowest order central difference form from Table 5.3. Clearly higher order versions could be devised including forward and backward difference versions if required.

It makes sense to use central difference formulae when possible, as they give the greatest accuracy in terms of the number of points included.

When using these central difference formulae, the *ij*th grid point always lies at the middle of the 'molecule'. This is the point at which a derivative is to be approximated, and the 'molecule' can be visualised as an overlay to the grid drawn on the two-dimensional solution domain.

It may be noted that the rectangular nature of the 'molecules' implies that the solution domains must also be rectangular. Problems with non-rectangular solution domains can be tackled using the 'molecule' approach, but some assumptions must be made when the boundary crosses between grid points. All the examples shown in this chapter will involve rectangular domains.

In summary, the finite difference approach to the solution of linear partial differential equations can be considered in four steps

(a) Divide the spatial solution domain into a grid. The shape of the grid should reflect the nature of the problem and the boundary conditions. The solution will be sought at the n grid points lying within and on the boundaries of the solution domain.
(b) Obtain the finite difference formula that approximately represents the governing PDE. This may be conveniently written as a computational 'molecule'.
(c) Overlay the 'molecule' over each grid point at which a solution is required taking account of boundary conditions. This will lead to n simultaneous equations.
(d) In the case of the elliptic system in Section 8.4 the equations will be algebraic equations which can be solved using the linear algebra techniques from Chapter 2. In the case of the parabolic and hyperbolic systems in Section 8.4 the equations will be ordinary differential equations in time which can be solved using the techniques from Chapter 7.

In more sophisticated computer-coded implementations of finite difference methods, all of these steps are combined in a single program allowing arbitrary boundary shapes. The present examples are only illustrative of what can be achieved.

8.6 Elliptic systems

Problems governed by elliptic systems of PDEs are characterised by closed solution domains (see Fig. 8.5). The boundary conditions must be known, and the problem amounts to finding the values of the dependent variable at internal locations. Numerically, after the finite difference 'discretisation' process has been completed, the solution of a set of (linear) simultaneous equations is involved.

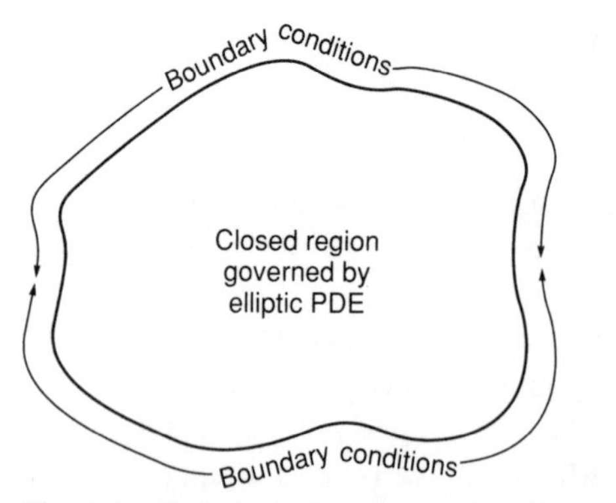

Fig. 8.5 Closed solution domain for elliptic problems

Example 8.3

The rectangular plate shown in Fig. 8.6 is subjected to the boundary temperatures as indicated. Use a finite difference scheme to estimate the steady state temperatures at the internal grid points.

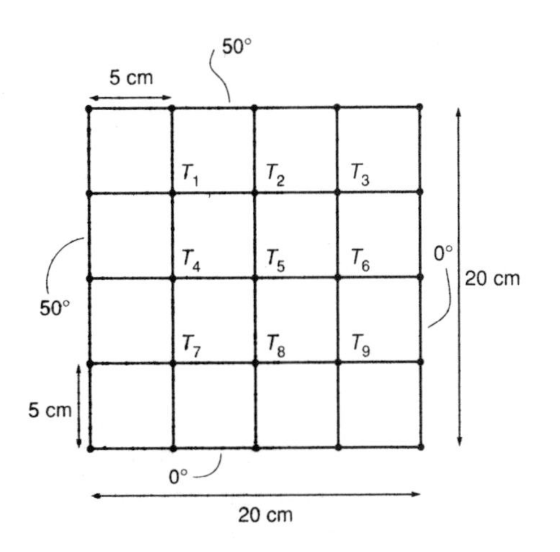

Fig. 8.6 Finite difference grid for Example 8.3

Solution 8.3

The distribution is governed by Laplace's equation

$$\frac{\partial^2 T}{\partial x^2}+\frac{\partial^2 T}{\partial y^2}=0$$

By symmetry, it can be stated that

$$T_2=T_4, \quad T_3=T_7, \quad T_6=T_8$$

The finite difference grid is square with a side length of 5 cm, hence the 'molecule' is given by

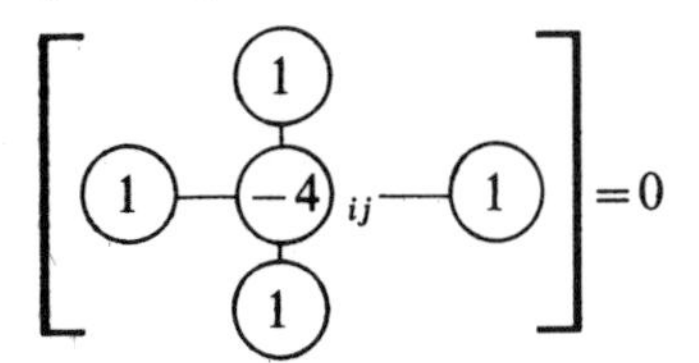

The centre of the molecule is now placed at each unknown grid point in turn to give

Point	
1	$50 + 50 + T_2 + T_4 - 4T_1 = 0$
4	$50 + T_1 + T_5 + T_7 - 4T_4 = 0$
5	$T_4 + T_2 + T_6 + T_8 - 4T_5 = 0$
7	$50 + T_4 + T_8 + 0 - 4T_7 = 0$
8	$T_7 + T_5 + T_9 + 0 - 4T_8 = 0$
9	$T_8 + T_6 + 0 + 0 - 4T_9 = 0$

Using symmetry, we can write this as a matrix product:

$$\begin{bmatrix} -4 & 2 & 0 & 0 & 0 & 0 \\ 1 & -4 & 1 & 1 & 0 & 0 \\ 0 & 2 & -4 & 0 & 2 & 0 \\ 0 & 1 & 0 & -4 & 1 & 0 \\ 0 & 0 & 1 & 1 & -4 & 1 \\ 0 & 0 & 0 & 0 & 2 & -4 \end{bmatrix} \begin{Bmatrix} T_1 \\ T_4 \\ T_5 \\ T_7 \\ T_8 \\ T_9 \end{Bmatrix} = \begin{Bmatrix} -100 \\ -50 \\ 0 \\ -50 \\ 0 \\ 0 \end{Bmatrix}$$

which is of the form $\mathbf{Ax}=\mathbf{b}$.

It is much more convenient for some computational purposes (see Chapter 2) if the leading diagonal terms in **A** are positive. At the same time, we notice that **A** can readily be made symmetric by, for example, multiplying the second, fourth and fifth equations by 2. It can also be noted that a feature of finite difference (and of all 'grid') methods is that the discretisation leads to a banded structure for **A**, in this case with a half bandwidth of 2.

An appropriate solution technique from Chapter 2 might therefore be Program 2.4 for symmetric, banded, positive definite systems. The input band matrix (in rectangular form) would be

$$\begin{bmatrix} 0 & 0 & 4 \\ 0 & -2 & 8 \\ 0 & -2 & 4 \\ -2 & 0 & 8 \\ -2 & -2 & 8 \\ 0 & -2 & 4 \end{bmatrix}$$

and the **b** vector would be

$\{100 \quad 100 \quad 0 \quad 100 \quad 0 \quad 0\}^{\mathrm{T}}$

leading to the solution

$$\begin{Bmatrix} T_1 \\ T_4 \\ T_5 \\ T_7 \\ T_8 \\ T_9 \end{Bmatrix} = \begin{Bmatrix} 42.86 \\ 35.71 \\ 25.00 \\ 25.00 \\ 14.29 \\ 7.14 \end{Bmatrix}$$

In the example just described, the value of T was known around the entire boundary. Derivative boundary conditions may also apply and can be tackled by the introduction of temporary grid points outside the solution domain. A problem of this type was described in Example 7.12.

Example 8.4

The plate shown in Fig. 8.7 has temperatures prescribed on three sides and a derivative boundary condition on the fourth side. Find the steady state temperature distribution using finite differences.

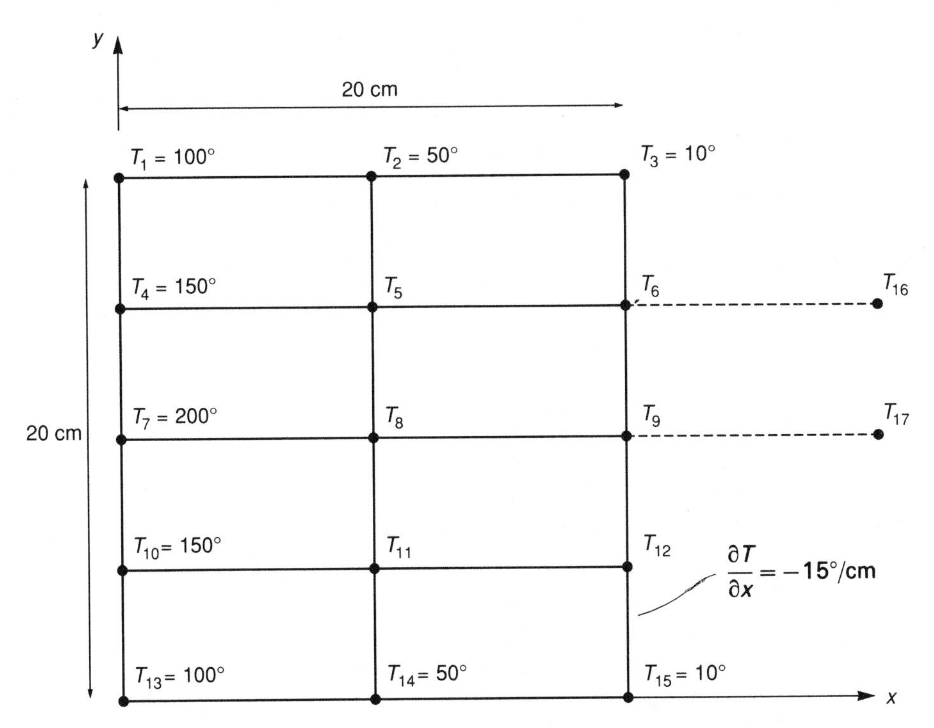

Fig. 8.7 Grid with derivative boundary condition in Example 8.4

Solution 8.4

By symmetry $T_5 = T_{11}$, $T_6 = T_{12}$. Introduce two temporary points at T_{16} and T_{17}. The distribution is governed by Laplace's equation, but in this case the grid is not square, with $h = 10$ cm, $k = 5$ cm.

Hence

$$\frac{\partial^2 T}{\partial x^2} + \frac{\partial^2 T}{\partial y^2} = 0$$

which in finite difference form becomes

$$\tfrac{1}{100}(T_{i-1,j} - 2T_{i,j} + T_{i+1,j}) + \tfrac{1}{25}(T_{i,j-1} - 2T_{i,j} + T_{i,j+1}) = 0$$

giving the computational 'molecule' for this example as

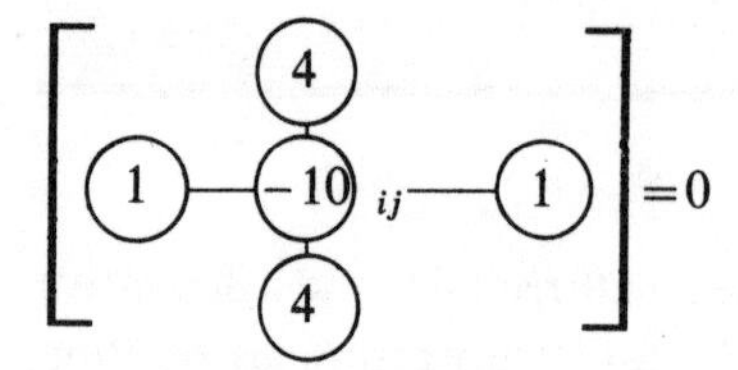

The centre of the molecule is now placed at each unknown grid point in turn to give

Point

5	$150 + 200 + T_6 + 4T_8 - 10T_5 = 0$
6	$T_5 + 40 + T_{16} + 4T_9 - 10T_6 = 0$
8	$200 + 4T_5 + T_9 + 4T_5 - 10T_8 = 0$
9	$T_8 + 4T_6 + T_{17} + 4T_6 - 10T_9 = 0$

Two more equations are required, so the boundary conditions are introduced in finite difference form.

Using central differences

$$\frac{T_{16} - T_5}{20} = -15, \qquad \frac{T_{17} - T_8}{20} = -15$$

Substituting for T_{16} and T_{17}, and changing the signs of all terms results in the matrix equation

$$\begin{bmatrix} 10 & -1 & -4 & 0 \\ -2 & 10 & 0 & -4 \\ -8 & 0 & 10 & -1 \\ 0 & -8 & -2 & 10 \end{bmatrix} \begin{Bmatrix} T_5 \\ T_6 \\ T_8 \\ T_9 \end{Bmatrix} = \begin{Bmatrix} 350 \\ -260 \\ 200 \\ -300 \end{Bmatrix}$$

Again this can be made symmetrical, for example by multiplying the first equation by 2 and dividing the last equation by 2. After these operations, the input matrix for Program 2.4 becomes

$$\begin{bmatrix} 0 & 0 & 20 \\ 0 & -2 & 10 \\ -8 & 0 & 10 \\ -4 & -1 & 5 \end{bmatrix}$$

with **b** vector

$\{700 \ -260 \ 200 \ -150\}^T$.

Solution of these equations gives

$$\begin{Bmatrix} T_5 \\ T_6 \\ T_8 \\ T_9 \end{Bmatrix} = \begin{Bmatrix} 55.9 \\ -32.3 \\ 60.3 \\ -43.8 \end{Bmatrix}$$

8.7 Parabolic systems

For typical parabolic equations, for example the 'conduction' or 'consolidation' equation, we require boundary conditions together with initial conditions in time. The solution then marches along in time for as long as required. Unlike elliptic problems, the solution domain is 'open', (see Fig. 8.8) in the sense that there is no limit to the time value at which a solution could be sought.

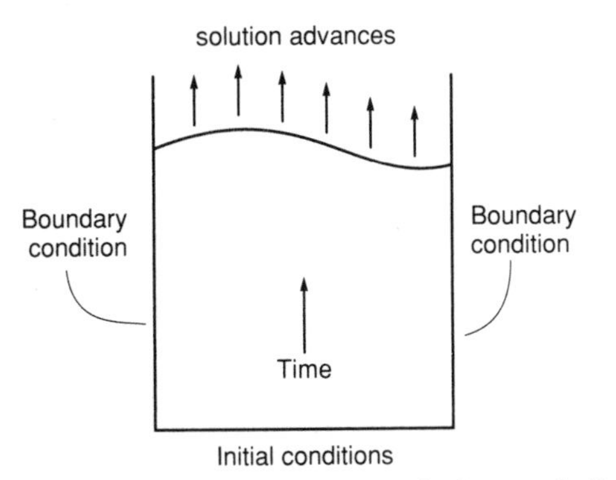

Fig. 8.8 Open solution domain for parabolic problems

A parabolic problem which often arises in civil engineering analysis is the consolidation equation in one dimension, where

$$c_v \frac{\partial^2 u}{\partial z^2} = \frac{\partial u}{\partial t} \tag{8.27}$$

c_v = the coefficient of consolidation (length^2/time); u = excess pore pressure (force/length^2); z = length coordinate; and t = time.

If c_v were replaced by a conductivity property, the same parabolic equation could be used to govern heat flow in a solid.

It is often convenient to non-dimensionalise eq. 8.27 so that a non-dimensional solution can be obtained which is more generally applicable.

Hence let

$$Z = \frac{z}{D}, \quad U = \frac{u}{U_0}, \quad T = \frac{c_v t}{D^2}$$

where D = reference length (drainage path length); U_0 = reference pressure (initial pressure at $t=0$); and T = dimensionless time known as the 'time factor'.

The derivatives can be written as

$$\frac{\partial u}{\partial z}=\frac{\partial u}{\partial Z}\frac{\partial Z}{\partial z}=\frac{1}{D}\frac{\partial u}{\partial Z}$$

$$\frac{\partial^2 u}{\partial z^2}=\frac{\partial}{\partial z}\left(\frac{\partial u}{\partial z}\right)=\frac{\partial}{\partial Z}\frac{\partial Z}{\partial z}\left(\frac{\partial u}{\partial z}\right)=\frac{1}{D^2}\frac{\partial^2 u}{\partial Z^2} \tag{8.28}$$

also $$\frac{\partial u}{\partial t}=\frac{\partial u}{\partial T}\frac{\partial T}{\partial t}=\frac{c_v}{D^2}\frac{\partial u}{\partial T} \tag{8.29}$$

and $$u=UU_0 \tag{8.30}$$

After substitution into eq. 8.27 we get

$$\frac{c_v}{D^2}\frac{\partial^2(UU_0)}{\partial Z^2}=\frac{c_v}{D^2}\frac{\partial(UU_0)}{\partial T}$$

hence $$\frac{\partial^2 U}{\partial Z^2}=\frac{\partial U}{\partial T} \tag{8.31}$$

which is the dimensionless form of eq. 8.27.

8.7.1 Stability of finite differences

Equation 8.27 is readily expressed in finite difference form. Using central differences for the second derivative in space, we get

$$\frac{\partial^2 u}{\partial z^2}\simeq\frac{1}{\Delta z^2}(u_{i-1,j}-2u_{i,j}+u_{i+1,j}) \tag{8.32}$$

where the subscript i refers to changes in z and subscript j refers to changes in t.

This 'semi-discretisation' of the space variables, applied over the whole space grid leads to a set of ordinary differential equations in the time variable. Any of the methods described in Chapter 7 could be used to integrate these sets of equations but in this introductory treatment we shall apply finite differences to the time dimension as well.

Using a simple forward difference scheme for the first time derivative, i.e. $\theta=0$ in Section 7.3.1.8, we get

$$\frac{\partial u}{\partial t}\simeq\frac{1}{\Delta t}(u_{i,j+1}-u_{i,j}) \tag{8.33}$$

Equations 8,32 and 8.33 can be substituted into eq. 8.27 and rearranged to give

$$u_{i,j+1}=u_{i,j}+\frac{c_v\Delta t}{\Delta z^2}(u_{i-1,j}-2u_{i,j}+u_{i+1,j}) \tag{8.34}$$

This type of relationship is termed 'explicit' because the value of u at the new time level is expressed solely in terms of values of u at the immediately preceding time level. However, as with all 'explicit' approaches, numerical stability is conditional on a satisfactory combination of spatial and temporal step lengths being employed. Numerical instability occurs when a perturbation (or error) introduced at a certain stage in the stepping procedure grows uncontrollably at subsequent steps.

It can be shown that numerical stability is only guaranteed in this explicit method if

$$\frac{c_v \Delta t}{\Delta z^2} \leq \frac{1}{2}. \tag{8.35}$$

Example 8.5

The insulated rod shown in Fig. 8.9 is initially at 0° at all points along its length except at the left-hand end which is maintained at 100°C. Use finite differences to compute the temperature variation as a function of position and time.

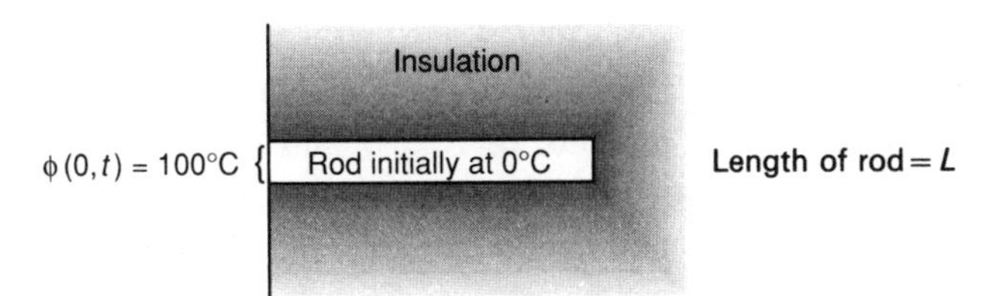

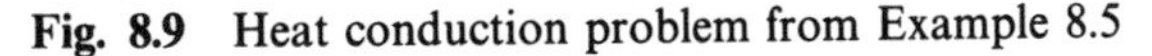

Fig. 8.9 Heat conduction problem from Example 8.5

Solution 8.5

Governing equation

$$k\frac{\partial^2\phi}{\partial x^2} = \frac{\partial\phi}{\partial t}$$

where k = conductivity; ϕ = temperature; x = space dimension; t = time.

See Fig. 8.9.

Boundary conditions $\qquad \phi(0, t) = 100°\text{C} = \phi_0$

$$\frac{\partial\phi}{\partial x}(L, t) = 0°\text{C/m (zero gradient at insulated end)}$$

Initial condition $\qquad \phi(x, 0) = 0°\text{C} \quad \text{for} \quad 0 < x \leq L$

After arranging the equation in non-dimensional form we get

$$\frac{\partial^2\Phi}{\partial X^2} = \frac{\partial\Phi}{\partial T}$$

where $X = \dfrac{x}{L}$, $\Phi = \dfrac{\phi}{\phi_0}$, $T = \dfrac{kt}{L^2}$

with boundary conditions $\Phi(0, T)=1, \quad \frac{\partial \Phi}{\partial X}(1, T)=0$

and initial condition $\Phi(X, 0)=0 \quad \text{for} \quad 0<X\leq 1.$

Expressing the dimensionless equation in finite difference form to give

$$\frac{1}{\Delta X^2}(\Phi_{i-1,j}-2\Phi_{ij}+\Phi_{i+1,j})=\frac{1}{\Delta T}(\Phi_{i,j+1}-\Phi_{i,j})$$

$$\therefore \quad \Phi_{i,j+1}=\Phi_{i,j}+\frac{\Delta T}{\Delta X^2}(\Phi_{i-1,j}-2\Phi_{i,j}+\Phi_{i+1,j})$$

The stability requirement is that $\Delta T/\Delta X^2\leq\frac{1}{2}$. If we let $\Delta T/\Delta X^2=\frac{1}{2}$ then the formula simplifies to

$$\Phi_{i,j+1}=\tfrac{1}{2}(\Phi_{i-1,j}+\Phi_{i+1,j})$$

Let $\Delta X=0.2$ and $\Delta T=0.02$ and set out results in tabular form as shown in Table 8.2. (Instead of applying the full temperature of $\Phi=1.0$ at $T=0.0$ and $X=0.0$, it is sometimes recommended that half is applied at $T=0.0$ and the remainder at $T=\Delta T$.)

The condition that $\partial\Phi/\partial X=0$ at $X=1$ is maintained by including the fictitious values at $X=1.2$ which by central differences, are maintained at the same values as those at $X=0.8$.

Table 8.2. Tabulated values of $\Phi=f(X, T)$

T \ X	0.0	0.2	0.4	0.6	0.8	1.0	1.2
0.0	0.5	0.0	0.0	0.0	0.0	0.0	0.0
0.02	1.0	0.25	0.0	0.0	0.0	0.0	0.0
0.04	1.0	0.50	0.125	0.0	0.0	0.0	0.0
0.06	1.0	0.563	0.25	0.063	0.0	0.0	0.0
0.08	1.0	0.625	0.313	0.125	0.031	0.0	0.031
0.10	1.0	0.656	0.375	0.172	0.063	0.031	0.063
0.12	1.0	0.688	0.414	0.219	0.102	0.063	0.102
0.14	1.0	0.707	0.453	0.258	0.141	0.102	0.141

Example 8.6

A specimen of uniform saturated clay 240 mm thick is placed in a conventional consolidation cell with drains at top and bottom. A sudden increment of vertical stress of 200 kN/m^2 is applied. If the coefficient of consolidation is $c_v=10$ m^2/yr estimate the excess pore pressure distribution after 1 hour.

Solution 8.6

Solve the dimensionless form (see p. 291)

$$\frac{\partial^2 U}{\partial Z^2} = \frac{\partial U}{\partial T}$$

The stability requirement is that $\Delta T/\Delta Z^2 \leq \frac{1}{2}$ so let $\Delta T = 0.02$, $\Delta Z = 0.2$ which leads to the finite difference equation

$$U_{i,j+1} = \tfrac{1}{2}(U_{i-1,j} + U_{i+1,j})$$

In order to find how many dimensionless time steps are required to reach one hour, we need the relationship between T and t, i.e.

$$\Delta T = \frac{c_v \Delta t}{D^2} \quad \text{where } D = \text{reference length}$$

$$= 0.12\text{ m due to symmetry}$$

$\Delta t = 0.02 \times 0.12^2 \times 365 \times 24/10 = 0.252$ hr, hence one hour is reached after approximately four dimensionless time steps. Using the finite difference equation, computed values of U are given in Table 8.3 assuming the boundary condition at the symmetrical 'mid-plane' is $\partial u/\partial z = 0$.

Table 8.3. Tabulated values of $U = f(Z, T)$ from Example 8.6

Z \ T	0.0	0.02	0.04	0.06	0.08
0.0	0.0	0.0	0.0	0.0	0.0
0.2	1.0	0.5	0.5	0.375	0.375
0.4	1.0	1.0	0.75	0.75	0.625
0.6	1.0	1.0	1.0	0.875	0.875
0.8	1.0	1.0	1.0	1.0	0.9375
1.0	1.0	1.0	1.0	1.0	1.0
1.2	1.0	1.0	1.0	1.0	0.9375

When $T = 0.08$, $t = 1.01$ hr, and the solution is plotted in Fig. 8.10 after multiplication by the initial load intensity 200 kN/m^2.

8.8 Hyperbolic systems

Hyperbolic systems usually involve propagation phenomena, for example as described by the wave equation. The displacements of a vibrating string are given by

$$c^2 \frac{\partial^2 v}{\partial x^2} = \frac{\partial^2 v}{\partial t^2} \tag{8.36}$$

where $c^2 = T/\rho$; T = tension in the string; and ρ = mass per unit length.

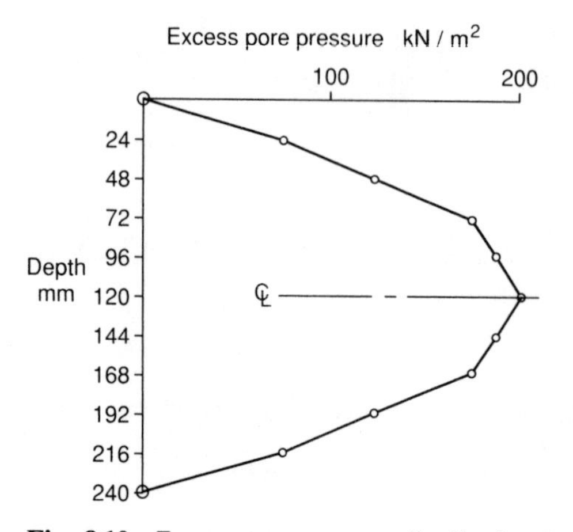

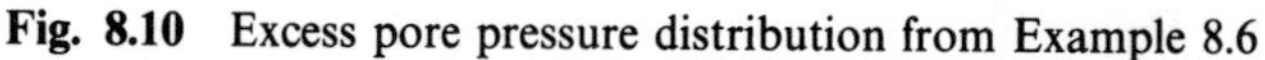

Fig. 8.10 Excess pore pressure distribution from Example 8.6

Longitudinal vibration displacements of a rod are given by

$$c^2\frac{\partial^2 u}{\partial x^2}=\frac{\partial^2 u}{\partial t^2} \tag{8.37}$$

where $c^2=E/\rho$; E=Young's modulus; and ρ=mass per unit volume.

Clearly these one-dimensional examples can be extended to two or three dimensions, in which case the Laplacian operator can be used, i.e.

$$c^2\nabla^2 u=\frac{\partial^2 u}{\partial t^2} \tag{8.38}$$

As with parabolic systems, it is often convenient to non-dimensionalise the one-dimensional wave equation. With reference to eq. 8.36, by making the substitutions

$$V=\frac{v}{v_0},\qquad X=\frac{x}{L}\quad\text{and}\quad T=\frac{ct}{L}$$

where v_0 is an initial displacement and L is a reference length; we can write

$$\frac{\partial^2 v}{\partial x^2}=\left(\frac{\partial X}{\partial x}\right)^2\frac{\partial^2 v}{\partial X^2}=\frac{1}{L^2}\frac{\partial^2 v}{\partial X^2}$$

$$\text{and}\qquad \frac{\partial^2 v}{\partial t^2}=\left(\frac{\partial T}{\partial t}\right)^2\frac{\partial^2 v}{\partial T^2}=\frac{c^2}{L^2}\frac{\partial^2 v}{\partial T^2} \tag{8.39}$$

$$\text{hence}\qquad \frac{\partial^2 V}{\partial X^2}=\frac{\partial^2 V}{\partial T^2} \tag{8.40}$$

The finite difference form of this equation can be written as

$$\frac{1}{\Delta X^2}(V_{i-1,j}-2V_{i,j}+V_{i+1,j})=\frac{1}{\Delta T^2}(V_{i,j-1}-2V_{i,j}+V_{i,j+1}) \tag{8.41}$$

where the subscript i refers to changes in X and subscript j refers to changes in T. ΔX and ΔT are the dimensionless step lengths in X and T respectively.

Rearrangement of eq. 8.41 gives

$$V_{i,j+1}=\frac{\Delta T^2}{\Delta X^2}(V_{i-1,j}-2V_{i,j}+V_{i+1,j})-V_{i,j-1}+2V_{i,j} \tag{8.42}$$

If we arrange for $\Delta X=\Delta T$ then eq. 8.42 simplifies considerably to become

$$V_{i,j+1}=V_{i-1,j}-V_{i,j-1}+V_{i+1,j} \tag{8.43}$$

which also turns out to be the exact solution! Other values of the ratio $\Delta T/\Delta X$ could be used, but stability is only guaranteed if it is less than one.

Example 8.7

Solve the wave equation

$$\frac{\partial^2 U}{\partial X^2}=\frac{\partial^2 U}{\partial T^2}$$

in the range $0 \leq X \leq 1$, $0 \leq T$ subject to the following initial and boundary conditions

At $X=0$ and $X=1$, $U=2\sin(\pi T/5)$ for all T

When $T=0$, $U=\partial U/\partial T=0$ for all X.

Solution 8.7

Choose $\Delta X=\Delta T=0.1$ and take account of symmetry. The finite difference equation 8.43 requires information from two preceding steps in order to proceed. In order to compute the result corresponding to $T=0.1$ the assumption was made that no change in the values of U had occurred. This is equivalent to a simple Euler step based on the initial condition that $\partial U/\partial T=0$ for all X. Application of eq. 8.43 leads to the results shown in Table 8.4.

Note that again this solution has involved 'explicit' finite differences in the time

Table 8.4. Tabulated values of $U=f(Z, T)$ from Example 8.7

T \ X	0.0	0.1	0.2	0.3	0.4	0.5	0.6
0.0	0.0	0.0	0.0	0.0	0.0	0.0	0.0
0.1	0.1256	0.0	0.0	0.0	0.0	0.0	0.0
0.2	0.2507	0.1256	0.0	0.0	0.0	0.0	0.0
0.3	0.3748	0.2507	0.1256	0.0	0.0	0.0	0.0
0.4	0.4974	0.3748	0.2507	0.1256	0.0	0.0	0.0
0.5	0.6180	0.4974	0.3748	0.2507	0.1256	0.0	0.1256
0.6	0.7362	0.6180	0.4974	0.3748	0.2507	0.2512	0.2507
0.7	0.8516	0.7362	0.6180	0.4974	0.5004	0.5014	0.5004

variable. The stability restrictions attached to this kind of method can be removed by 'binding together' the solutions at two (or more) time levels in what are called 'implicit' difference schemes. Numerically these mean that a set of simultaneous equations has to be solved on each time step, involving again the linear algebra library subroutines from Chapter 2.

8.9 Exercises

1 Given the equation

$$\sqrt{x}\frac{\partial z}{\partial x}+z\frac{\partial z}{\partial y}=-z^2,$$

use a numerical method to estimate $z(3.5, y)$ on the characteristic through $(3, 0)$ given that $z=1$ at all points on the x-axis.
Answer: 0.783.

2 Given

$$z(xp-yzq)=y^2-x^2$$

where

$$p=\frac{\partial z}{\partial x} \quad \text{and} \quad q=\frac{\partial z}{\partial y}$$

For the characteristic passing through $(1, 1)$ estimate the value of y and z when $x=1.5$ given that $z(1, 1)=1$.
Answer: 0.702, 0.562.

3 Find the steady state temperature distribution in the square plate shown:

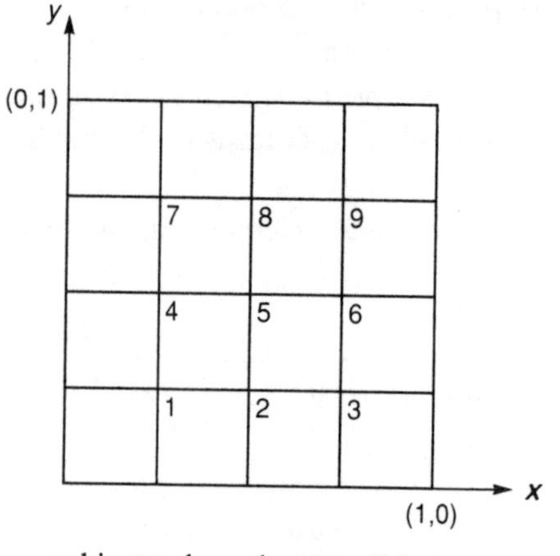

subject to boundary conditions:

At $x=0$ $T=100$
$x=1$ $T=0$
$y=0$ $T=100$
$y=1$ $T=0$

Laplace's equation governs the behaviour.

Use a two-dimensional finite difference grid of side length 0.25.

Answer: $T_1=85.71$ $T_2=T_4=71.43$ $T_5=50.00$ $T_3=T_7=50.00$ $T_9=14.29$ $T_6=T_8=28.57$

4 The figure below represents the cross-section of an 8 cm square bar subjected to pure torsion

	7	8	9
	4	5	6
	1	2	3

The stress function ϕ is found to be distributed across the bar according to Poisson's equation

$$\frac{\partial^2\phi}{\partial x^2}+\frac{\partial^2\phi}{\partial y^2}+2=0$$

Given that $\phi=0$ on the boundaries, find its value at the internal grid points.

Answer: $\phi_1=\phi_7=\phi_9=\phi_3=5.5\text{ cm}^2$ $\phi_2=\phi_4=\phi_6=\phi_8=7\text{ cm}^2$ $\phi_5=9\text{ cm}^2$

5 The variation of a quantity T, with respect to Cartesian directions x and y, is defined by the fourth order differential equation

$$\frac{\partial^4 T}{\partial x^4}+\frac{\partial^4 T}{\partial y^4}=0.$$

Starting from first principles, derive the finite difference form of this equation at the point (x_i, y_j) assuming a square mesh.

Using the finite difference approximation you have derived, find the value for T at the centre of the square area ABCD shown in the figure using the boundary conditions given and a mesh spacing of unity.

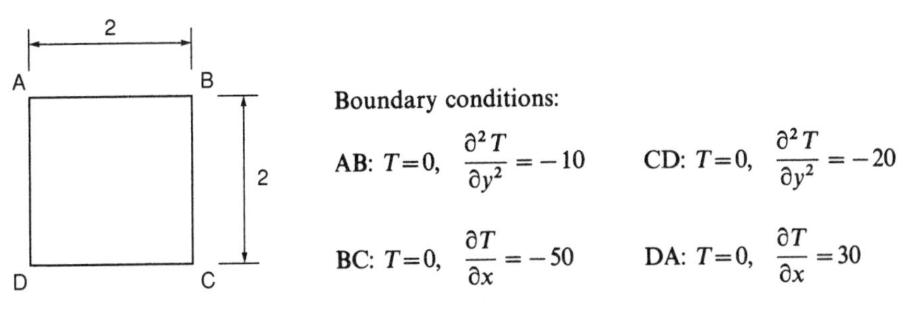

Answer: 15.833

6 The steady two-dimensional distribution in a heat conducting material is given by Laplace's equation

$$\frac{\partial^2 T}{\partial x^2}+\frac{\partial^2 T}{\partial y^2}=0,$$

where T is temperature and x, y represent the Cartesian coordinate system. Using a finite difference form of this equation, find temperatures T_1, T_2, T_3 and T_4 for the metal plate shown in the figure.

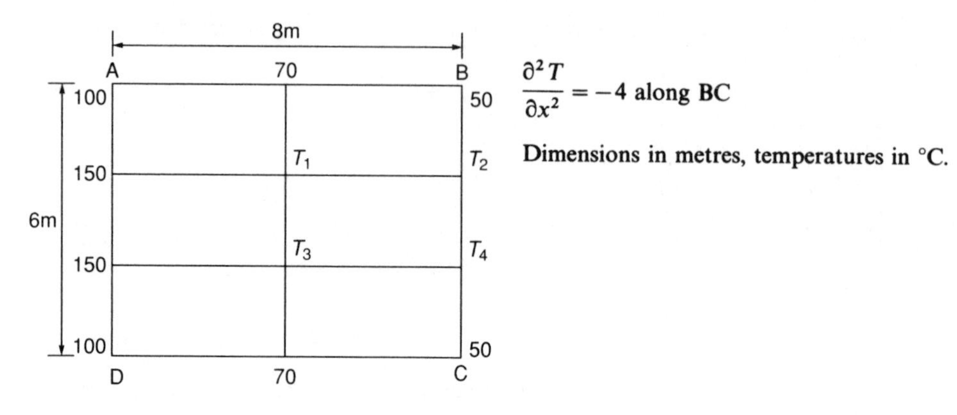

The thermal energy, Q, stored in the plate is calculated from

$$Q = \rho c \int T \, dV,$$

where Q is in joules (J), ρ is density (=8000 kg/m^3), c is specific heat capacity of the material (=1000 J/kg/°C), and dV is an increment of volume (m^3). If the thickness of the plate is 50 mm, use the repeated trapezium rule to estimate Q.
Answer: $T_1 = T_3 = 77.33$, $T_2 = T_4 = 34$, $Q = 1.55$ gJ

7 The steady three-dimensional temperature distribution in a heat conducting material may be described by Laplace's equation

$$\nabla^2 T = \frac{\partial^2 T}{\partial x^2} + \frac{\partial^2 T}{\partial y^2} + \frac{\partial^2 T}{\partial z^2} = 0,$$

where T is temperature and x, y, and z represent the Cartesian coordinate system. Derive a finite difference form of this equation at the point (x_i, y_j, z_k) assuming a cubic mesh.

Use this approximation, and a grid spacing of 0.5, to determine the temperature at the centre of a cube of unit dimensions, given that

$$\nabla^2 T = 0, \quad 0 \le x, y, z \le 1;$$

$$T(0, y, z) = 100, \quad \frac{\partial T}{\partial x}(1, y, z) = -20, \quad 0 \le y, z \le 1;$$

$$T(x, 0, z) = 40, \quad T(x, 1, z) = 70, \quad 0 \le x, z \le 1;$$

$$T(x, y, 0) = 50, \quad T(x, y, 1) = 50, \quad 0 \le x, y \le 1.$$

If the latter boundary conditions were replaced by

$$\frac{\partial T}{\partial z}(x, y, 0) = \frac{\partial T}{\partial z}(x, y, 1) = 0, \quad 0 \le x, y \le 1$$

how could the analysis be simplified?
Answer: $T_1 = 60.29$, change in boundary conditions leads to two-dimensional analysis.

8 A horizontal clay stratum of thickness 5 m is subjected to a loading which produces a pressure distribution which varies from $p\,\text{kN/m}^2$ at the top to $0.5\,p\,\text{kN/m}^2$ at the bottom.

By subdividing the soil into five layers and using steps of one month, use a finite difference approximation to estimate the excess pore pressures after five months.

What is the percentage degree of consolidation at this time?

Let $c_v = 3\,\text{m}^2/\text{yr}$, assume double drainage.

N.B. The degree of consolidation at time t is given by

$$U_t = \frac{(A_0 - A_t)}{A_0}\%$$

where A_0 and A_t are the areas of the pore pressure distributions at time $t=0$ and t respectively.

Answer: 54%

9 Classify each of the following equations as elliptic, hyperbolic or parabolic

(a) $\dfrac{\partial^2 u}{\partial x^2} - \dfrac{\partial^2 u}{\partial y^2} = 0$ (b) $\dfrac{\partial u}{\partial x} + \dfrac{\partial^2 u}{\partial x \partial y} = 8$

(c) $\dfrac{\partial^2 u}{\partial x^2} - 2\dfrac{\partial^2 u}{\partial x \partial y} + 2\dfrac{\partial^2 u}{\partial y^2} = x + 3y$

(d) $\dfrac{\partial^2 u}{\partial x^2} + 3\dfrac{\partial^2 u}{\partial x \partial y} + 4\dfrac{\partial^2 u}{\partial y^2} + 5\dfrac{\partial u}{\partial x} - 2\dfrac{\partial u}{\partial y} + 4u = 2x - 6y$

(e) $\dfrac{\partial^2 u}{\partial x^2} - 7\dfrac{\partial^2 u}{\partial x \partial y} + \dfrac{\partial^2 u}{\partial y^2} = 0$ (f) $\dfrac{\partial^2 u}{\partial x^2} + \dfrac{\partial^2 u}{\partial x \partial y} - 6\dfrac{\partial^2 u}{\partial y^2} = 0$

(g) $\dfrac{\partial^2 u}{\partial x^2} + 6\dfrac{\partial^2 u}{\partial x \partial y} + 9\dfrac{\partial^2 u}{\partial y^2} = 0$

Answer: H, H, E, E, H, H, P

10 A function $u(x, t)$ is to satisfy the differential equation

$$\frac{\partial^2 u}{\partial x^2} = \frac{\partial^2 u}{\partial t^2}$$

in the domain $0 \le x \le 10$

$$0 \le t < \infty$$

The following boundary/initial conditions apply:

$u(0, t) = \dfrac{\partial u}{\partial x}(10, t) = 0$ for all t

$u(x, 0) = x(20 - x)$ for $0 \le x \le 10$

$\dfrac{\partial u}{\partial t}(x, 0) = 0$ for $0 \le x \le 10$

Take $\Delta x = \Delta t = 2$ and find $u(x, 20)$ for $x = 2, 4, 6, 8$.

Answer: $u(2, 20) = -36$, $u(4, 20) = -64$, $u(6, 20) = -84$, $u(8, 20) = -96$.

8.10 Further reading

Burden, R.L. and Faires, J.D. (1989). *Numerical Analysis*, 4th edn, PWS-Kent, Boston, Massachusetts.

Chapra, S.C. and Canale, R.P. (1988). *Numerical Methods for Engineers*, 2nd edn, McGraw-Hill, New York.

Cheney, W. and Kincaid, D. (1985). *Numerical Mathematics and Computing*, 2nd edn, Brooks/Cole, Monterey, California.

Forsythe, G. and Wasow, W. (1960). *Finite Difference Methods for Partial Differential Equations*, Wiley, New York.

Froberg, C.E. (1985). *Numerical Mathematics*, Benjamin/Cummings, Menlo Park, California.

Gladwell, I. and Wait, R. (eds) (1979). *A Survey of Numerical Methods for Partial Differential Equations*, Oxford University Press, Oxford.

Lapidus, L. and Pinder, G. (1982). *Numerical Solution of Partial Differential Equations in Science and Engineering*, Wiley, New York.

Mitchell, A.R. (1969). *Computational Methods in Partial Differential Equations*, Wiley, New York.

Ralston, A. and Wilf, H.S. (1967). *Mathematical Methods for Digital Computers*, Wiley, New York.

Richtmyer, R. and Morton, K. (1967). *Difference Methods for Initial Value Problems*, 2nd edn, Wiley, New York.

Salvadori, M.G. and Baron, M.L. (1961). *Numerical Methods in Engineering*, Prentice-Hall, Englewood Cliffs, New Jersey.

Schwarz, H.R. (1989). *Numerical Analysis*, Wiley, New York.

Shoup, T.E. (1979). *A Practical Guide to Computer Methods for Engineers*, Prentice-Hall, Englewood Cliffs, New Jersey.

Smith, G.D. (1985). *Numerical Solution of Partial Differential Equations*, 3rd edn, Oxford University Press, London.

Smith, I.M. and Griffiths, D.V. (1988). *Programming the Finite Element Method*, 2nd edn, Wiley, New York.

Zienkiewicz, O.C. (1989). *The Finite Element Method*, 4th edn, McGraw-Hill, Maidenhead.

A1

Alphabetic Description of Subroutine Library

All subroutines called by the programs in this book appear either at the end of the main programs or are held in a subroutine library. In this appendix, all the library subroutines are described together with their argument lists. Underlined arguments are those returned by the routines as output. Other arguments are provided to the routines as input, although in some cases the same arguments are used for both input and output. Names in parentheses indicate subroutines normally used in conjunction with the ones they follow.

Subroutine	*Description*
BILIN	Computes values of the bilinear shape functions and their derivatives with respect to local coordinates over a quadrilateral, at Gaussian integration points
Arguments	
<u>DER</u>	Shape function derivatives in local coordinates
<u>FUN</u>	Shape functions
SAMP	Array holding sampling points (column 1) and weighting coefficients (column 2) for Gauss–Legendre quadrature
ISAMP	Size of SAMP in main program
I, J	Counters identifying Gaussian integration points

Subroutine	*Description*
CHECK	Checks the change in a scalar from one iteration to the next to see if convergence is achieved. X0 is updated to X1
Arguments	
X1	New value
X0	Old value
TOL	Convergence tolerance
<u>ICON</u>	Convergence satisfied (= 1), not satisfied (= 0)

Subroutine	*Description*
CHECON	Checks the change in a vector (column array) from one iteration to the next to see if convergence is achieved. OLDLDS is updated to LOADS
Arguments	
LOADS	Vector of new values
OLDLDS	Vector of old values
N	Length of vectors to be compared
TOL	Convergence tolerance
ICON	Convergence satisfied (=1), not satisfied (=0).

Subroutine	*Description*
CHOFAC (CHOSUB)	Factorises a lower triangular symmetrical banded matrix (stored as a rectangle) using Cholesky's method
Arguments	
LB	Symmetrical banded matrix stored as a rectangle (N rows, IW + 1 columns). Holds Cholesky factors on output
ILB	Size of LB in main program
N	Number of equations
IW	Half bandwidth

Subroutine	*Description*
CHOSUB (CHOFAC)	Performs forward and back substitution using Cholesky factors produced by CHOFAC
Arguments	
LB	Array holding Cholesky factors on which backsubstitution is to be performed
ILB	Size of LB in main program
B	Right hand side vector, overwritten by solution vector
N	Number of equations
IW	Half bandwidth

Subroutine	*Description*
GAULAG	Generates weights and sampling points for Gauss–Laguerre quadrature
Arguments	
SAMP	Array holding sampling points (column 1) and weighting coefficients (column 2) for Gauss–Laguerre quadrature.
ISAMP	Size of SAMP in main program
NGL	Required number of Gauss–Laguerre points

Subroutine	*Description*
GAULEG	Generates weights and sampling points for Gauss–Legendre quadrature.
Arguments	
SAMP	Array holding sampling points (column 1) and weighting coefficients (column 2) for Gauss–Legendre quadrature
ISAMP	Size of SAMP in main program
NGP	Required number of Gauss–Legendre points

Subroutine	*Description*
LDLFOR (SUBBAC)	Performs forward substitution on factors in **LDL**^{T} method
Arguments	
A	Array holding upper triangular factors stored as a lower triangle on which forward substitution is to be performed
IA	Size of A in main program
B	Right-hand side vector, overwritten by factored values
N	Number of equations

Subroutine	*Description*
LDLT (LDLFOR) (SUBBAC)	Factorises a symmetrical matrix using **LDL**^{T} method
Arguments	
A	Array holding symmetrical matrix on input. Holds **U** and **U**^{T} on output.
IA	Size of A in main program
D	Vector containing diagonals of **D**
N	Number of equations

Subroutine	*Description*
LUFAC (SUBFOR) (SUBBAC)	Factorises a matrix using **LU** method
Arguments	
A	Array holding equation coefficients
UPTRI	Array holding upper triangle factors **U**
LOWTRI	Array holding lower triangle factors **L**
IA	Size of A, UPTRI and LOWTRI in main program
N	Number of equations

Subroutine	*Description*
LUPFAC (LUPSOL)	Factorises a matrix using **LU** method with pivoting
Arguments	
A	Array holding equation coefficients on input. Holds **LU** factors after pivoting on output
IA	Size of A in main program
N	Number of equations
ROW	Vector of integers holding reordered row numbers

Subroutine	*Description*
LUPSOL (LUPFAC)	Performs forward and back substitution on factors in **LU** method with pivoting
Arguments	
A	Array holding **LU** factors
IA	Size of A in main program
B	Right-hand side vector
SOL	Solution vector
N	Number of equation
ROW	Reordered row numbers

Subroutine	*Description*
MATINV	Inverts a non-singular matrix
Arguments	
A	Array on input and the inverse on output
IA	Size of A in main program
N	Number of rows and columns in A

Subroutine	*Description*
MATMUL	Multiplies matrices **A** and **B** to give **C**
Arguments	
A	First array (L rows, M columns)
IA	Size of A in main program
B	Second array (M rows, N columns)
IB	Size of B in main program
C	Product of A and B (L rows, N columns)
IC	Size of C in main program
L, M, N	Number of rows and columns of A, B and C

Subroutine	*Description*
MATRAN	Forms matrix **A** which is the transpose of matrix **B**
Arguments	
A	Transpose of array B
IA	Size of A in main program
B	Array whose transpose is required
IB	Size of B in main program
M	Number of rows in B
N	Number of columns in B

Subroutine	*Description*
MSMULT	Multiplies matrix **A** by scalar *C*
Arguments	
A	Array on input and after multiplication by scalar C on output
IA	Size of A in main program
C	Scalar multiplier
M	Number of rows in A
N	Number of columns in A

Subroutine	*Description*
MVMULT	Multiplies matrix **M** by vector **V** to give vector **Y**
Arguments	
M	Array
IM	Size of M in main program
V	Vector (length L)
K	Number of rows in M
L	Number of columns in M
Y	Product of M and V (length K)

Subroutine	*Description*
NEWCOT	Provides weighting coefficients for Newton–Cotes quadrature
Arguments	
WEIGHT	Vector of weighting coefficients for Newton–Cotes integration rules
NNC	Required number of points

Subroutine	*Description*
NULL	Nulls a two-dimensional array
Arguments	
A	Array on input and nulled array on output
IA	Size of A in main program
M	Number of rows in A
N	Number of columns in A

Subroutine	*Description*
NULVEC	Nulls a vector (column array)
Arguments	
VEC	Vector on input and nulled vector on output
N	Length of vector

Subroutine	*Description*
PRINTA	Prints an array to the required output channel
Arguments	
A	Array to be printed to output channel ICH, with format 6EI2.4
IA	Size of A in main program
M	Number of rows in A
N	Number of columns in A
ICH	Output channel

Subroutine	*Description*
PRINTV	Prints a vector (column array) to the required output channel
Arguments	
VEC	Vector to be printed to output channel ICH, with format 6E12.4
N	Length of vector
ICH	Output channel

Subroutine	*Description*
SKYFAC (SKYSUB)	Factorises a symmetrical banded matrix stored as a vector using a 'skyline' storage approach and Cholesky's method
Arguments	
A	Vector of 'skyline' coefficients on input and Cholesky factors on output
N	Number of equations
KDIAG	Vector of integers holding locations of diagonal terms in A

Subroutine	*Description*
SKYSUB (SKYFAC)	Performs forward and back substitution using Cholesky factors from 'skyline' storage approach
Arguments	
A	Vector of Cholesky factors
B̲	Right-hand side vector on input overwritten by solution vector on output
N	Number of equations
KDIAG	Vector of integers holding locations of diagonal terms in A

Subroutine	*Description*
SOLVE	Solves a system of equations using Gaussian elimination with partial pivoting
Arguments	
K	Array of equation coefficients
IK	Size of K in main program
U̲	Solution vector
F	Right-hand side vector
N	Number of equations

Subroutine	*Description*
SUBBAC (SUBFOR) (LDLFOR)	Performs back substitution on factors following forward substitution in $\mathbf{LDL}^T$ or **LU** methods
Arguments	
A	Array of upper triangular factors for back substitution
IA	Size of A in main program
B̲	Right-hand side vector from SUBFOR on input, solution vector on output
N	Number of equations

Subroutine	*Description*
SUBFOR (SUBBAC)	Performs forward substitution on factors from **LU** method
Arguments	
A	Array of lower triangular factors for forward substitution
IA	Size of A in main program
B	Right-hand side vector on input, factored vector on output
N	Number of equations

Subroutine	*Description*
VDOTV	Forms dot product DOTPR of vectors **V1** and **V2**
Arguments	
V1	First vector
V2	Second vector
DOTPR	Dot product
N	Length of V1 and V2

Subroutine	*Description*
VECADD	Adds vectors **A** and **B** to give **C**
Arguments	
A	First vector
B	Second vector
C	Sum of first two vectors
N	Length of vectors

Subroutine	*Description*
VECCOP	Forms vector copy **B** of vector **A**
Arguments	
A	Vector to be copied
B	Copy of A
N	Length of vectors

Subroutine	*Description*
VECSUB	Forms vector **C** by subtracting **B** from **A**
Arguments	
A	First vector
B	Second vector
C	Result of subtracting B from A
N	Length of vectors

Subroutine	*Description*
VSMULT	Multiplies vector **V** by scalar SCAL
Arguments	
V	Vector on input and after multiplication by scalar SCAL on output
SCAL	Scalar multiplier
N	Length of vector

Subroutine	*Description*
VVMULT	Forms vector product of vectors **V1** and **V2** to give matrix **A**
Arguments	
V1	First vector (length M)
V2	Second vector (length N)
A	Array formed by vector product of V1 and V2 (M rows, N columns)
IA	Size of A in main program
M	Number of rows in A
N	Number of columns in A

Alphabetic Listing of Subroutine Library in FORTRAN 77

```
C
C
C
C
C
C
      SUBROUTINE BILIN(DER,FUN,SAMP,ISAMP,I,J)
C
C      SHAPE FUNCTIONS AND DERIVATIVES
C      FOR QUADRILATERAL TRANSFORMATION
C
      REAL DER(2,4),FUN(4),SAMP(ISAMP,2)
      ETA = SAMP(I,1)
      XI = SAMP(J,1)
      ETAM = .25* (1.-ETA)
      ETAP = .25* (1.+ETA)
      XIM = .25* (1.-XI)
      XIP = .25* (1.+XI)
      FUN(1) = 4.*XIM*ETAM
      FUN(2) = 4.*XIM*ETAP
      FUN(3) = 4.*XIP*ETAP
      FUN(4) = 4.*XIP*ETAM
      DER(1,1) = -ETAM
      DER(1,2) = -ETAP
      DER(1,3) = ETAP
      DER(1,4) = ETAM
      DER(2,1) = -XIM
      DER(2,2) = XIM
      DER(2,3) = XIP
      DER(2,4) = -XIP
      RETURN
      END
C
C
C
      SUBROUTINE CHECK(X1,X0,TOL,ICON)
C
C         SETS ICON TO 0 IF RELATIVE CHANGE IN X
C         IS GREATER THAN TOL
C
      ICON = 1
      IF (ABS(X1-X0)/ABS(X1).GT.TOL) ICON = 0
      X0 = X1
      RETURN
      END
C
C
C
      SUBROUTINE CHECON(LOADS,OLDLDS,N,TOL,ICON)
C
C      SETS ICON TO ZERO IF THE RELATIVE CHANGE
```

```
C       IN VECTORS 'LOADS' AND 'OLDLDS' IS GREATER THAN 'TOL'
C
      REAL LOADS(*),OLDLDS(*)
      ICON = 1
      BIG = 0.
      DO 1 I = 1,N
    1 IF (ABS(LOADS(I)).GT.BIG) BIG = ABS(LOADS(I))
      DO 2 I = 1,N
          IF (ABS(LOADS(I)-OLDLDS(I))/BIG.GT.TOL) ICON = 0
    2 OLDLDS(I) = LOADS(I)
      RETURN
      END
C
C
C
      SUBROUTINE CHOFAC(LB,ILB,N,IW)
C
C       CHOLESKI FACTORISATION OF BANDED LOWER TRIANGLE LB
C
      REAL LB(ILB,*)
      DO 1 I = 1,N
          X = 0.0
          DO 2 J = 1,IW
              X = X + LB(I,J)**2
    2     CONTINUE
          LB(I,IW+1) = SQRT(LB(I,IW+1)-X)
          DO 3 K = 1,IW
              X = 0.0
              IF (I+K.LE.N) THEN
                  IF (K.NE.W) THEN
                      DO 4 L = IW - K,1,-1
                          X = X + LB(I+K,L)*LB(I,L+K)
    4                 CONTINUE
                  END IF
                  IA = I + K
                  IB = IW - K + 1
                  LB(IA,IB) = (LB(IA,IB)-X)/LB(I,IW+1)
              END IF
    3     CONTINUE
    1 CONTINUE
      RETURN
      END
C
C
C
      SUBROUTINE CHOSUB(LB,ILB,B,N,IW)
C
C       CHOLESKI FORWARD AND BACKWARD SUBSTITUTION COMBINED
C
      REAL LB(ILB,*),B(*)
      B(1) = B(1)/LB(1,IW+1)
      DO 1 I = 2,N
          X = 0.0
          K = 1
          IF (I.LE.IW+1) K = IW - I + 2
          DO 2 J = K,IW
              X = X + LB(I,J)*B(I+J-IW-1)
    2     CONTINUE
          B(I) = (B(I)-X)/LB(I,IW+1)
    1 CONTINUE
      B(N) = B(N)/LB(N,IW+1)
      DO 3 I = N - 1,1,-1
          X = 0.0
          L = I + IW
          IF (I.GT.N-IW) L = N
          M = I + 1
```

```
          DO 4 J = M,L
             X = X + LB(J,IW+I-J+1)*B(J)
    4     CONTINUE
          B(I) = (B(I)-X)/LB(I,IW+1)
    3 CONTINUE
      RETURN
      END
C
C
C
      SUBROUTINE GAULAG(SAMP,ISAMP,NGL)
C
C      WEIGHTS AND SAMPLING POINTS
C      FOR GAUSS-LAGUERRE QUADRATURE
C
      REAL SAMP(ISAMP,*)
      GO TO (1,2,3,4,5),NGL
    1 SAMP(1,1) = 1.
      SAMP(1,2) = 1.
      GO TO 100
    2 SAMP(1,1) = 0.585786437627
      SAMP(2,1) = 3.414213562373
      SAMP(1,2) = 0.853553390593
      SAMP(2,2) = 0.146446609407
      GO TO 100
    3 SAMP(1,1) = 0.415774556783
      SAMP(2,1) = 2.294280360279
      SAMP(3,1) = 6.289945082937
      SAMP(1,2) = 0.711093009929
      SAMP(2,2) = 0.278517733569
      SAMP(3,2) = 0.0103892565016
      GO TO 100
    4 SAMP(1,1) = 0.322547689619
      SAMP(2,1) = 1.745761101158
      SAMP(3,1) = 4.536620296921
      SAMP(4,1) = 9.395070912301
      SAMP(1,2) = 0.603154104342
      SAMP(2,2) = 0.357418692438
      SAMP(3,2) = 0.038887908515
      SAMP(4,2) = 0.000539294705561
      GO TO 100
    5 SAMP(1,1) = 0.263560319718
      SAMP(2,1) = 1.413403059107
      SAMP(3,1) = 3.596425771041
      SAMP(4,1) = 7.085810005859
      SAMP(5,1) = 12.640800844276
      SAMP(1,2) = 0.521755610583
      SAMP(2,2) = 0.398666811083
      SAMP(3,2) = 0.0759424496817
      SAMP(4,2) = 0.00361175867992
      SAMP(5,2) = 0.0000233699723858
  100 CONTINUE
      RETURN
      END
C
C
C
      SUBROUTINE GAULEG(SAMP,ISAMP,NGP)
C
C      WEIGHTS AND SAMPLING POINTS
C      FOR GAUSS-LEGENDRE QUADRATURE
C
      REAL SAMP(ISAMP,*)
      GO TO (1,2,3,4,5,6,7),NGP
    1 SAMP(1,1) = 0.
      SAMP(1,2) = 2.
```

```
      GO TO 100
    2 SAMP(1,1) = 1./SQRT(3.)
      SAMP(2,1) = -SAMP(1,1)
      SAMP(1,2) = 1.
      SAMP(2,2) = 1.
      GO TO 100
    3 SAMP(1,1) = .2*SQRT(15.)
      SAMP(2,1) = .0
      SAMP(3,1) = -SAMP(1,1)
      SAMP(1,2) = 5./9.
      SAMP(2,2) = 8./9.
      SAMP(3,2) = SAMP(1,2)
      GO TO 100
    4 SAMP(1,1) = .861136311594053
      SAMP(2,1) = .339981043584856
      SAMP(3,1) = -SAMP(2,1)
      SAMP(4,1) = -SAMP(1,1)
      SAMP(1,2) = .347854845137454
      SAMP(2,2) = .652145154862546
      SAMP(3,2) = SAMP(2,2)
      SAMP(4,2) = SAMP(1,2)
      GO TO 100
    5 SAMP(1,1) = .906179845938664
      SAMP(2,1) = .538469310105683
      SAMP(3,1) = .0
      SAMP(4,1) = -SAMP(2,1)
      SAMP(5,1) = -SAMP(1,1)
      SAMP(1,2) = .236926885056189
      SAMP(2,2) = .478628670499366
      SAMP(3,2) = .568888888888889
      SAMP(4,2) = SAMP(2,2)
      SAMP(5,2) = SAMP(1,2)
      GO TO 100
    6 SAMP(1,1) = .932469514203152
      SAMP(2,1) = .661209386466265
      SAMP(3,1) = .238619186083197
      SAMP(4,1) = -SAMP(3,1)
      SAMP(5,1) = -SAMP(2,1)
      SAMP(6,1) = -SAMP(1,1)
      SAMP(1,2) = .171324492379170
      SAMP(2,2) = .360761573048139
      SAMP(3,2) = .467913934572691
      SAMP(4,2) = SAMP(3,2)
      SAMP(5,2) = SAMP(2,2)
      SAMP(6,2) = SAMP(1,2)
      GO TO 100
    7 SAMP(1,1) = .949107912342759
      SAMP(2,1) = .741531185599394
      SAMP(3,1) = .405845151377397
      SAMP(4,1) = .0
      SAMP(5,1) = -SAMP(3,1)
      SAMP(6,1) = -SAMP(2,1)
      SAMP(7,1) = -SAMP(1,1)
      SAMP(1,2) = .129484966168870
      SAMP(2,2) = .279705391489277
      SAMP(3,2) = .381830050505119
      SAMP(4,2) = .417959183673469
      SAMP(5,2) = SAMP(3,2)
      SAMP(6,2) = SAMP(2,2)
      SAMP(7,2) = SAMP(1,2)
  100 CONTINUE
      RETURN
      END
C
C
C
```

```
      SUBROUTINE LDLFOR(A,IA,B,N)
C
C      FORWARD SUBSTITUTION ON A LOWER TRIANGLE
C      STORED AS AN UPPER TRIANGLE
C
      REAL A(IA,*),B(*)
      DO 1 I = 1,N
          SUM = B(I)
          IF (I.GT.1) THEN
              DO 2 J = 1,I - 1
                  SUM = SUM - A(J,I)*B(J)
    2         CONTINUE
          END IF
          B(I) = SUM/A(I,I)
    1 CONTINUE
      RETURN
      END
C
C
C
      SUBROUTINE LDLT(A,IA,D,N)
C
C      L*D*LT FACTORISATION OF A SQUARE MATRIX A
C      LT OVERWRITES A
C
      REAL A(IA,*),D(*)
      DO 1 K = 1,N - 1
          D(1) = A(1,1)
          IF (ABS(A(K,K)).GT.1.E-6) THEN
              DO 2 I = K + 1,N
                  X = A(I,K)/A(K,K)
                  DO 3 J = K + 1,N
                      A(I,J) = A(I,J) - A(K,J)*X
    3             CONTINUE
                  D(I) = A(I,I)
    2         CONTINUE
          ELSE
              WRITE (6,1000)
              WRITE (6,*) K
          END IF
    1 CONTINUE
 1000 FORMAT ('ZERO PIVOT FOUND IN THIS LINE:')
      RETURN
      END
C
C
C
      SUBROUTINE LUFAC(A,UPTRI,LOWTRI,IA,N)
C
C          L*U FACTORISATION OF A SQUARE MATRIX A
C
      REAL A(IA,*),UPTRI(IA,*),LOWTRI(IA,*)
      CALL NULL(UPTRI,IA,N,N)
      CALL NULL(LOWTRI,IA,N,N)
      DO 10 I=1,N
         UPTRI(1,I)=A(1,I)
   10    LOWTRI(I,I)=1.0
      DO 1 K=1,N-1
         IF(ABS(UPTRI(K,K)).GT.1.E-6)THEN
            DO 2 I=K+1,N
C      LOWER TRIANGULAR COMPONENTS
               DO 3 J=1,I-1
                  SUM=0.0
                     DO 4 L=1,J-1
                        SUM=SUM-LOWTRI(I,L)*UPTRI(L,J)
    4                CONTINUE
```

```
                LOWTRI(I,J)=(A(I,J)+SUM)/UPTRI(J,J)
    3        CONTINUE
C      UPPER TRIANGULAR COMPONENTS
             DO 5 J=I,N
                SUM=0.0
                DO 6 L=1,I-1
                   SUM=SUM-LOWTRI(I,L)*UPTRI(L,J)
    6           CONTINUE
                UPTRI(I,J)=A(I,J)+SUM
    5        CONTINUE
    2     CONTINUE
       ELSE
       WRITE(6,1000)
       WRITE(6,*)K
       STOP
       ENDIF
    1 CONTINUE
 1000 FORMAT('ZERO PIVOT FOUND IN THIS ROW')
      RETURN
      END
C
C
C
      SUBROUTINE LUPFAC(A,IA,N,ROW)
C
C      L*U FACTORISATION OF A SQUARE MATRIX WITH PIVOTS
C
      REAL A(IA,*)
      INTEGER ROW(*)
      DO 1 I = 1,N
    1 ROW(I) = I
      DO 2 I = 1,N - 1
          IP = I
          PVAL = A(ROW(IP),IP)
          DO 3 J = I + 1,N
              IF (ABS(A(ROW(J),I)).GT.ABS(PVAL)) THEN
                  IP = J
                  PVAL = A(ROW(J),I)
              END IF
    3     CONTINUE
          IF (ABS(PVAL).LT.1.E-10) THEN
              WRITE (6,1000)
              STOP
          END IF
          IH = ROW(IP)
          ROW(IP) = ROW(I)
          ROW(I) = IH
          DO 4 J = I + 1,N
              IE = ROW(J)
              PIVOT = A(IE,I)/PVAL
              A(IE,I) = PIVOT
              IROW = ROW(I)
              DO 5 K = I + 1,N
                  A(IE,K) = A(IE,K) - A(IROW,K)*PIVOT
    5         CONTINUE
    4     CONTINUE
    2 CONTINUE
      IF (ABS(A(ROW(N),N)).LT.1.E-10) THEN
          WRITE (6,1000)
          STOP
      END IF
 1000 FORMAT ('SINGULAR EQUATIONS DETECTED')
      RETURN
      END
C
C
```

```
C
      SUBROUTINE LUPSOL(A,IA,B,SOL,N,ROW)
C
C      FORWARD AND BACKWARD SUBSTITUTION COMBINED WITH PIVOTS
C
      REAL A(IA,*),B(*),SOL(*)
      INTEGER ROW(*)
      DO 1 I = 1,N
          IROW = ROW(I)
          SUM = B(IROW)
          IF (I.GT.1) THEN
              DO 2 J = 1,I - 1
                  SUM = SUM - A(IROW,J)*B(ROW(J))
    2         CONTINUE
              B(IROW) = SUM
          END IF
    1 CONTINUE
      DO 3 I = N,1,-1
          IROW = ROW(I)
          SUM = B(IROW)
          IF (I.LT.N) THEN
              DO 4 J = I + 1,N
                  SUM = SUM - A(IROW,J)*B(ROW(J))
    4         CONTINUE
          END IF
          B(IROW) = SUM/A(IROW,I)
    3 CONTINUE
      DO 5 I = 1,N
    5 SOL(I) = B(ROW(I))
      RETURN
      END
C
C
C
      SUBROUTINE MATINV(A,IA,N)
C
C      FORMS THE INVERSE OF A MATRIX
C      USING GAUSS-JORDAN TRANSFORMATION
C
      REAL A(IA,*)
      DO 1 K = 1,N
          CON = A(K,K)
          A(K,K) = 1.
          DO 2 J = 1,N
    2     A(K,J) = A(K,J)/CON
          DO 1 I = 1,N
              IF (I.EQ.K) GO TO 1
              CON = A(I,K)
              A(I,K) = 0.
              DO 3 J = 1,N
    3         A(I,J) = A(I,J) - A(K,J)*CON
    1 CONTINUE
      RETURN
      END
C
C
C
      SUBROUTINE MATMUL(A,IA,B,IB,C,IC,L,M,N)
C
C      PRODUCT OF TWO MATRICES
C
      REAL A(IA,*),B(IB,*),C(IC,*)
      DO 1 I = 1,L
          DO 1 J = 1,N
              X = 0.0
              DO 2 K = 1,M
```

```
    2          X = X + A(I,K)*B(K,J)
               C(I,J) = X
    1 CONTINUE
      RETURN
      END
C
C
C
      SUBROUTINE MATRAN(A,IA,B,IB,M,N)
C
C      FORMS THE TRANSPOSE OF A MATRIX
C
      REAL A(IA,*),B(IB,*)
      DO 1 I = 1,M
          DO 1 J = 1,N
    1 A(J,I) = B(I,J)
      RETURN
      END
C
C
C
      SUBROUTINE MSMULT(A,IA,C,M,N)
C
C      MULTIPLIES A MATRIX BY A SCALAR
C
      REAL A(IA,*)
      DO 1 I = 1,M
          DO 1 J = 1,N
    1 A(I,J) = A(I,J)*C
      RETURN
      END
C
C
C
      SUBROUTINE MVMULT(M,IM,V,K,L,Y)
C
C      MULTIPLIES A MATRIX BY A VECTOR
C
      REAL M(IM,*),V(*),Y(*)
      DO 1 I = 1,K
          X = 0.
          DO 2 J = 1,L
    2     X = X + M(I,J)*V(J)
          Y(I) = X
    1 CONTINUE
      RETURN
      END
C
C
C
      SUBROUTINE NEWCOT(WEIGHT,NNC)
C
C      WEIGHTING COEFFICIENTS
C      FOR NEWTON-COTES QUADRATURE
C
      REAL WEIGHT(*)
      GO TO (1,2,3,4,5) NNC
    1 WEIGHT(1) = 1.
      GO TO 100
    2 WEIGHT(1) = .5
      WEIGHT(2) = .5
      GO TO 100
    3 WEIGHT(1) = 0.333333333333333
      WEIGHT(2) = 1.333333333333333
      WEIGHT(3) = 0.333333333333333
      GO TO 100
```

```
    4 WEIGHT(1) = 0.375
      WEIGHT(2) = 1.125
      WEIGHT(3) = 1.125
      WEIGHT(4) = 0.375
      GO TO 100
    5 WEIGHT(1) = 0.311111111111111
      WEIGHT(2) = 1.422222222222222
      WEIGHT(3) = 0.533333333333333
      WEIGHT(4) = 1.422222222222222
      WEIGHT(5) = 0.311111111111111
  100 CONTINUE
      RETURN
      END
C
C
C
      SUBROUTINE NULL(A,IA,M,N)
C
C      NULLS A 2-D ARRAY
C
      REAL A(IA,*)
      DO 1 I = 1,M
          DO 1 J = 1,N
    1 A(I,J) = 0.0
      RETURN
      END
C
C
C
      SUBROUTINE NULVEC(VEC,N)
C
C      NULLS A COLUMN VECTOR
C
      REAL VEC(*)
      DO 1 I = 1,N
    1 VEC(I) = 0.
      RETURN
      END
C
C
C
      SUBROUTINE PRINTA(A,IA,M,N,ICH)
C
C      WRITES A 2-D ARRAY TO OUTPUT CHANNEL 'ICH'
C
      REAL A(IA,*)
      DO 1 I = 1,M
    1 WRITE (ICH,2) (A(I,J),J=1,N)
    2 FORMAT (1X,6E12.4)
      RETURN
      END
C
C
C
      SUBROUTINE PRINTV(VEC,N,ICH)
C
C      WRITES A COLUMN VECTOR TO OUTPUT CHANNEL 'ICH'
C
      REAL VEC(*)
      WRITE (ICH,1) (VEC(I),I=1,N)
    1 FORMAT (1X,6E12.4)
      RETURN
      END
C
C
C
```

```
      SUBROUTINE SKYFAC(A,N,KDIAG)
C
C      CHOLESKI FACTORISATION OF VARIABLE BANDWIDTH
C      A STORED AS A VECTOR AND OVERWRITTEN
C
      REAL A(*)
      INTEGER KDIAG(*)
      A(1) = SQRT(A(1))
      DO 1 I = 2,N
          KI = KDIAG(I) - I
          L = KDIAG(I-1) - KI + 1
          DO 2 J = L,I
              X = A(KI+J)
              KJ = KDIAG(J) - J
              IF (J.NE.1) THEN
                  LL = KDIAG(J-1) - KJ + 1
                  LL = MAX0(L,LL)
                  IF (LL.NE.J) THEN
                      M = J - 1
                      DO 3 K = LL,M
                          X = X - A(KI+K)*A(KJ+K)
    3                 CONTINUE
                  END IF
              END IF
              A(KI+J) = X/A(KJ+J)
    2     CONTINUE
          A(KI+I) = SQRT(X)
    1 CONTINUE
      RETURN
      END
C
C
C
      SUBROUTINE SKYSUB(A,B,N,KDIAG)
C
C      CHOLESKI FORWARD AND BACKWARD SUBSTITUTION COMBINED
C      VARIABLE BANDWIDTH FACTORED A STORED AS A VECTOR
C
      REAL A(*),B(*)
      INTEGER KDIAG(*)
      B(1) = B(1)/A(1)
      DO 1 I = 2,N
          KI = KDIAG(I) - I
          L = KDIAG(I-1) - KI + 1
          X = B(I)
          IF (L.NE.I) THEN
              M = I - 1
              DO 2 J = L,M
                  X = X - A(KI+J)*B(J)
    2         CONTINUE
          END IF
          B(I) = X/A(KI+I)
    1 CONTINUE
      DO 3 IT = 2,N
          I = N + 2 - IT
          KI = KDIAG(I) - I
          X = B(I)/A(KI+I)
          B(I) = X
          L = KDIAG(I-1) - KI + 1
          IF (L.NE.I) THEN
              M = I - 1
              DO 4 K = L,M
                  B(K) = B(K) - X*A(KI+K)
    4         CONTINUE
          END IF
    3 CONTINUE
```

```
      B(1) = B(1)/A(1)
      RETURN
      END
C
C
C
      SUBROUTINE SOLVE(K,IK,U,F,N)
C
C      PERFORMS GAUSSIAN ELIMINATION WITH
C      PARTIAL PIVOTING ON A FULL N*N MATRIX
C
      REAL K(IK,*),F(*),U(*)
C
C      PIVOTING STAGE
C
      DO 1 I = 1,N - 1
          BIG = ABS(K(I,I))
          IHOLD = I
          DO 10 J = I + 1,N
              IF (ABS(K(J,I)).GT.BIG) THEN
                  BIG = ABS(K(J,I))
                  IHOLD = J
              END IF
   10     CONTINUE
          IF (IHOLD.NE.I) THEN
              DO 12 J = I,N
                  HOLD = K(I,J)
                  K(I,J) = K(IHOLD,J)
                  K(IHOLD,J) = HOLD
   12         CONTINUE
              HOLD = F(I)
              F(I) = F(IHOLD)
              F(IHOLD) = HOLD
          END IF
C
C      ELIMINATION STAGE
C
          DO 3 J = I + 1,N
              FAC = K(J,I)/K(I,I)
              DO 4 L = I,N
    4         K(J,L) = K(J,L) - K(I,L)*FAC
              F(J) = F(J) - F(I)*FAC
    3     CONTINUE
    1 CONTINUE
C
C      BACK-SUBSTITUTION STAGE
C
      DO 9 I = N,1,-1
          SUM = 0.
          DO 6 L = I + 1,N
    6     SUM = SUM + K(I,L)*U(L)
          U(I) = (F(I)-SUM)/K(I,I)
    9 CONTINUE
      RETURN
      END
C
C
C
      SUBROUTINE SUBBAC(A,IA,B,N)
C
C      BACKWARD SUBSTITUTION ON AN UPPER TRIANGLE
C
      REAL A(IA,*),B(*)
      DO 1 I = N,1,-1
          SUM = B(I)
          IF (I.LT.N) THEN
```

```
            DO 2 J = I + 1,N
               SUM = SUM - A(I,J)*B(J)
    2       CONTINUE
         END IF
         B(I) = SUM/A(I,I)
    1 CONTINUE
      RETURN
      END
C
C
C
      SUBROUTINE SUBFOR(A,IA,B,N)
C
C      FORWARD SUBSTITUTION ON A LOWER TRIANGLE
C
      REAL A(IA,*),B(*)
      DO 1 I = 1,N
         SUM = B(I)
         IF (I.GT.1) THEN
            DO 2 J = 1,I - 1
               SUM = SUM - A(I,J)*B(J)
    2       CONTINUE
         END IF
         B(I) = SUM/A(I,I)
    1 CONTINUE
      RETURN
      END
C
C
C
      SUBROUTINE VDOTV(V1,V2,DOTPR,N)
C
C             DOT PRODUCT V1*V2
C
      REAL V1(*),V2(*)
      DOTPR = 0.0
      DO 1 I = 1,N
    1 DOTPR = DOTPR + V1(I)*V2(I)
      RETURN
      END
C
C
C
      SUBROUTINE VECADD(A,B,C,N)
C
C      ADDS VECTORS  A+B=C
C
      REAL A(*),B(*),C(*)
      DO 1 I = 1,N
    1 C(I) = A(I) + B(I)
      RETURN
      END
C
C
C
      SUBROUTINE VECCOP(A,B,N)
C
C      COPIES VECTOR A INTO VECTOR B
C
      REAL A(*),B(*)
      DO 1 I = 1,N
    1 B(I) = A(I)
      RETURN
      END
C
C
```

```
C
      SUBROUTINE VECSUB(A,B,C,N)
C
C      VECTOR SUBTRACT C:=A-B
C
      REAL A(*),B(*),C(*)
      DO 1 I = 1,N
    1 C(I) = A(I) - B(I)
      RETURN
      END
C
C
C
      SUBROUTINE VSMULT(V,SCAL,N)
C
C             MULTIPLY A VECTOR BY A SCALAR
C
      REAL V(*)
      DO 1 I = 1,N
    1 V(I) = V(I)*SCAL
      RETURN
      END
C
C
C
      SUBROUTINE VVMULT(V1,V2,A,IA,M,N)
C
C      FORMS A VECTOR PRODUCT
C
      REAL V1(*),V2(*),A(IA,*)
      DO 1 I = 1,M
          DO 1 J = 1,N
    1 A(I,J) = V1(I)*V2(J)
      RETURN
      END
```

Index